THE HANDBOOK OF ENVIRONMENTAL HEALTH

Frank R. Spellman
Melissa L. Stoudt

THE SCARECROW PRESS, INC.
Lanham • Toronto • Plymouth, UK
2013

Published by Scarecrow Press, Inc.
A wholly owned subsidiary of The Rowman & Littlefield Publishing Group, Inc.
4501 Forbes Boulevard, Suite 200, Lanham, Maryland 20706
www.rowman.com

10 Thornbury Road, Plymouth PL6 7PP, United Kingdom

British Library Cataloguing in Publication Information Available

Library of Congress Cataloging-in-Publication Data

Spellman, Frank R.

The handbook of environmental health / Frank R. Spellman, Melissa L. Stoudt.
p. cm.
Includes bibliographical references and index.
ISBN 978-0-8108-8685-8 (cloth : alk. paper) — ISBN 978-0-8108-8686-5 (ebook)
1. Environmental health—Handbooks, manuals, etc. I. Stoudt, Melissa L. II. Title.
RA566.22.S64 2013
362.1—dc23 2012041369

∞™ The paper used in this publication meets the minimum requirements of American National Standard for Information Sciences—Permanence of Paper for Printed Library Materials, ANSI/NISO Z39.48-1992.

Printed in the United States of America

For
Suzanne Wilson
and
Nancy Velasquez

Contents

Preface

THIS BOOK IS ABOUT OUR WORLD: the sun, the air, the water, the soil, the dust, the plants and animals, the chemicals, and the metals. They support life. They make it beautiful and fun. They allow us to live and maintain the so-called good life. This book points out, however, that as wonderful as they are . . . they can also make some people sick, and damage plants and animals—the environment in which we live.

This book is a comprehensive treatment of a very broad field. The chapters are timely and factual, well balanced, referenced, and written in plain English for general readers, college students, environmental health practitioners, lawyers, public administrators, utility directors (water and wastewater treatment), public service managers, non-environmental managers, regulators, and non-environmental health professionals in any field who desire to obtain understanding of everyday environmental health issues.

Also, this book examines health issues, scientific understanding of causes, and possible future approaches to control the major environmental problems in industrialized and developing countries. This book offers an overview of the methodology and paradigms of the dynamic, evolving field, ranging from ecology (environment and aquatic and terrestrial) to epidemiology, from environmental psychology to toxicology, and from genetics to ethics. Topics include how the body reacts to environmental pollutants and stressors, and they cover the many effects of environmental factors: physical, chemical, and biological agents of environmental contamination; vectors for dissemination (air, water, soil); solid and hazardous waste; susceptible populations; biomarkers and risk analysis; the scientific basis for policy decision; occupational health and safety issues; and emerging global environmental problems.

Moreover, this book not only covers the environmental sciences but also the human population. It emphasizes the environmental practices that support human life as well as the need to control factors that are harmful to human life. With chapters providing a judicious use of figures and tables, each sector includes an appropriate amount of material for an overview of the topics presented.

The bottom line is that this book is not just a *book*; it is a handbook. While it can be used anywhere—in the library, in the classroom, in the office, and/or in the laboratory—it is designed to be used primarily in the field. It is designed to be a "handheld" reference and guide providing guidelines about environmental health and how to maintain it in the environmental world.

Frank R. Spellman and Melissa L. Stoudt

> The real problem is the shock of severe, dangerous illness, its unexpectedness and surprise. Most of us, patients and doctors alike, can ride almost all the way through life with no experience of real peril and when it does come, it seems an outrage, a piece of unfairness. We are not used to disease as we used to be, and we are not at all used to being incorporated into a high technology.
>
> —Lewis Thomas (1983), *The Youngest Science*

Prologue

On Tuesday afternoon, a fifty-two-year-old man with previously diagnosed coronary artery disease controlled by nitroglycerin describes episodes of recurring headache for the past three weeks. Mild nausea often accompanies the headache; there is no vomiting. He describes a dull frontal ache that is not relieved by aspirin. The patient states that the headaches are sometimes severe; at other times they are a nagging annoyance. The durations ran from half an hour to a full day.

His visit was also prompted by a mild angina attack that he suffered this past weekend shortly after he awoke on Sunday morning. He has experienced no further cardiac symptoms since that episode.

History of previous illness indicates that the patient was diagnosed with angina pectoris three years ago. He has been taking 0.4 milligrams (mg) nitroglycerin sublingually (placed beneath the tongue) prophylactically before vigorous exercise. He also takes one aspirin every other day. He has been symptom-free for the past two and a half years.

Sublingual nitroglycerin relieved the pain of the Sunday morning angina attack within several minutes.

The patient does not smoke and rarely drinks alcohol. He is a trim man with a slightly ruddy complexion.

Information in this chapter from CDC-ATSDR, *Case Studies in Environmental Medicine*. Accessed 02/08/12 at http://www.atsdr.cdc.gov/csem/csem.html.

At present, he is afebrile (i.e., not having a fever), and his vital signs are as follows:

- blood pressure 120/85
- pulse 80
- respirations 20

Physical exam is normal.

The results of an electrocardiogram (ECG) with a rhythm strip performed in your office are unremarkable.

Subsequent laboratory testing reveals normal blood lipids, cardiac enzymes, complete blood cell count (CBC), sedimentation rate, glucose, creatinine, and thyroid function.

Initial Check Questions

1. What should be included in the patient's problem list?
2. What would you include in the differential diagnosis?
3. What additional information would you seek to assist in the diagnosis?

Did You Know?

Basically, differential diagnosis, sometimes abbreviated DDx, ddx DD, D/Dx, or ΔΔ, is a systematic process of elimination to determine actual probability of having or not having disease.

Environmental factors rarely enter into the clinician's differential diagnosis. Consequently, most environmental and occupational diseases either manifest as common medical problems or have nonspecific symptoms. As a result, clinicians miss the opportunity to make a correct diagnosis that might influence the course of disease in some afflicted individuals (by stopping exposure) and that might prevent disease in others (by avoiding exposure).[1]

The obvious question should be: what can the clinician do to improve recognition of disease related to current or past exposures?

- Suppose this patient lived near a hazardous waste site.
- Suppose the patient is an accountant who has had the same job and same residence for many years.
- Suppose the patient owns a commercial cleaning service and uses cleaning products at various industrial and commercial sites.

- Suppose this patient is a retired advertising copywriter who lives in the vicinity of an abandoned industrial complex.

These are examples of the work and lifestyles that the clinician needs to know to make an accurate diagnosis.

Note

1. Goldman, R.H., and J.M. Peters. (1981). The Occupational and Environmental Health History. *JAMA* 246 (24): 2831–2836.

1

Introduction

If you want one year of prosperity, plant corn. If you want ten years of prosperity, plant trees. If you want one hundred years of prosperity, educate people.

—Chinese Proverb

Growth for the sake of growth is the ideology of the cancer cell.

—Edward Abbey

What Is Environmental Health?

There is an ensemble of definitions of environmental health. Though the exact content of each definition varies, it is possible, without rising to the level of parochial hyperbole and formal analysis, to observe some features that are prevalent across the definitions (EHPC, 1998).

- All definitions mention human health, public health, humans, or similar words.
- In addition to mentioning human health, some definitions mention ecologic health or ecological balances.
- A few definitions mention specific environmental stressors, such as physical, chemical, and biologic agents.

Before presenting a few example definitions (and the one used in this text) of environmental health, it is useful to cite the World Health Organization's (WHO, 2012) definition of health—*Health is a complete state of physical, mental, and social well-being, not just the absence of infirmity or disease.*

Definitions of Environmental Health

Dade W. Moeller (2011)—In its broadest sense, *environmental health* is the segment of public health that is concerned with assessing understanding, and controlling the impacts of people on their environment and the impacts of the environment on them.

Robert H. Friis (2012)—*Environmental health* comprises those aspects of human health, including quality of life, that are determined by physical, chemical, biological, social, and psychosocial factors in the environment. It also refers to the theory and practice of assessing, correcting, controlling and preventing those facts in the environment that potentially can affect adversely the health of present and future generations.

Herman Koren (2005)—*Environmental health* is the art and science of the protection of good health, the promotion of aesthetic values, the prevention of disease and injury through control of positive environmental factors, and the reduction of potential physical, biological, chemical, and radiological hazards.

National Environmental Health Association (NEHA, 1996)—*Environmental health* and protection refers to protection against environmental factors that may adversely impact human health or the ecological balances essential to long-term human health and environmental quality, whether in the natural or [human]-made environment.

In this book, NEHA's altered definition (man-made altered to human-made) is the preferred usage. Moreover, the author shares Moeller's (2011) view that environmental health is better defined by the problems it faces than by the approaches it uses.

Environmental Health Concerns

There are numerous environmental health concerns. In this book we focus on the health concerns shown in figure 1.1. Each of these concerns is also addressed in this book and is explained in detail. Again, keep in mind that there are many more environmental health concerns than those shown in figure 1.1; the focus in the text is on those health concerns that the public might be expected to come into contact with or experience around the office, farm, industrial facility, recreational facilities, or the home.

The major chemicals of concern to the environmental health practitioner and/or occupational health clinician are shown in figure 1.2. There are literally thousands of chemicals and chemical compounds used in modern society, but again, we focus on those that are most commonly encountered at work and in the home.

The environmental health practitioner is also concerned with those diseases related to environmental exposure. Many of these environmental diseases, causative factors, and health sectors of concern and listed by the National Institutes of Health (NIH, 2012) are shown in figure 1.3. In addition, the most common of these environmental diseases are briefly discussed in this chapter.

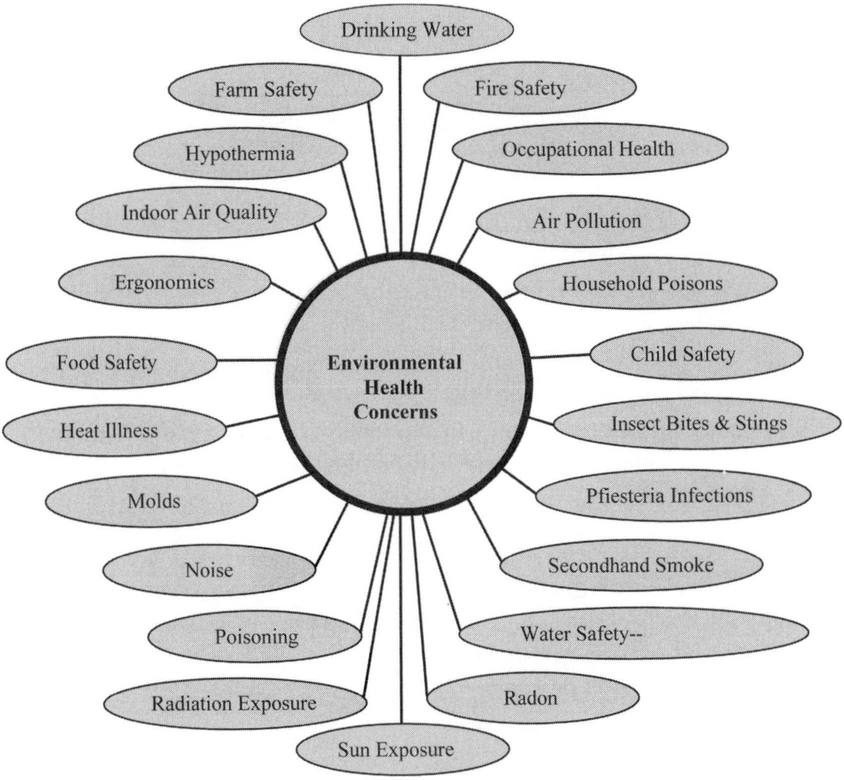

FIGURE 1.1
Environmental health concerns.

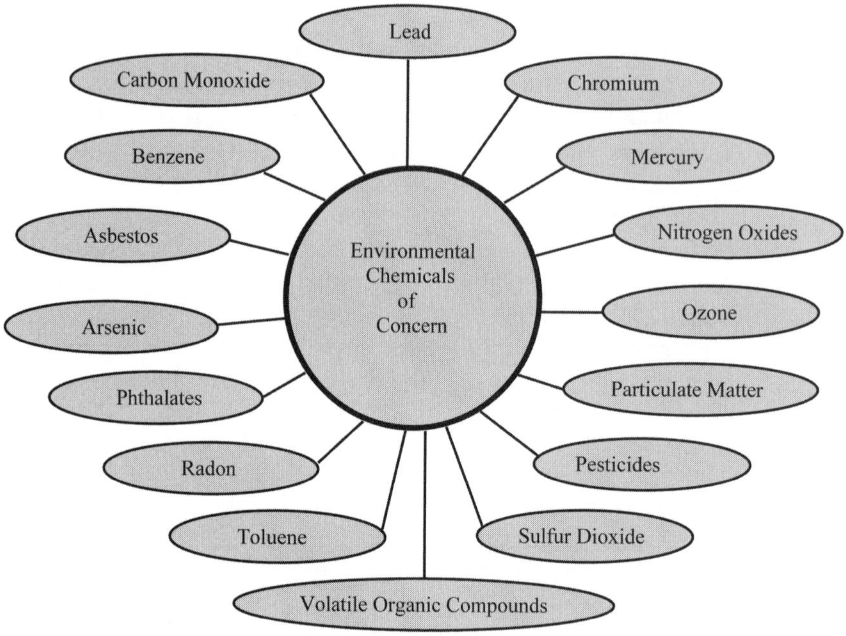

FIGURE 1.2
Major environmental chemicals of concern to the environmental practitioner.

Did You Know?

Public Health is the approach to medicine that is concerned with the health of the community as a whole. Public health is community health. It has been said that: "Health care is vital to all of us some of the time, but public health is vital to all of us all of the time" (MedTerms, 2012).

Did You Know?

Occupational Health is the promotion and maintenance of the highest degree of physical, mental, and social well-being of workers in all occupations by preventing departures from health, controlling risks, and the adaptation of work to people, and people to their jobs (ILO/WHO, 1950).

Environmental Diseases from A to Z

(Information in this section is based on NIEHS-NIH 2007. *Environmental Diseases from A to Z.* NIH Publication No. 96-4145 at http://www.niehs.nih

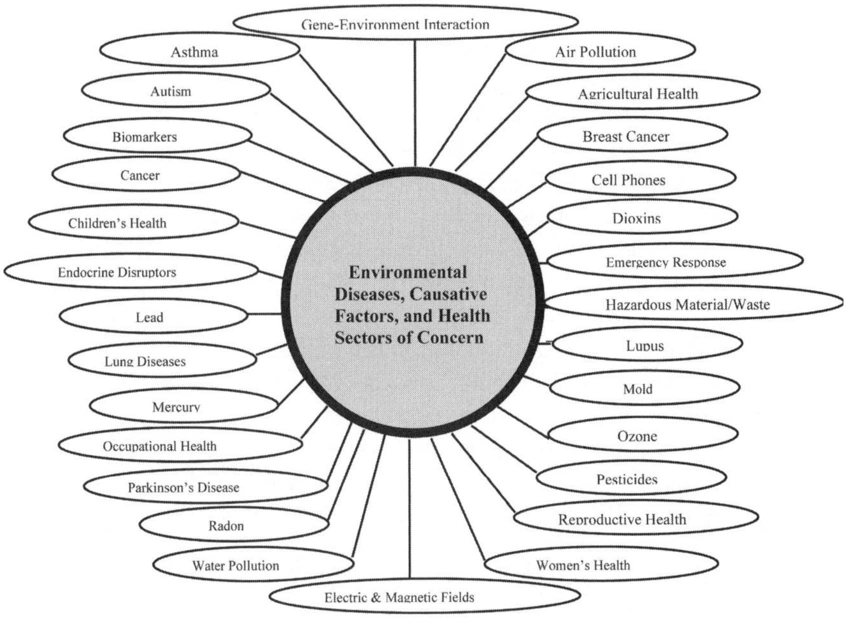

FIGURE 1.3
Environmental diseases, causative factors, and health sectors of concerns. Based on information from NIH, 2012, accessed 02/10/12 at http://www.neihs.nih.gov/health/.Topics/atoz/index.

.gov.) The National Institute of Environmental Health Sciences (NIEHS) and National Institute of Health (2007) have published a second edition of their informative *Environmental Diseases from A to Z*. The information contained therein is based on data obtained from various national agencies, such as NIH, CDC, National Institute of Child Health and Human Development, National Library of Medicine's *Medline Plus*, and others, and prestigious environmental journals, such as *Journal of Allergy Clinical Immunology*. In the following section, based on the information from these organizations we present environmental diseases and some ideas of preventing and caring for them.

Allergies and Asthma—Slightly more than half of the 300-plus million people living in the United States are sensitive to one or more allergens.[1] They sneeze, their noses run, and their eyes itch from pollen, dust, and other substances. Some suffer sudden attacks that leave them breathless and gasping for air. This is allergic asthma. Asthma attacks often occur after periods of heavy exercise or during sudden changes in the weather. Some can be triggered by pollutants and other chemicals in the air and in the home. Doctors can test to find out which substances are causing reactions. They can also prescribe drugs to relieve the symptoms.

Birth Defects—Sometimes, when pregnant women are exposed to chemicals or drink a lot of alcohol, harmful substances reach the fetus. Some of these babies are born with an organ, tissue, or body part that has not developed in a normal way. Aspirin and cigarette smoke can also cause birth problems. Birth defects are the leading cause of death for infants during the first year of life (NIH, 2007). Many of these could be prevented.

Cancer—Cancer occurs when a cell or group of cells begins to multiply more rapidly than normal. As the cancer cells spread, they affect nearby organs and tissues in the body. Eventually, the organs are not able to perform their normal functions. Cancer is the second leading cause of death in the United States, causing more than 500,000 deaths each year (NIH, 2007). Some cancers are caused by substances in the environment: cigarette smoke, asbestos, radiation, natural and man-made chemicals, alcohol, and sunlight. People can reduce their risk of getting cancer by limiting their exposure to these harmful agents.

Dermatitis—Dermatitis derives from Greek *derma* ("skin") + *-itis* ("inflammation"). Many of us have experienced the oozing bumps and itching caused by poison ivy, oak, and sumac. There are several different types of dermatitis. The different kinds usually have in common an allergic reaction to specific allergens (e.g., rubber, metal [nickel]), jewelry, cosmetics, fragrances and perfume, weeds [poison ivy], or a common ingredient found in topical antibiotic creams [neomycin]). Fabrics, foods, and certain medications can cause unusual reactions in some individuals. The term may describe eczema, which is also called dermatitis eczema and eczematous dermatitis. An eczema diagnosis often implies atopic dermatitis (which is very common in children and teenagers) but, without proper context, may refer to any kind of dermatitis. Cleaning products like skin soaps, detergents, laundry soap, or bleach may cause contact dermatitis. Some chemicals found in paints, dyes, cosmetics, and detergents can also cause rashes and blisters (Ring et al., 2006).

Did You Know?

Irritant contact dermatitis accounts for 80 percent of all cases of contact dermatitis. As the name implies, contact dermatitis is caused by an allergen or an irritating substance like soap or solvent (Bigby et al., 2000).

Emphysema—Emphysema is a chronic response caused by air pollution and cigarette smoke that breaks down sensitive tissue in the lungs. Once this happens, the lungs cannot expand and contract properly; damage to the alveoli (air sacs) results in air trapping—the lungs lose their normal elasticity.

Fertility Problems—Fertility is the ability to produce children. However, one in eight couples has a problem. Moreover, more than 10 percent of couples cannot conceive after one year of trying to become pregnant (NIH, 2007). Infertility can be caused by infections that come from sexual diseases or from exposure to chemicals on the job or elsewhere in the environment. Researchers at the National Institute of Environmental Health Sciences (NIEHS) have shown that too much caffeine in the diet can temporarily reduce a women's fertility. NIEHS scientists have also pinpointed the days when a woman is likely to be fertile.

Goiter—A goiter is a swelling in the thyroid gland, which can lead to a swelling of the neck or larynx (voice box) (*Dorland's Medical Dictionary*, 2007). This condition usually occurs in people who do not get enough iodine from the foods they eat. This can cause the thyroid gland to grow larger. The thyroid can become so large that it looks like a baseball sticking out of the front of your neck. Since the thyroid controls basic functions like growth and energy, goiter can produce a wide range of effects. Some goiter patients are unusually restless and nervous. Others tend to be sluggish and lethargic. Goiter became rare after public health officials decided that iodine should be added to salt (NIH, 2007).

Heart Disease—Heart disease and cardiovascular disease are a class of diseases that involve the heart or blood vessels (arteries and veins) (Maton, 1993). Heart disease is the leading cause of death in the United States and is a major cause of disability. Almost 700,000 Americans die of heart disease each year (NIH, 2007). While this may be due in part to poor eating habits and/or lack of exercise, environmental chemicals also play a role. While most chemicals that enter the body are broken down into harmless substances by the liver, some are converted into particles called free radicals that can react with proteins in the blood to form fatty deposits called plaques, which can clog blood vessels. A blockage can cut off the flow of blood to the heart, causing a heart attack.

Immune Deficiency Diseases—The immune system fights germs, viruses, and poisons that attack the body. It is composed of white blood cells and other warrior cells. When a foreign particle enters the body, these cells surround and destroy this "enemy." We have all heard of AIDS and the harm it does to the immune system. Some chemicals and drugs can also weaken the immune system by damaging its specialized cells. When this occurs, the body is more vulnerable to disease and infections (NIH, 2007).

Job-Related Illnesses—Every job has certain hazards. Even a writer can get a paper cut or an electrical shock from his or her laptop computer. But did you know that about 137 workers die from job-related disease every day? This is more than eight times the number of people who die from job-related accidents. Many of these illnesses are caused by chemicals and other agents present in the workplace. Factories and scientific laboratories can contain poisonous chemicals, dyes, and metals. Doctors and other health workers have to work with radiation. People who work in airports or play in rock concerts can suffer hearing loss from loud noise. Some jobs involve extreme heat or cold. Workers can protect themselves from hazards by wearing special suits and using goggles, gloves, earplugs, and other protective gear. Note that in this text we have dedicated a large volume of space to addressing hazards and illnesses related to the work environment.

Did You Know?

Employers who cannot reach a hospital infirmary or clinic within a reasonable amount of time must be prepared to provide first aid to workers who experience injuries or illnesses on the job. OSHA requires that adequate first-aid supplies must be readily available and that someone must be adequately trained to render first aid. OSHA also encourages employers to consider acquiring automated external defibrillators (AEDs)—medical devices designed to revive victims of sudden cardiac arrest.

Kidney Diseases—About 7.5 million adults have some evidence of chronic kidney disease (NIH, 2007). These diseases range from simple infections to total kidney failure. People with kidney failure cannot remove wastes and poisons from their blood. They depend on expensive kidney machines in order to stay alive. Some chemicals found in the environment can produce kidney damage. Some nonprescription drugs, when taken too often, can also cause kidney problems.

Lead Poisoning—Sometimes, infants and children will pick up and eat paint chips and other objects that contain lead. Lead dust, fumes, and lead-contaminated water can also introduce lead into the body. Lead can damage the brain, kidneys, liver, and other organs. Severe lead poisoning can produce headaches, cramps, convulsions, and even death. Even small amounts can cause learning problems and changes in behavior. Doctors can test for lead in the blood and recommend ways to reduce further exposure.

Mercury Poisoning—Mercury poisoning (also known as hydrargyria or mercurialism) is a disease caused by exposure to mercury or its compounds. Mercury (chemical symbol Hg) is a silvery heavy metal that is extremely poisonous. Very small amounts can damage the kidneys, liver, and brain. Years ago, workers in hat factories were poisoned by breathing fumes from mercury used to shape the hats. Today mercury exposure usually results from eating contaminated fish and other foods that contain small amounts of mercury compounds. Since the body cannot get rid of mercury, it gradually builds up inside the tissues. If it is not treated, mercury poisoning can eventually cause pain, numbness, weak muscles, loss of vision, paralysis, and even death (Clifton, 2007).

Did You Know?

The Hatter is often referred to as the Mad Hatter because of Lewis Carroll's classic 1865 tale of *Alice's Adventures in Wonderland*. However, this term was never used by Carroll. The phrase "mad as a hatter" predates Carroll's works and the characters the Hatter and March Hare are initially referred to as "both mad" by the Cheshire Cat with both first appearing in *Alice's Adventures in Wonderland*, in the seventh chapter titled "A Mad Tea-Party."

Nervous System Disorders—The nervous system includes the brain, spinal cord, and nerves, and it commands and controls our thoughts, feelings, movements, and behavior. The nervous system consists of billions of nerve cells. They carry messages and instructions from the brain and spinal cord to other parts of the body. When these cells are damaged by toxic chemicals, injury, or disease, this information system breaks down. This can result in disorders ranging from mood changes and memory loss to blindness, paralysis, and death. Proper use of safety devices such as seat belts, child restraints, and bike helmets can prevent injuries and save lives (NIH, 2007).

Osteoporosis—NIH (2007) points out that over 10 million Americans have osteoporosis, while 18 million others have lost bone mass and are likely to develop osteoporosis in the future. About 25 million Americans suffer from some kind of bone thinning. As people get older, back problems become more common, and bones in the spine, hip, and wrists break more easily. Young people can lower their chances of getting osteoporosis in later years by exercising and eating calcium-rich foods like milk and yogurt.

Pneumoconiosis—Ordinary house and yard dusts do not pose a serious health hazard. But some airborne particles can be very dangerous and can result in occupational lung disease and a restrictive lung disease caused by the inhalation of dust. Depending upon the type of dust, the disease is given different names:

- Asbestosis—asbestos
- Silicosis—silica

- Bauxite fibrosis—bauxite
- Berylliosis—beryllium
- Siderosis—iron
- Byssinosis—cotton
- Silicosiderosis—mixed dust containing silica and iron
- Black lung (anthracosis)—coal worker's pneumoconiosis
- Labrador Lung—a type of asbestos-disease found in miners in Labrador, Canada

Queensland Fever—Q fever is caused by a bacterium called *Coxiella burnetii*. People get infected by inhaling the bacteria usually while in contact with infected animals, animal tissues, or animal products. The main carriers of the disease are farm animals such as cattle, sheep, and goats, but in rural areas kangaroos are also important. A wide range of other animals can be infected including camels, llamas, alpacas, rodents, cats, dogs, birds, wallabies, and other marsupials. The bacteria can survive harsh conditions and remain in the environment for long periods of time, so hay, dust, and other small particles may also carry the bacteria. Symptoms include fever, chills, and muscle aches and pains (NIH, 2007; Queensland Health, 2011).

Reproductive Disorders—NIH (2007) points out that beginning in the later 1940s, many women who were in danger of losing their unborn babies were prescribed a synthetic female hormone called DES (diethylstilbestrol). In 1971, scientists discovered that some of the daughters of these women were developing a very high rate of cancer of the reproductive organs. Since then, the use of DES and other synthetic hormones during pregnancy has been discontinued. NIEHS and other agencies are studying the possibility that some natural chemicals and man-made pesticides may cause similar problems. They are finding that some of these chemicals are so similar to female estrogen that they may actually "mimic" this important hormone. As a result, they may interfere with the development of male and female reproductive organs. This can lead to an increased risk of early puberty, low sperm counts, ovarian cysts, and cancer of the breast or testicles.

Sunburn and Skin Cancer—Too much sunlight can also produce the most common type of cancer—skin cancer. In the United States in 2007, 58,094 people were diagnosed with melanomas of the skin, and 8,461 people died from it (USCS, 2008).

Tooth Decay—In the 1930s, health experts noticed that people who lived in areas where the water contained natural chemicals called fluorides had few cavities. Today, all U.S. residents are exposed to fluoride to some degree, and its use has resulted in a significant decline in tooth decay. National surveys report that the incidence of tooth decay among children twelve to seventeen years of age has declined from 90 percent in 1971 to 67 percent in 1988. Dentists can also protect young teeth by applying special coatings called sealants (NIH, 2007).

Uranium Poisoning—Uranium is a dangerous element because it is radioactive. This means it gives off high-energy particles that can go through the body and damage living tissue. A single high dose of radiation can kill. Small doses over a long period can also be harmful. For example, miners who are exposed to uranium dust are more likely to get lung cancer. Uranium poisoning can also damage the kidneys and interfere with the body's ability to fight infection. While most people will never come in contact with uranium, those who work with medical x-rays or radioactive components are at risk. They should wear lead shields and follow recommended safety guidelines to protect themselves from unnecessary exposures.

Vision Problems—Our eyes are especially sensitive to the environment. Gases found in polluted air can irritate the eyes and produce a burning sensation. Tiny particles from smoke and soot can also cause redness and itching of the eyes. Airborne organisms like molds and fungus can cause infections of the eyes and eyelids. Too much exposure to the sun's rays can eventually produce a clouding of the lens called a cataract.

Waterborne Diseases—Even our cleanest streams, rivers, and lakes can contain chemical pollutants. Heavy metals like lead and mercury can produce severe organ damage. Some chemicals can interfere with the development of organs and tissues, causing birth defects. Others can cause normal cells to become cancerous. Some of our waterways also contain human and animal wastes. The bacteria in the waste can cause high fever, cramps, vomiting, and diarrhea.

Xeroderma Pigmentosa—Xeroderma is a rare condition that people inherit from their parents. When these people are exposed to direct sunlight, their skin breaks out into tiny dark spots that look like freckles. If this condition is not treated, the spots can become cancerous. These areas must then be removed by a surgeon.

***Yusho Poisoning* (Japanese for *oil disease*)**—In 1968, more than one thousand people in western Japan became seriously ill. They suffered from fatigue, headache, cough, numbness in the arms and legs, and unusual skin sores. Pregnant women later delivered babies with birth defects. These people had eaten food that was cooked in contaminated rice oil. Toxic chemicals called PCBs (polychlorinated biphenyls) had accidentally leaked into the oil during the manufacturing process.

For years, PCBs were widely used in the manufacturing of paints, plastics, and electrical equipment. When scientists discovered that low levels of PCBs could kill fish and other wildlife, their use was dramatically reduced. By this time, PCBs were already leaking into the environment from waste disposal sites and other sources. Today, small amounts of these compounds can still be found in our air, water, soil, and some of the foods we eat.

Zinc Deficiency/Poisoning—Zinc is a mineral that the body needs to function properly. In rare cases, people can be poisoned if there is too much zinc in the food or water. However, most people can take in large quantities without any harmful effects. In areas where nutrition is a problem, people may not get enough zinc from their diet. This can lead to retarded growth, hair loss, delayed sexual maturation, eye and skin lesions, and loss of appetite (NIH, 2007).

The Toxic Environment and Stressors

Later, in chapter 2, we discuss the basics of our physical and ecological environment. Prior to this discussion, however, it is important to list those specific and pertinent areas of environmental concern addressed in this book. In accomplishing this, we found the best source of guidance (for our purposes) is provided by NIH. Specifically, NIH's Tox Town (http://toxtown.nlm.nih.gov) is an interactive guide to commonly encountered toxic substances, your health, and the environment.

In particular regard to toxic substances, your health, and the environment, it is the environmental stressors and their impact on people and other organisms based on location and chemical toxicants that are the focus of concern.

In the work environment, typically, it is the industrial hygienist who focuses on evaluating the healthfulness of the workplace environment, either for short periods or for a work-life of exposure. When required, the industrial hygienist recommends corrective procedures to protect health, based on solid quantitative data, experience, and knowledge. The control measures he or she often recommends include: isolation of a work process, substitution of a less harmful chemical or material, and/or other measures designed solely to increase the healthfulness of the work environment. Along with the workplace environment, we also must focus on stressors that are around us 24/7: those that are in our towns, cities, farms, the U.S.-Mexico border, homes, places of recreation, and other locations. Public health, occupational health, and environmental practitioners are the specialists tasked with monitoring environmental health and implementing proper methods and techniques to mitigate environmental hazards, whatever those hazards might be.

To ensure a healthy workplace environment and associated environs, the environmental professional focuses on the recognition, evaluation, and control of chemical, physical, or biological and ergonomic stressors that can cause sickness, impaired health, or significant discomfort to humans, animals, and plants.

The key word just mentioned was *stressors*, or simply, *stress*—the stress caused by the external or internal environmental demands placed upon us. Increases in external stressors beyond a person's tolerance level affect his or her overall health and/or on-the-job performance.

The environmental practitioner must understand not only that environmental stressors exist, but also that they are sometimes cumulative (additive). For example, in the workplace studies have shown that some assembly line processes are little affected by either low illumination or vibration; however, when these two stressors are combined, assembly line performance deteriorates.

Other cases have shown just the opposite effect. For example, the worker who has had little sleep and then is exposed to a work area where noise levels are high actually benefits (to a degree, depending on the intensity of the noise level and the worker's exhaustion level) from increased arousal level; a lack of sleep combined with a high noise level is compensatory.

In order to recognize environmental stressors and other factors that influence an individual's health, the environmental health practitioner must be familiar with work operations, home environment, lifestyle factors, and other processes. In the workplace, for example, an essential part of the new industrial hygienist's employee orientation process should include an overview of all pertinent company work operations and processes. Obviously, the newly hired industrial hygienist who has not been fully indoctrinated on company work operations and processes not only is not qualified to study the environmental effects of such processes, but also suffers from another disability—lack of credibility with supervisors and workers. This point cannot be emphasized strongly enough—if your intention is to correct or remove environmental stressors, you must know your organization and what it is all about.

Did You Know?

Information on the risk to workers from chemical hazards can be obtained from the Material Safety Data Sheet (MSDS) that OSHA's Hazard Communication Standard require be supplied by the manufacturer or importer to the purchaser of all hazardous materials. The MSDS is a summary of the important health, safety, and toxicological information on the chemical or the mixture's ingredients. Other provisions of the Hazard Communication Standard require that all containers of hazardous substances in the workplace have appropriate warning and identification labels.

What are the workplace and general environmental stressors the industrial hygienist and environmental professional should be concerned with? The stressors of concern should be those that are likely to accelerate the aging process, cause significant discomfort and inefficiency, cause chronic illness, or may be immediately dangerous to life and health (Spellman, 1998). Several stressors fall into these categories; the most important work-related health stressors include:

Chemical Stressors—Harmful chemical compounds in the form of solids, liquids, gases, mists, dusts, fumes, and vapors exert toxic effects by inhalation (breathing), absorption (through direct contact with the skin), or ingestion (eating or drinking). Airborne chemical hazards exist as concentrations of mists, vapors, gases, fumes, or solids. Some are toxic through inhalation, and some of them irritate the skin on contact; some can be toxic by absorption through the skin or through ingestion, and some are corrosive to living tissue. The degree of individual risk from exposure to any given substance depends on the nature and potency of the toxic effects and the magnitude and duration of exposure.

Physical Stressors—These include excessive levels of ionizing and non-ionizing electromagnetic radiation, noise, vibration, illumination, and temperature. In occupations where there is exposure to *ionizing* radiation, time, distance, and shielding are important tools in ensuring worker safety. Danger from radiation increases with the amount of time one is exposed to it; hence, the shorter the time of exposure, the smaller the radiation danger. Distance also is a valuable tool in controlling exposure to both ionizing and non-ionizing radiation. Radiation levels from some sources can be estimated by comparing the squares of the distances between the work and the source. For example, at a reference point of ten feet from a source, the radiation is 1/100 of the intensity at one foot from the source. Shielding is also a way to protect against radiation. The greater the protective mass between a radioactive source and the worker, the lower the radiation exposure. *Non-ionizing* radiation also is dealt with by shielding workers from the source. However, sometimes limiting exposure times to non-ionizing radiation or increasing the distance is not effective. Laser radiation, for example, cannot be controlled effectively by imposing time limits. An exposure that is hazardous can happen faster than the blinking of an eye. Increasing the distance from a laser source may require miles before the energy level reaches a point where the exposure would not be harmful.

Noise, another significant physical hazard, can be controlled by various measures. Noise can be reduced by installing equipment and systems that have been engineered, designed, and built to operate quietly; by enclosing or shielding noisy equipment; by making certain that equipment is in good repair and properly maintained with all worn or unbalanced parts replaced;

by mounting noisy equipment on special mounts to reduce vibration; and by installing silencers, mufflers, or baffles. Substituting quiet work methods for noisy ones is another significant way to reduce noise, for example, welding parts rather than riveting them. Also, treating floors, ceilings, and walls with acoustical material can reduce reflected or reverberant noise. In addition, erecting sound barriers at adjacent work stations around noisy operations will reduce worker exposure to noise generated at adjacent work stations.

It is also possible to reduce noise exposure by increasing the distance between the source and the receiver, by isolating workers in acoustical booths, by limiting workers' exposure time to noise, and by providing hearing protection. OSHA requires that workers in noisy surroundings be periodically tested as a precaution against hearing loss.

Another physical hazard, radiant heat exposure in factories such as steel mills, can be controlled by installing reflective shields and by providing protective clothing.

Biological Stressors—These include bacteria, viruses, fungi, and other living organisms that can cause acute and chronic infections by entering the body either directly or through breaks in the skin. Occupations that deal with plants or animals or their products or with food and food processing may expose workers to biological hazards. Laboratory and medical personnel also can be exposed to biological hazards. Any occupations that result in contact with bodily fluids pose a risk to workers from biological hazards.

In occupations where animals are involved, biological hazards are dealt with by preventing and controlling diseases in the animal population as well as proper care and handling of infected animals. Also, effective personal hygiene, particularly proper attention to minor cuts and scratches, especially those on the hands and forearms, helps keep worker risks to a minimum.

In occupations where there is potential exposure to biological hazards, workers should practice proper personal hygiene, particularly hand washing. Hospitals should provide proper ventilation, proper personal protective equipment such as gloves and respirators, adequate infectious waste disposal systems, and appropriate controls including isolation in instances of particularly contagious diseases such as tuberculosis.

Ergonomic Stressors—The science of ergonomics studies and evaluates a full range of tasks including, but not limited to, lifting, holding, pushing, walking, and reaching. Many ergonomic problems result from technological changes such as increased assembly line speeds, adding specialized tasks, and increased repetition; some problems arise from poorly designed job tasks. Any of those conditions can cause ergonomic hazards such as excessive vibration and noise, eye strain, repetitive motion, and heavy lifting problems. Improperly designed tools or work areas also can be ergonomic hazards.

Repetitive motions or repeated shocks over prolonged periods of time, as in jobs involving sorting, assembling, and data entry, can often cause irritation and inflammation of the tendon sheath of the hands and arms, a condition known as carpal tunnel syndrome.

Ergonomic hazards are avoided primarily by the effective design of a job or job site and better designed tools or equipment that meet workers' needs in terms of physical environment and job tasks. Through thorough work-site analyses, employers can set up procedures to correct or control ergonomic hazards by using the appropriate engineering controls (e.g., designing or redesigning work stations, lighting, tools, and equipment); teaching correct work practices (e.g., proper lifting methods); employing proper administrative controls (e.g., shifting workers among several different tasks, reducing production demand, and increasing rest breaks); and, if necessary, providing and mandating personal protective equipment. Evaluating working conditions from an ergonomics standpoint involves looking at the total physiological and psychological demands of the job on the worker.

Overall, environmental health professionals point out that the benefits of a well-designed, ergonomic work environment can include increased efficiency, fewer accidents, lower operating costs, and more effective use of personnel.

In the workplace, the environmental health professional should review the following to anticipate potential health stressors:

- Raw materials
- Support materials
- Chemical reactions
- Chemical interactions
- Products
- By-products
- Waste products
- Equipment
- Operating procedures

Note: Stressors are discussed in further detail in chapter 10.

Environmental Toxins

An individual does not have to be in the workplace alone to be exposed to harmful chemical toxins. Chemical toxins are found everywhere; they are found in airplanes, dental offices, drinking water, food service locations, funeral homes, hair and nail salons, homes, hospitals, parks, pharmacies, rivers, lakes, schools, vehicles, and many more. If you live and work on a farm, you may be exposed to many of the same chemical toxins and also animal waste, agricultural runoff, barns and silos, crop fields, farm ponds, feeding operations, hydraulic fracturing, landfills, meat processing, off-road vehicles, pests, pets, tree farm and logging, urban sprawl, and others. If you reside near the U.S.-Mexico border, you may be exposed to Colonia, crop fields, drinking water, illegal dumps and tire piles, maquiladora, pests, storm water sewage, trienda, trash burning, wildfires, and others. If you live or work near a seaport, you could be exposed to chemical toxins from algae blooms, cesspools, chemical storage tanks, cruise ships, fish farms, fuel pipelines, marinas and boats, nuclear power plants, septic systems, shellfishing, shipyards, storms and floods, vehicles, wastewater treatment plants, and others.

Discussion Questions

1. What is the difference between public health, occupational health, and environmental health? Should they be combined into one discipline? Explain.
2. Are there any locations on earth that are free of natural or human-made toxins?
3. Do you believe that the physical and psychological stressors in the workplace have the same effect on human well-being as other aspects outside of it? Explain.

Note

1. Prevalence of Positive Skin Test Responses to 10 Common Allergens in the U.S. Population: Results from the Third National Health and Nutrition Examination Survey." *Jour. Allergy Clinical Immunol.* 2005.

References and Recommended Reading

Bigby, M.E., K.A. Arndt, and S.A. Coopman, 2000. Skin Disorders. *In* B.S. Levy and D.H. Wegman Occupational Health, 4th ed. Philadelphia: Lippincott Williams & Wilkins.

Clifton, J.C., 2007. Mercury exposure and public health. *Pediatr Clin North Am* 54 (2): 237–69.

Dorland's Medical Dictionary. 2007. Philadelphia: Saunders Publishing.

EHPC, 1998. An Ensemble of Definitions of Environmental Health. Accessed 02/09/12 at http://www.health.gov/environemtnt/DefinitionsofEnvHeatlth/edh-der2.htm.

Friis, R.H., 2012. *Essentials of Environmental Health*, 2nd ed. Sudbury, MA: Jones & Bartlett.

ILO/WHO, 1950. International Labour Organization/World Health Organization. *UN Proclamation*. New York: United Nations.

Koren, H., 1995. *Illustrated Dictionary and Resource Directory of Environmental & Occupational Health*, 2nd ed. Boca Raton, FL: CRC Press.

Maton, A., 1993. *Human Biology and Health*. Englewood Cliffs, New Jersey: Prentice Hall.

MedTerms, 2012. Definition of Public Health. Accessed 02/10/12 at http://www.med-terms.com/script/main/art.asp?articlekey=5120.

Moeller, D.W., 2011. *Environmental Health*, 4th ed. Cambridge, MA: Harvard University Press.

NEHA, 1996. Environmental Health. Accessed 02/09/12 at http://www.neha.org/position_papers/def_env_health.html.

NIH, 2007. NIH Publication 96-4145. http://www.niehs.nih.gov.

NIH, 2012. Environmental Diseases from A to Z. Accessed 02/10/12 at http://www.niehs.nih.goc/health/topics/atoz/index.cfm.

Queensland Health, 2011. Fact Sheet. Accessed 02/12/12 at gtto://www.health.gld.gov.au.

Ring, J., B. Przybilla, and T. Ruzicka, 2006. *Handbook of atopic eczema*, 2nd ed. New York: Springer.

USCS, 2008. Skin Cancer. Accessed 02/12/12 at http://apps.nccd.cdc.gov/uscs/.

WHO, 2012. Definition of Health. World Health Organization. Accessed 02/09/12 at http://apps.who.inst/aboutwho/en/defintion.html.

2

The Environment and Ecology

Our environment affects our health. Our average life span has almost doubled over the past century or so mainly because we have clean, safe drinking water. If parts of the environment, like air and water or soil become polluted, it can lead to health problems. For example, asthma attacks can result from pollutants and other chemicals in the air and in the home.

—NIH: National Institute of Environmental Health Sciences, 2012

Notice: USDA Food Safety and Inspection Service Texas Farm Recalls Cobb Salads Due to Possible Listeria Contamination.

GH Foods SW, a Houston, Texas, establishment, is recalling approximately 515 pounds of Cobb salad products. The salads contain eggs that are the subject of a Food and Drug Administration (FDA) recall due to concerns about contamination with *Listeria monocytogenes*.

Killing the Wolf

We were eating lunch on a high rimrock, at the foot of which a turbulent river elbowed its way. We saw what we thought was a doe fording the torrent, her breast awash in white water. When she climbed the bank toward us and shook out her tail, we realized our error: it was a wolf. A half-dozen others, evidently grown pups, sprang from the willows and all joined in a welcoming melee of wagging tails and playful maulings. What was literally a pile of wolves writhed and tumbled in the center of an open flat at the foot of our rimrock.

> In those days we had never heard of passing up a chance to kill a wolf. In a second we were pumping lead into the pack, but with more excitement than accuracy; how to aim a steep downhill shot is always confusing. When our rifles were empty, the old wolf was down, and a pup was dragging a leg into impassable side-rocks.
>
> We reached the old wolf in time to watch a fierce green fire dying in her eyes. I realized then, and have known ever since, that there was something new to me in those eyes—something known only to her and the mountain. I was young then, and full of trigger-itch; I thought that because fewer wolves mean more deer, that no wolves would mean hunters' paradise. But after seeing the green fire die, I sensed that neither the wolf nor the mountain agreed with such a view. (Aldo Leopold, 1948)

Leopold's account of killing wolves, and by extension of the entire species, has a profound impact on our surroundings, on our environment, on our river of life; we—as representatives of but a few of the organisms who depend on the river for life, and for survival—are tied to it, component and portion of the environment that we influence and are influenced by. The point is that what we do to one part of our environment has a profound effect on another part of the environment. The killing off of wolves, a natural predator, affects the balance of nature. Leopold made this point clear in his later statement: "I have watched the face of many a newly wolf-less mountain, and seen the south-facing slopes wrinkle with a maze of new deer trails. I have seen every edible bush and seedling browsed, first to anemic desuetude, and then to death" (Leopold, 1948, p. 132).

This text examines four specific environmental areas: air, water, soil, and biota. However, our principal focus is on the first three—because without safe air, water, or soil, no viable biota can exist. Without them the planet would be a sterile hunk of orbiting rock. Without air, water, and soil, nothing is left that we can—or could—relate to.

We damage the environment with every breath we take; with every glass of water we drink; with every food and non-food (e.g., forest trees) crop we plant. We damage the environment through living, through use, misuse, and abuse of technology.

In order to ensure good health and safety we must avoid human illness and injury not only by learning to live in our environment but also by learning to efficiently use it. Accordingly, it is helpful to subscribe to and follow the sage advice of Koren and Bisesi (2003): "Environmental health and safety is the art and science of protecting human function; promoting aesthetic values; and preventing illness and injury through the control of positive environmental facts and the reduction of potential physical, biological, and chemical hazards" (p. 1).

Technology and the Environment

Frequently we use technological advances before we fully understand their long-term effects on the environment—we dam a river before we think about the salmon. We weigh the advantages a technological advance can give us against the environment and discount the importance of the environment, through greed, through hubris, or simply through lack of knowledge of what that new technology will do to our environment. We often only examine short-term plans without fully developing how problems may be handled years later. We assume that later, when the situation becomes critical, technology will be there to fix it. Scientists will be able to figure it out; we believe this, and in turn, we ignore the immediate and long-term consequences of our technological abuse.

Consider this: While technological advances have provided us with nuclear power, the lightbulb and its energy source, plastics, the internal combustion engine, air-conditioning and refrigeration, genetic engineering, stem cell treatment, DNA analysis, artificial intelligence, machine vision, computers, LEDs (light emitting diodes), digital cameras, iPods, cell phones, BlackBerry phones, GPS devices, flat screen TVs, and scores of other advances that make our modern lives pleasant and comfortable, many of these advances, especially when not properly used and/or disposed of (e.g., nuclear waste), have affected the earth's environment in ways we did not expect, in ways we deplore, and in ways we may not be able to live with. Note, however, that technology is a double-edged sword. That is, on one side it can do great damage to the environment; on the other edge it can remediate or prevent environmental damage. It is only when we presume that technology is the answer to all problems—including human-made problems—that we let the genie out of the bottle and let her go in whichever direction she chooses.

In this chapter, we set out on a journey that begins with the basics, the building blocks of environmental health that enable us to pursue and understand difficult concepts. We develop clearer perceptions about ideas that some people view as controversial concerns (although almost everyone will agree that maintaining our well-being has impacted our environment in such a manner as to make it not as pristine and unspoiled as we want—and/or expect it to be) that impact us all. Most importantly, we offer insights that will enable you to make up your mind on what needs to be done, on what needs to be undone, and on what needs to be mankind's focus on maintaining life as we know, as we want, as we deserve it to be. That is, maintaining our health and safety while at the same time maintaining a healthy and relatively safe environment (remember, we do not control Mother Nature). Have a smooth, healthy, safe, and enlightening journey.

What Is the Environment?

When we say the "environment," what do we mean exactly? Think about it. Environment can mean many different things to many different people. For example, some may view "environment" as the office environment, the creative environment, the learning environment, the corporate environment, the virtual environment, the aquatic environment, the tropical environment, the social environment, or the conservation environment. Or, in this digital age, maybe we are referring to the desktop environment, the integrated development environment, the runtime environment, and so forth. Obviously, when we use the term "environment" we need to be more specific. In this text, we are specific, actually more specific by defining the *environment* as the natural environment, which encompasses all living and nonliving (such as air, soil, and water) things that influence organisms.

Did You Know?

The terms "environmental science" and "ecology" are often used interchangeably, but technically "ecology" refers only to the study of organisms and their interactions with each other and their environment. Ecology could be considered a subset of environmental science. In practice, there is considerable overlap between the work of ecologists and other environmental scientists.

The Environment and Environmental Health Connections

Environmental health addresses all human-health-related aspects of both the natural environment and the human-made environment. Environmental health concerns are shown in figure 2.1.

Many environmental health specialists have widely varying, diverse backgrounds, training, and experience. A well-trained environmentalist is a generalist, trained as a biologist, an epidemiologist, toxicologist, ecologist (formerly known as natural scientist), geologist, environmental engineer—and in many related areas. While practitioners of environmental health should be generalists (i.e., they have a broad view based on knowledge gained from several subject areas, including the arts), and while they may have concentrated their study on a particular specialty, solidly trained environmental health specialists have one thing in common: they are well grounded in several different branches of science.

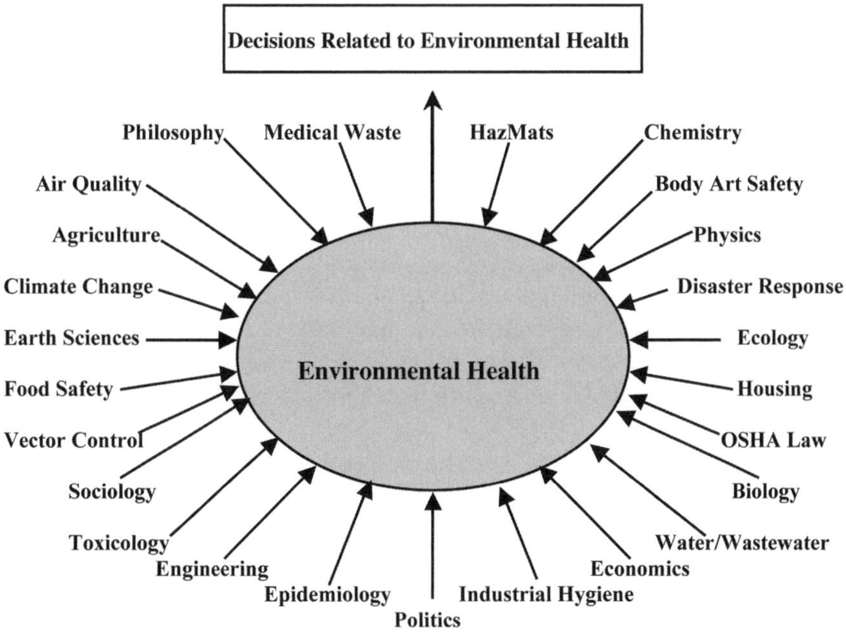

FIGURE 2.1
Components of environmental health.

Did You Know?

An academic program in environmental health focuses on the application of biological, chemical, and physical principles to the study of the physical environment and its interface with human health.

In its broadest sense, environmental health also encompasses the social and cultural aspects of the environment. As a mixture of several traditional sciences, political awareness, and societal values, environmental health demands examination of more than the concrete physical aspects of the world around us—and many of those political, societal, and cultural aspects are far more slippery than what we can prove as scientific fact.

Did You Know?

In the early 1980s the California Waste Management Board was concerned about siting municipal landfills. It contracted with a consulting firm to help it determine how to successfully site landfills. The consultant's report suggested siting landfills in minority neighborhoods because these individuals were less apt to know how to stop the building of landfills in their neighborhoods.

In short, we can accurately say that environmental health, like medicine itself, is a pure science, because it includes study of all the mechanisms of environmental and medical processes: *the study of the air, water, soil, and health.* But it is also an applied science, because it examines problems with the goal of contributing to their solution: *the study of the effects of technology thereon.* As mentioned, to solve environmental problems and understand the issues, environmental health practitioners need a broad base of information from which to draw.

The environment in which we live has been irreversibly affected by advancements in technology—and has been affected for as long as humans have wielded tools to alter their circumstances. We will continue to alter our environment to suit ourselves as long as we remain a viable species, but to do so wisely, we need to closely examine what we do and how we do it. We must build a bridge between science and technology: on one side, science, and on the other side, technology.

Why is the study and practice of environmental health necessary? In regard to the environment, we can all see the signs of decay around us. You need not study nor be a neurosurgeon to see the air we breathe is filthy; the water we drink has a foul odor and taste; the landfill (not in my backyard!—NIMBY) is overflowing; the trees in the forest are diseased and dying or dead and decayed; the local beach is a trash heap; a composite of sounds range from blasting stereo systems to the roar of supersonic transport jets—why do we need environmental health to tell us that? We know the basic requirements for a healthy environment: clean air, safe and sufficient water, safe and adequate food, safe and peaceful settlements, and a stable global environment (Yassi et al., 2001).

We need environmental health (and science in general), first for quantitative analysis. We use the precepts of environmental health and science to obtain basic information of existing conditions of air, water, and soil. We need to know what causes the problem and to define its severity. It's a matter of using the six step problem-solving paradigm (used in Risk Assessment/Management Practice; figure 2.2).

We also need the practice of environmental health and science to show us the hidden problems—the ones we can't see: the lake that appears normal on the surface but, in fact, is sterile from acid rain (and yet may remain a source of drinking water). We rely on scientific measurements and computer models

Define → Measure → Understand → Intervene → Set Policy → Implement and Evaluate

FIGURE 2.2
Problem-solving paradigm.

to help us define, understand, and affect change on these problems. The environmental challenges we face today are often less visible to the unaware, and more global in scope—and have longer response times. We can solve these problems only through scientific methods.

Let's pause a moment. Are we making too big of a deal about the influence of the environment on our health? No, not really. The physical environment, our habitat, is the most important determinant of human health. Moreover, protection of the environment and preservation of ecosystems are the most fundamental steps in preventing human illness. Keep in mind that environmental problems are global and long term. And a huge contributor to the problem is human belief systems—our belief systems.

Are you now convinced that environment influences our health? No, maybe our health is intrinsically genetic? Maybe it is age and time that influence our health and the onset of disease? Well, all these factors play a role in the status of our health, of course. But it would be prudent of us to remember those words of Judith Stern: "Genetics loads the gun, but environment pulls the trigger."

Fast-forward to the present. We use technology to address environmental problems of air and water quality, of soil contamination, and of solid and hazardous waste—to clean up the environment.

Technology? Didn't we say that technology caused many of our present environmental problems in the first place? Why do we want to make a bad situation worse by throwing more technology at it?

For better or worse, technology has changed our environment. And while technology has contributed to our environmental problems, remember that the human element must bear the brunt of the blame. People must use technology to repair the damage, as well. For example, water and wastewater treatment technologies have made enormous strides in the task of purifying the water we drink (water treatment technology) and treating the water we waste (wastewater treatment technology). Moreover, like it or dislike it, it is technology and science that will develop the cure for diabetes, heart disease, Parkinson's disease, cancer, and the rest of the deadly diseases.

Whether we agree or disagree with the advantages or disadvantages of technology, to sustain life on our planet, we must learn that the marriage between environmental health/science and technology is not only compatible—but also critically important. In short, we simply can't mitigate the anthropogenic effects on the natural environment, which, in turn, can lead to health problems, without using technology. What are the issues environmental scientists use technology and technological advances to mitigate? The following is a short list:

- Anoxic waters
- Climate change
- Energy
- Environmental degradation
- Environmental health
- Genetic engineering
- Land degradation
- Nanotechnology
- Nuclear issues
- Overpopulation
- Ozone depletion
- Pollution
- Resource depletion

The Good Life versus Environmental Health

(Note: Information in this section is based on information from F. R. Spellman and N. Whiting [2006]. *Environmental Science & Technology: Concepts and Applications,* 2nd ed., Lanham, MD: Government Institutes Press.) As long as capitalism drives most modern economies, people will desire material things—precipitating a high level of consumption. For better or for worse, the human desire to lead the "good life" (which Americans may interpret as a life enriched by material possessions) is a fact of life. Arguing against someone who wants to purchase a new, modern home with all the amenities, or who wants to purchase the latest, greatest automobile, is difficult. Arguing against the person wanting to make a better life for his or her children, by making sure they have all they need and want to succeed in their chosen pursuit, is even harder. How do you argue against such goals with someone who earns his or her own way and spends his or her hard-earned money at will? Look at the trade-offs, though. The trade-off often affects the environment. That new house purchased with hard-earned money may sit in a field of radon-rich soil or on formerly undeveloped land. That new SUV may get only eight miles to the gallon. The boat they use on weekends gets even worse gas mileage, and it exudes wastes into the local lake, river, or stream. Their weekend retreat on the five wooded acres is part of the watershed of the local community, and it disturbs breeding and migration habitats for several species.

The environmental trade-offs never enter the average person's mind. Most of us don't think much about the environment until we damage it, until it becomes unsightly, until it is so fouled that it not only offends us but literally reaches out and impacts our health. People can put up with a lot of environmental abuse, especially with our surroundings—until the surroundings no longer please us; until it makes us sick or worse. We treat our resources the same way. How often do we think about the air we breathe, the water we drink, the soil in which our agribusiness conglomerates plant our vegetables? Not often enough.

Resource utilization and environmental degradation are tied together. While people depend on resources and must use them, this use impacts the environment. A *resource* is usually defined as anything obtained from the physical environment that is of use to man. Some resources, such as edible growing plants, water (in many places), and fresh air, are directly available to man. But most resources, like coal, iron, oil, groundwater, game animals, and fish, are not. They become resources only when man uses science and technology to find them, extract them, process them, and convert them, at a reasonable cost, into usable and acceptable forms. Natural gas, found deep below the earth's surface, was not a resource until the technology for drilling a well and installing pipes to bring it to the surface became available. For centuries, man stumbled across stinky, messy pools of petroleum and had no idea of its potential uses or benefits. When its potential was realized, man exploited petroleum by learning how to extract it and convert (refine) it into heating oil, gasoline, sulfur extract, road tar, and other products.

Earth's natural resources and processes that sustain other species and us are known as Earth's Natural Capital. This includes air, water, soil, forests, grasslands, wildlife, minerals, and natural cycles. Societies are the primary engines of resource use, converting materials and energy into wealth, delivering goods and services, and creating waste or pollution. This provision of necessities and luxuries is often conducted in ways that systematically degrade the earth's natural capital—the ecosystems that support all life.

Excluding *perpetual resources* (solar energy, tides, wind, and flowing water), two different classes (types) of resources are available to us: renewable and nonrenewable (see figure 2.3). *Renewable resources* (fresh air, fresh water, fertile soil, plants, and animals, via genetic diversity) can be depleted in the short run if used or contaminated too rapidly, but normally will be replaced through natural processes in the long run. Because renewable resources are relatively plentiful, we often ignore, overlook, destroy, contaminate, and/or mismanage them.

Mismanage? Yes. Classifying anything as "renewable" is a double-edged sword. Renewable resources are renewable only to a point. Timber or grass used for grazing must be managed for *maximum sustainable yield* (the

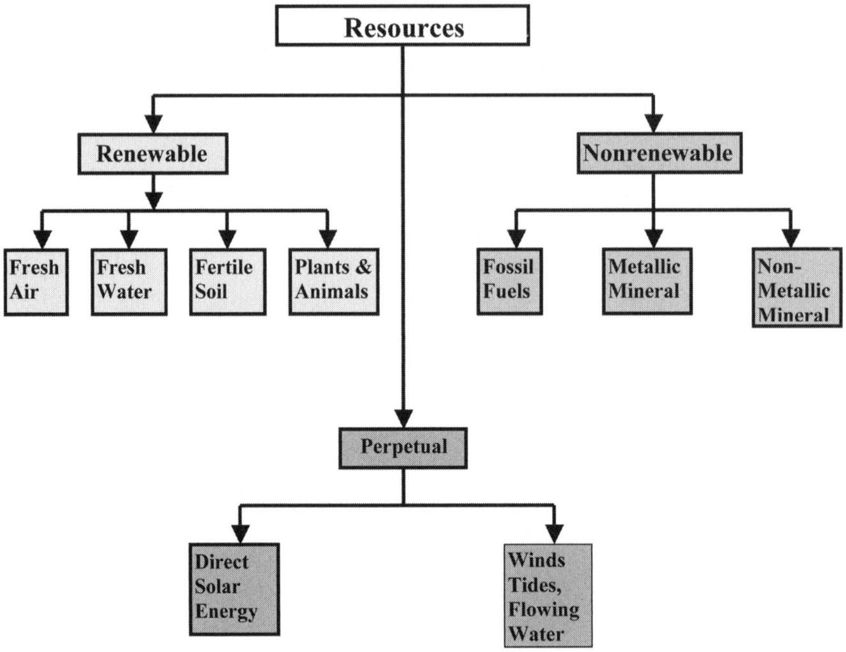

FIGURE 2.3
Major types of material resources.

highest rate at which a renewable resource can be used without impairing or damaging its ability to be fully renewed). If timber or grass yields exceed this rate, the system gives ever-diminishing returns. Recovery is complicated by the time-factor, which is life-cycle dependent. Grass can renew itself in a season or two. Timber takes decades. Any length of time is problematic when people get impatient.

Exceeding maximum sustainable yield is only the tip of the iceberg—other environmental, social, and economic problems may develop. Let's look at *overgrazing* (depleting) grass on livestock lands. The initial problem occurs when the grass and other grazing cover is depleted. But secondary problems kick in fast. Without grass, the soil erodes quickly. In very little time, so much soil is gone that the land is no longer capable of growing grass—or anything else. Productive land converted to non-productive deserts (*desertification*) is a process of *environmental degradation*—and it impacts social and economic factors. Those who depend on the grasslands must move on, and moving on costs time, energy, and money—and puts more land at risk. Should the same level of poor stewardship of land resources continue on more acreage?

Land is often environmentally degraded when a metropolitan area expands. In high-growth areas, productive land is covered with concrete, asphalt, buildings, water, or silt to such an extent that agricultural productivity declines and wildlife habitat is lost.

Nonrenewable resources (copper, coal, tin, and oil, among many others) have built up or evolved in a geological timespan. They can't be replaced at will—only over a similar time-scale. In this age of advanced technology, we often hear that (for example) when high-grade tin ore runs out (when 80 percent of its total estimated supply has been removed and used), low-grade tin ore (the other 20 percent) will become economically workable. This erroneous view neglects the facts of energy resource depletion and increasing pollution with lower-grade burdens. In short, to find, to extract, and to process the remaining 20 percent generally costs more than the result is worth. Even with unlimited supplies of energy (impossible according to the *Laws of Thermodynamics* [discussed later]), what if we could extract that last 20 percent? When it is gone, nothing is going to bring it back except time measured in centuries and millennia, paired with the elements that produce the resource.

Advances in technology have allowed us to make great strides in creating the "good life." These same technological advances have increased environmental degradation. But not all the news is bad. Technological advances have also let us (via recycling and reuse) conserve finite resources—aluminum, copper, iron, plastics, and glass, for example. *Recycling* involves collecting household waste items (aluminum beverage cans, for example) and reprocessing usable portions. *Reuse* involves using a resource over and over in the same form (refillable beverage bottles, water).

We discussed the so-called good life earlier—modern homes, luxury cars and boats, the second home in the woods. With the continuing depletion of natural resources, prices must be forced upward until economically, attaining the "good life," or even gaining a foothold toward it, becomes difficult or impossible—and maintaining it becomes precarious. Ruthless exploitation of natural resources and the environment—overfishing a diminishing species (look at countless marine species populations, for example), intense exploitation of energy and mineral resources, cultivation of marginal land without proper conservation practices, degradation of habitat by unbalanced populations or introduced species, and the problems posed by further technological advances—will result in environmental degradation that will turn the "good life" into something we don't want to even think about.

So—what's the answer? What are we to do? What should we do? Can we do anything? Some would have us all "return to nature." Those people suggest returning to Thoreau's Walden Pond on a large scale, to give up the "good life" to which we have become accustomed. They think that giving up the cars, the boats, the fancy homes, the bulldozers that make construction and farming easier, the pesticides that protect our crops, the medicines that improve our health and save our lives—the myriad material improvements that make our lives comfortable and productive—will solve the problem. Is this approach the answer—or even realistic? To a small minority, it is—although for those who realize how urban Walden Pond was, the idea is amusing.

To the rest of us? It cannot. It should not. And it will not happen. We can't abandon ship—we must prevent the need for abandoning our society from ever happening. Technological development is a boon to civilization, and will continue to be. Technological development isn't the problem—improper use of technology is. But we must continue to make advances in technology, we must find further uses for technology, and we must learn to use technology for the benefit of mankind and the environment. Technology and the environment must work hand in hand, not stand opposed. We must also foster respect for, and care for, what we have left.

Just how bad are the problems of technology's influence on environment?

Major advances in technology have provided us with enormous landscape transformation and pollution of the environment. While transformation is generally glaringly obvious (damming a river system, for example), "polluting" or "pollution" is not always as clear. What do we mean by pollution? To *pollute* means to impair the purity of some substance or environment. *Air pollution* and *water pollution* refer to alteration of the normal compositions of air and water (their environmental quality) by the addition of foreign matter (gasoline, sewage). Ways technology has contributed to environmental transformation and pollution include:

- Extraction, production, and processing of raw natural resources, such as minerals, with accompanying environmental disruption.
- Manufacturing enormous quantities of industrial products that consume huge amounts of natural resources and produce large quantities of hazardous waste and water/air pollutants.
- Agricultural practices resulting in intensive cultivation of land, irrigation of arid lands, drainage of wetlands, and application of chemicals.
- Energy production and use accompanied by disruption and contamination of soil by strip mining, emission of air pollutants, and pollution of water by release of contaminants from petroleum production, and the effects of acid rain.
- Transportation practices (particularly reliance on automobiles) that cause loss of land by road and storage construction, emission of air pollutants, and increased demand for fuel (energy) resources.

- Transportation practices (particularly reliance on the airplane) that cause scarring of land surfaces from airport construction, emission of air pollutants, and greatly increased demands for fuel (energy) resources.

Environmental health encompasses an array of pollutants, or, more properly, agents that may cause acute or chronic health effects in the population. Thus, it is important to recognize pollutant sources and pathways to human receptors. First, we must recognize those agents that may affect health. For example:

- Ambient air pollutants (e.g., toxic chemicals, particulate matter, and ozone)
- Lead-based paint
- Foodborne pathogens (e.g., *Escherichia coli* 0157:H7)
- Indoor air pollutants (e.g., carbon monoxide, tobacco smoke, and molds)
- Disinfection by-products in drinking water
- Hazards in the work setting
- Stressors that cause injury (e.g., automobiles)
- Pesticide residues in food

Pollutant source pathways emanate from the pollutant source itself; enter air, water, plant, and soil; and are passed on to human receptors via inhalation, ingestion, and dermal exposure (absorption).

Did You Know?

When you throw a stone into a pool of quiet water, the ensuing ripples move out in concentric circles from the point of impact. Eventually, those ripples, much dissipated, reach the edge of the pond, where they break, disturbing the shore environment. When we alter our environment, even in the most subtle fashion, similar repercussions affect the world around us—and some of these actions can—or will be—felt across the world. We use technology to alter our environment to suit our needs. That same technology can be put into effect so that our environment is protected from unrecoverable losses. Environmental health practitioners must maintain an acute sense of awareness concerning the global repercussions of problems we create for the environment—to extend the boundaries of the problem beyond our own backyard.

Ecological Concepts

(Note: Information from this section is from F. R. Spellman [2008]. *Ecology for Non-Ecologists*. Lanham, MD: Government Institutes Press.) Now we shift gears to discuss fundamental concepts foundational to more complex material that follows in subsequent chapters. We also provide understanding of the relationship of the environment to humans and how to protect humans from illness and injury. The first area discussed deals with the basics involved with the circulation of matter through the ecosystem—the *biogeochemical cycles.* Because biogeochemical cycles (and most other processes on earth) are driven by energy from the sun, we present energy and energy transfer next. We also provide a brief discussion of the basics of ecology, including productivity, population, and succession.

Biogeochemical Cycles

To live, grow, and reproduce, the nutrient atoms, ions, and molecules that organisms need are continuously cycled from the nonliving (abiotic) environment to living organisms (biotic), then back again. This takes place in what are called biogeochemical cycles (nutrient cycles)—literally, life-earth-chemical cycles. The cycle generally describes the physical state, chemical form, and biogeochemical processes affected by the substance at each point in the cycle in an undisturbed ecosystem. Many of these processes are influenced by microbial populations that are naturally adapted to life in aerobic (oxygenated) or anaerobic (oxygen-free) conditions. Because both of these conditions are readily created by varied and fluctuating water levels, wetlands support a greater variety of these processes than other ecosystems (USDA FS, 1995).

To understand our physical world, you must understand the natural biogeochemical cycles that take place in our environment. Biogeochemical cycles are categorized into two types, the *gaseous* and the *sedimentary.* Gaseous cycles include the carbon and nitrogen cycles. The atmosphere and the ocean are the main sinks (storage sites, or reservoirs) of nutrients in the gaseous cycle. The sedimentary cycles include the sulfur and phosphorous cycles. Soil and the rocks of the earth's crust are the main sinks for sedimentary cycles. These cycles are ultimately powered by the sun, and fine-tuned and directed by energy expended by organisms. Another important cycle, the *hydrological cycle* (discussed later), is also solar-powered and acts like a continuous conveyor system that moves materials essential for life through the ecosystem.

Between twenty and forty elements of the earth's ninety-two naturally occurring elements are ingredients that make up living organisms. The chemical elements carbon, hydrogen, oxygen, nitrogen, and phosphorous are critical in maintaining life as we know it on earth. Of the elements needed by living organisms to survive, oxygen, hydrogen, carbon, and nitrogen are needed in larger quantities than some of the other elements. The point is—no matter what elements are needed to sustain life, these elements exhibit definite biogeochemical cycles. For now, let's cover the life-sustaining elements in greater detail.

The elements needed to sustain life are products of the global environment. The global environment consists of three main subdivisions:

Hydrosphere—includes all the components formed of water bodies on the earth's surface.

Lithosphere—comprises the solid components, such as rocks.

Atmosphere—the gaseous mantle that envelops the hydrosphere and lithosphere.

To survive, organisms require inorganic metabolites from all three parts of the biosphere. For example, the hydrosphere supplies water as the exclusive source of needed hydrogen. The lithosphere provides essential elements (calcium, sulfur, and phosphorus). Finally, the atmosphere provides oxygen, nitrogen, and carbon dioxide.

Within the biogeochemical cycles, all the essential elements circulate from the environment to organisms and back to the environment. Because these elements are critically important for sustaining life, you can easily understand why the biogeochemical cycles are readily and realistically labeled *nutrient cycles*.

Through these biogeochemical (or nutrient) cycles, nature processes and reprocesses the critical life-sustaining elements in definite inorganic-organic phases. Some cycles (the carbon cycle, for example) are more perfect than others—that is, the cycle loses no material in the process for long periods of time. Others are less perfect, but one essential point to keep in mind is that energy flows through an ecosystem (we'll explain how later), but nutrients are cycled and recycled.

Because humans need almost all the elements in our complex culture, we have sped up the movement of many materials so that the cycles tend to become imperfect, or what Odum (1971) calls *acyclic*. One example of a somewhat imperfect (acyclic) cycle is demonstrated by man's use of phosphate, which, of course, affects the phosphorus cycle. Phosphate rock is mined and processed with careless abandon, which leads to severe local pollution near mines and phosphate mills. We also increase the input of phosphate fertilizers in agricultural systems without controlling in any way the inevitable increase in runoff output. This severely stresses our waterways and reduces water quality through *eutrophication*, the natural aging of a land-locked body of water.

In agricultural ecosystems, we often supply necessary nutrients in the form of fertilizer to increase plant growth and yield. In natural ecosystems, however, these nutrients are recycled naturally through each trophic level (feeding level). Plants take up elemental forms. The consumers ingest these elements in the form of organic plant material. They cycle through the food chain from producer to consumer, and eventually, the nutrients are degraded back to the inorganic form again.

The following sections present and discuss the nutrient cycles for carbon, nitrogen, phosphorus, and sulfur.

Carbon Cycle

Carbon, an essential ingredient for all living things and the basic building block of the large organic molecules necessary for life (carbohydrates, fats, proteins, DNA, and others), is cycled into food webs from the atmosphere (see Figure 2.4).

Green plants obtain carbon dioxide (CO2) from the air (figure 2.4), and through photosynthesis—probably the most important chemical process on earth—produce the food and oxygen upon which all organisms live. Part of the carbon produced remains in living matter; the other part is released as CO2 in cellular respiration, where it is then returned to the atmosphere.

Some carbon is contained in buried dead animal and plant materials. Over the course of eons, much of these buried plant and animal materials was transformed into fossil fuels (coal, oil, and natural gas), which contain large amounts of carbon. When fossil fuels are burned, stored carbon combines with oxygen in the air to form carbon dioxide, which enters the atmosphere.

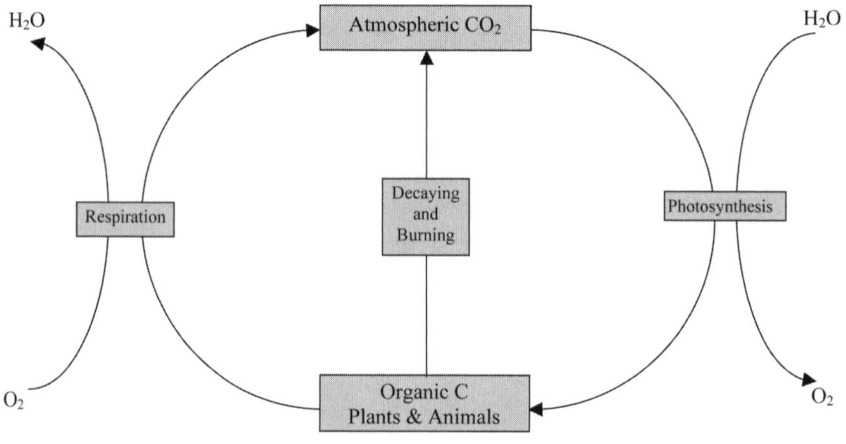

FIGURE 2.4
Carbon cycle.

Did You Know?

Major sinks are the atmosphere (CO_2, actually only a small amount of carbon is stored here), the lithosphere (carbonate rocks and fossil fuels, the largest sink), the hydrosphere (CO_2, carbonate, and bicarbonate molecules in water), and as biomass (carbon-based organic molecules) in the ecosphere.

Earth's heat is radiated into space. This balance is important. As more carbon dioxide is released into the atmosphere, that balance can and is altered. Massive increases of carbon dioxide into the atmosphere tend to increase the possibility of global warming. The consequences of global warming might be catastrophic, and the resulting climate change may be irreversible.

Nitrogen Cycle

The atmosphere contains 78 percent by volume of nitrogen. Nitrogen, an essential element for all living matter, constitutes 1 to 3 percent of the dry weight of cells, yet nitrogen is not a common element on earth. Although it is an essential ingredient for plant growth, nitrogen is chemically very inactive, and before the vast majority of the biomass can incorporate it, it must be fixed.

Though nitrogen gas does make up about 78 percent of the volume of the earth's atmosphere, in that form it is useless to most plants and animals. Fortunately, nitrogen gas is converted into compounds containing nitrate ions, which are taken up by plant roots as part of the nitrogen cycle (shown in simplified form in figure 2.5).

Aerial nitrogen is converted into nitrates mainly by microorganisms, bacteria, and blue-green algae. Lightning also converts some aerial nitrogen gas into forms that return to the earth as nitrate ions in rainfall and other types of precipitation. Ammonia plays a major role in the nitrogen cycle (see Figure 2.5). Excretion by animals and aerobic decomposition of dead organic matter by bacteria produces ammonia. Ammonia, in turn, is converted by nitrification bacteria into nitrites, then into nitrates. This process is known as *nitrification*. Nitrification bacteria are *aerobic*. Bacteria that convert ammonia into nitrites are known as *nitrite bacteria* (*Nitrosococcus* and *Nitrosomonas*); they convert nitrites into nitrates and nitrate bacteria (*Nitrobacter*).

Did You Know?

Most nitrogen in the atmosphere is in the form of N_2. This gaseous molecule is not reactive and, therefore, needs to be converted to a usable,

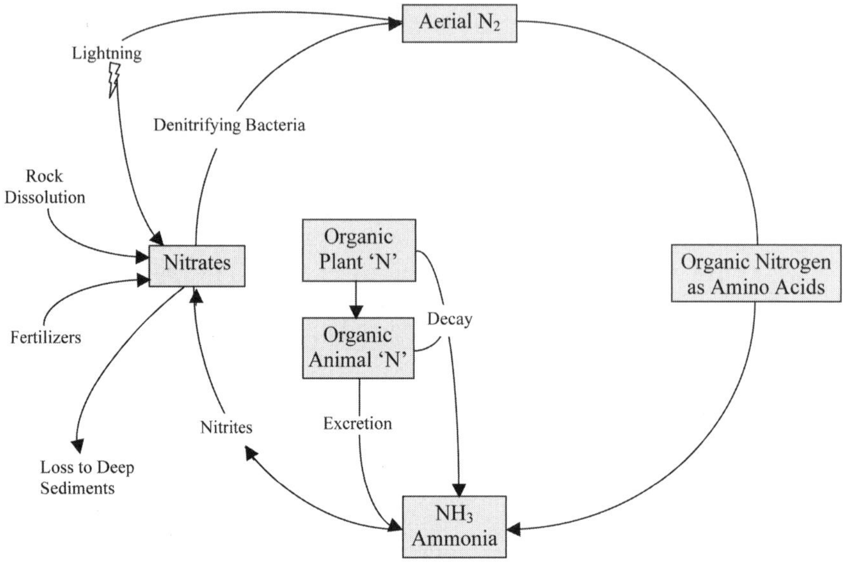

FIGURE 2.5
The nitrogen cycle.

> reactive form. This is accomplished by nitrogen fixation (by lightning and bacteria). In turn, ammonia (NH_3) and ammonium (NH_4) are formed; then they become nitrite (NO_2) and nitrate (NO_3). Nitrate is easily used by plants and then by consumers and returned to the atmosphere in the form of N_2 by ammonification (by decomposers) and then denitrification (by bacteria) processes.

Because nitrogen is often a *limiting factor* in naturally occurring soil, it can inhibit plant growth. Nitrogen is removed from topsoil when we harvest nitrogen-rich crops, irrigate crops, and burn or clear grasslands and forests before planting crops. To increase yields, farmers often provide extra sources of nitrogen by applying inorganic fertilizers or by spreading manure on the field and relying on the soil bacteria to decompose the organic matter and release the nitrogen for plant use.

Phosphorus Cycle

Phosphorus is another element (P) common in the structure of living organisms. Phosphorus circulates through water, the earth's crust, and living organisms in the phosphorus cycle (see figure 2.6). The ultimate source of

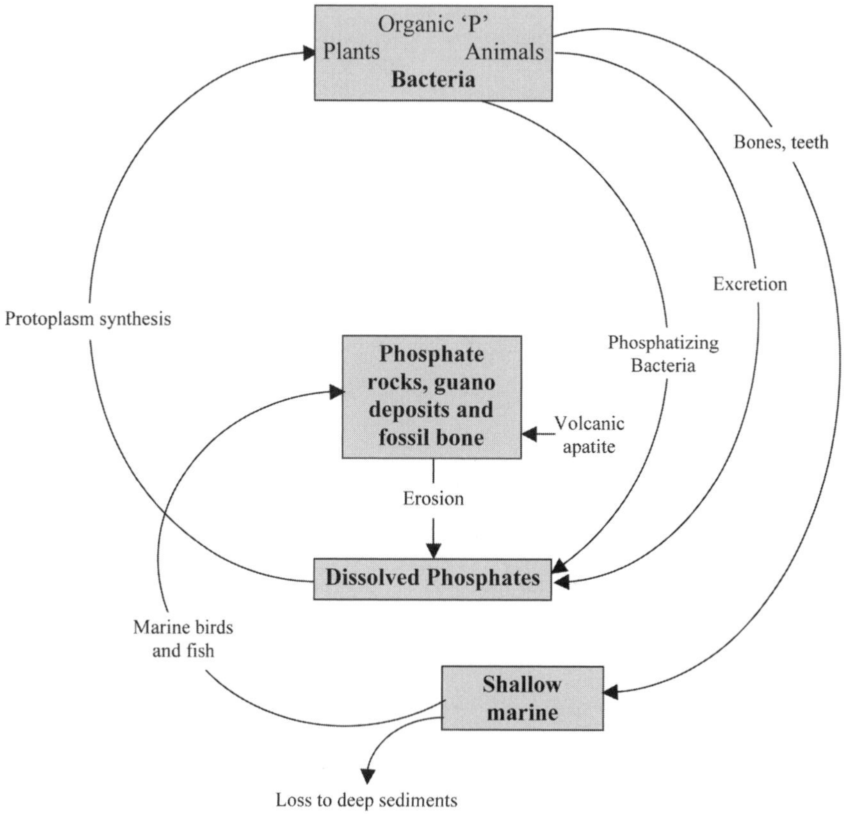

FIGURE 2.6
The phosphorus cycle.

phosphorus is rock (figure 2.6). Phosphorus occurs as phosphate or other minerals formed in past geological ages. It is often stored for long periods of time (millions of years) in phosphate rocks. These massive deposits are gradually eroding, providing phosphorus to various ecosystems. A large amount of eroded phosphorus ends up in deep sediments in the oceans and in lesser amounts in shallow sediments. Some phosphorus reaches land when marine animals are brought out. Birds also play a role in phosphorus recovery. The great guano deposit (bird excreta) of the Peruvian coast is an example. Humans have hastened the rate of phosphorus loss through mining and the production of fertilizers, which are washed away and lost.

Phosphorus has become very important in water quality studies, since it is often a limiting factor. Phosphates, upon entering a stream, act as fertilizer, which promotes the growth of undesirable algae blooms. As the organic matter decays, dissolved oxygen levels decrease, and fish and other aquatic species die, limiting producer populations in freshwater systems.

Sulfur Cycle

Sulfur, like nitrogen, is characteristic of organic compounds. Much of it is stored in the lithosphere as sulfide and sulfate minerals. The sulfur cycle (see Figure 2.7) is both sedimentary and gaseous. Bacteria play a major role in the conversion of sulfur from one form to another. In an *anaerobic* environment, bacteria break down organic matter, thereby producing hydrogen sulfide with its characteristic rotten-egg odor. Bacteria called *Beggiatoa* converts hydrogen sulfide (H_2S) into elemental sulfur (S). An aerobic sulfur bacterium, *Thiobacillus thiooxidans*, converts sulfur into sulfates. Other sulfates are contributed by the dissolving of rocks and some sulfur dioxide (SO_2) during volcanic eruptions. Sulfur is incorporated by plants into proteins. Some of these plants are then consumed by organisms. Sulfur from proteins is liberated by many heterotrophic anaerobic bacteria, as hydrogen sulfide. In the atmosphere, SO_2 reacts with water to form sulfuric acid (H_2SO_4) and acid rain.

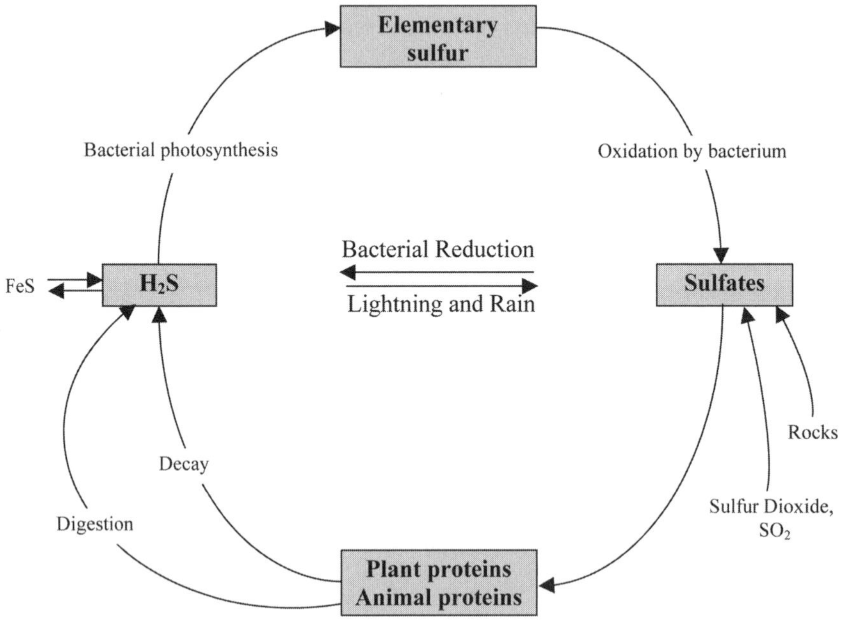

FIGURE 2.7
The sulfur cycle.

Energy Flow through an Ecosystem and the Biosphere

We often take energy for granted through a deceptive familiarity, because we think of it in so many different ways: oil-derived energy, atomic energy, alternative energy, renewable energy, food energy, cheap energy, expensive energy, abundant energy, energy shortages, and so on. This presents a huge double irony, because on one hand most people know that without energy our energy-dependent industrialized society (i.e., residential, commercial, industrial, and transportation energy use) would grind to a halt. On the other hand, energy is more than just the force that powers our machines, our civilization—it powers hurricanes, the movement of the planets—the entire universe. Despite its pervasiveness and its familiarity, energy is a complex and puzzling concept. It cannot be seen, tasted, smelled, or touched.

Key Terms Defined

Energy—is the ability or capacity to do work. Energy is degraded from a higher to a lower state.

First Law of Thermodynamics—states that energy is transformed from one form to another, but is neither created nor destroyed. Given this principle, we should be able to account for all the energy in a system in an energy budget, a diagrammatic representation of the energy flows through an ecosystem.

Second Law of Thermodynamics—asserts that energy is only available due to degradation of energy from a concentrated to a dispersed form. This indicates that energy becomes more and more dissipated (randomly arranged) as it is transformed from one form to another or moved from one place to another. It also suggests that any transformation of energy will be less than 100 percent efficient (i.e., the transfers of energy from one trophic level to another are not perfect); some energy is dissipated during each transfer.

Population density—is the number of a particular species in an area. This is affected by natality (birth and reproduction), immigration (moving into), mortality (death), and emigration (moving out of).

Ultimate carrying capacity—is the maximum number of a species an area can support; the environmental carrying capacity is the actual maximum capacity a species maintains in an area. Ultimate capacity is always greater than the environmental capacity.

Systems and Throughputs

A *system* is a set of components that operate in a connected and predictable way. It is a defined, physical part of the universe (e.g., the atmosphere, a pond or lake, the human body). Large systems are often made up of many smaller systems. All environmental systems have *inputs* (matter, energy, and information), *throughputs* (flows), and *outputs* (wastes; figure 2.8). The latter may become inputs for other systems. Consider food and wastewater treatment, for example; we put food in our bodies (input), metabolize it (throughput), and expel wastes (output). This waste then becomes the input to a wastewater treatment plant or bacteria.

Materials Balance

(Material in this section from USEPA 2010, Material Balance—Air Pollution Control Module 1.) Probably the simplest way to express *materials balance*, one of the most important and fundamental scientific principles, is to point

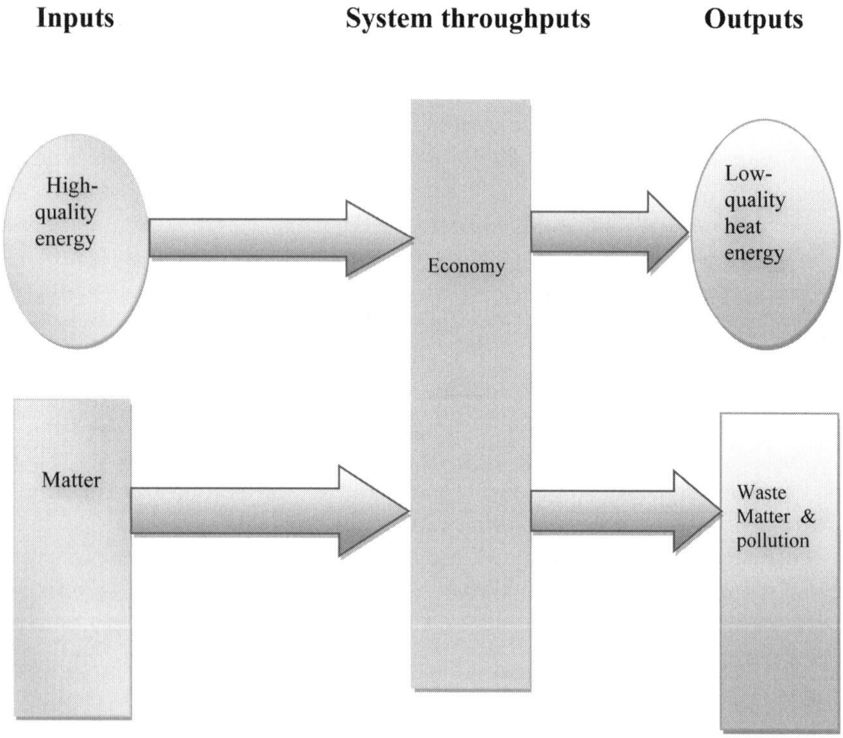

FIGURE 2.8
Any system requires inputs of energy and matter from the environment. These resources flowing through the economy are converted to low-quality heat energy, wastes, and pollutants.

out that *everything has to go somewhere*. Stated simply, a material balance means "what goes in, must come out." According to the *law of conservation of mass or matter*, when chemical reactions take place, matter is neither created nor destroyed (exception: in a nuclear reaction, mass can be converted to energy). The importance of this concept in environmental science is that it allows us to track pollutants from one location to another using *mass balance equations*. For example, in a wide variety of air pollution control calculations, material balance equations can be used to evaluate the following:

- Formation of combustion products in boilers
- Rates of air infiltration into air pollution control systems
- Material requirements for process operations
- Rate of ash collection in air pollution control systems
- Humidities of exhaust gas streams
- Exhaust gas flow rates from multiple sources controlled by a single air pollution control system
- Gas flow rates from combustion processes

Material balance, or conservation of matter, can be applied in solving problems involving the quantities of matter moving in various parts of a process, and is illustrated in example problem 2.1.

Example 2.1

Problem: This problem illustrates how a mass balance calculation can be used to check the results of an air emission test.

During an air emission test, the inlet gas stream to a fabric filter is 100,000 actual ft^3/min (ACFM) and the particulate loading is 2 grains/actual feet (ACF). The outlet gas stream from the fabric filter is 109,000 ACFM and the particulate loading is 0.025 grains/ACF (see figure 2.9). What is the

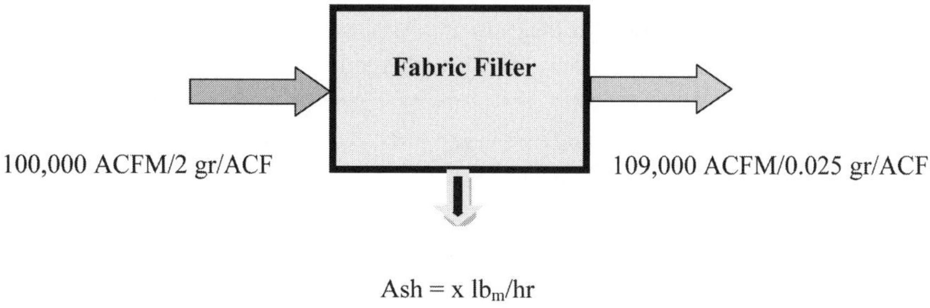

FIGURE 2.9
For example 2.1.

maximum quantity of ash that will have to be removed per hour from the fabric filter hopper based on these test results?

Solution: Based on particulate mass balance, $Mass_{(in)} = Mass_{(out)}$ / Inlet gas stream particulate = Outlet gas steam particulate + Hopper Ash.

1. Calculate the inlet and outlet particulate quantities in pounds mass per hour.

$$\text{Inlet particulate quantity} = 100{,}000\ \frac{\text{ACF}}{\text{min}} \times \left(2\ \frac{\text{gr}}{\text{ACF}}\right) \times \left(\frac{1\ \text{lb}_m}{7000\ \text{gr}}\right) \times \left(\frac{60\ \text{min}}{1\ \text{hr}}\right)$$

$$= 1{,}714.3\ \text{lb}_m/\text{hr}$$

$$\text{Outlet particulate quantity} = 109{,}000\ \text{ACF/min} \times \left(0.025\ \frac{\text{gr}}{\text{ACF}}\right) \times \left(\frac{1\ \text{lb}_m}{7000\ \text{gr}}\right) \times \left(\frac{60\ \text{min}}{1\ \text{hr}}\right)$$

$$= 23.4\ \text{lb}_m/\text{hr}$$

2. Calculate the quantity of ash that will have to be removed from the hopper per hour.

Hopper ash = Inlet gas stream particulate – Outlet gas
Stream particulate
$= 1{,}714.3\ \text{lb}_m/\text{hr} - 23.4\ \text{lb}_m/\text{hr}$
$= 1{,}690.9\ \text{lb}_m/\text{hr}$

To perform mass balance analysis, you must first define the particular region to be analyzed. The region you select could include anything—the fabric filter in example 2.1, a lake, a stretch of river or stream, an air basin above a city or factory, a chemical mixing vat, a coal-fired power plant, or the earth itself. Whatever region you select for analysis, you must confine the region with an imaginary boundary (see figure 2.10). From such a region we can begin to identify the flow of materials across the boundary as well as the accumulation of materials within the region.

When a material enters the region, it has three possible fates: Some of it may enter and slip through the region unchanged; some of it may accumulate within the boundary; and some of it may be converted—for example, CO to CO_2—to some other material. If we use figure 2.6 as a guide, a materials balance equation (equation 2.1) can be written:

$$\text{Input rate} = \text{Output rate} + \text{Decay rate} + \text{Accumulation rate} \quad (2.1)$$

Note that the decay rate in equation 2.1 does not imply a violation of the law of conservation of mass. No constraints occur on the change of one substance to another (chemical reactions), but atoms are conserved.

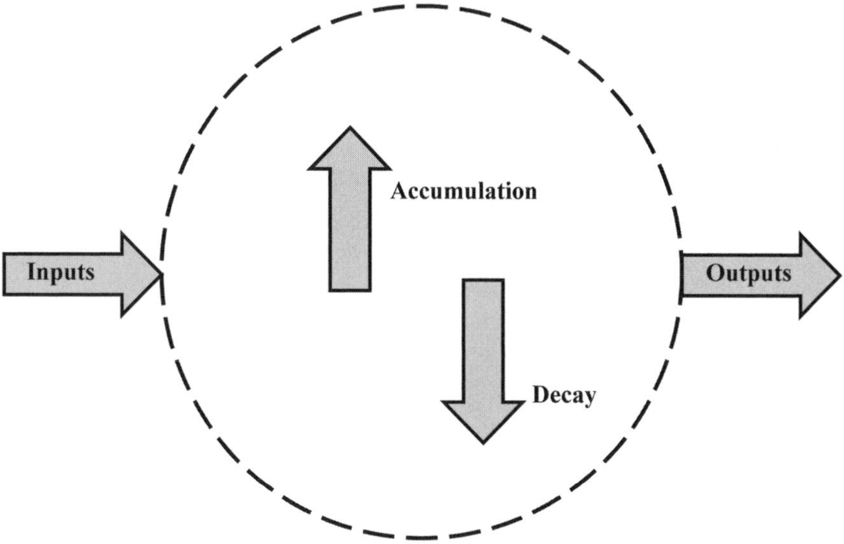

FIGURE 2.10
Materials balance diagram.

Note: In practice, equation 2.1 can be (and often is) simplified by assuming steady state-of-equilibrium conditions (that nothing changes with time), but discussion of this practice is beyond the scope of this text, and is generally presented in environmental engineering studies.

Let's get back to our discussion of energy. First of all, what is energy? *Energy* is often defined as the capacity for doing work, and work is often described as the product of force and the displacement of some object caused by that force.

Along with understanding and analyzing the flow of materials through a particular region, we can also determine and analyze the flow of energy. Using the First Law of Thermodynamics, we can write *energy balance equations*. The First Law of Thermodynamics states that energy cannot be created nor destroyed. In short, energy may change forms in a given process, but we should be able to account for every bit of energy as it takes part in the process. In simplified form, this relationship is shown in equation 2.2.

$$\text{Energy in} = \text{Energy out} \tag{2.2}$$

Equation 2.2 may give you the false impression that the transfer of energy in a process is 100 percent efficient. This is, of course, not the case. In a coal-fired electrical power generating plant, for example, only a portion of the energy from the burned coal is converted directly into electricity. A large portion of the coal-fired energy ends up as waste heat given off to the environment, because of the Second Law of Thermodynamics, which states that every process generates some waste heat; devising a process or machine that can convert heat to work with 100 percent efficiency is impossible.

Heat can be transferred in three ways: by conduction, by convection, and by radiation. When direct contact between two physical objects at different temperatures occurs, heat is transferred via *conduction* from the hotter object to the colder one. When a gas or liquid is placed between two solid objects, heat is transferred by *convection*. Heat is also transferred when no physical medium exists by *radiation* (for example, radiant energy from the sun).

Did You Know?

"The Second Law of Thermodynamics holds, I think, the supreme position among laws of nature. . . . If your theory is found to be against the Second Law of Thermodynamics, I can give you no hope."—Arthur S. Eddington

Energy Flow in the Biosphere

Energy flow in the biosphere all starts with the sun. The sun's radiant energy sustains all life on earth. The sun not only lights and warms the earth, but it provides energy used by green plants to synthesize the compounds that keep them alive. These compounds serve as food for almost all other organisms. The sun's solar energy also powers the biochemical cycles and drives climate systems that distribute heat and fresh water over the earth's surface.

Figure 2.11 reflects an important point: not all solar radiant energy reaches the earth. Approximately 34 percent of incoming solar radiation is reflected back to space by clouds, dust, and chemicals in the atmosphere and by the earth's surface. Most of the remaining 66 percent warms the atmosphere and land, evaporates water and cycles it through the biosphere, and generates winds. Surprisingly, only a small percentage (about 0.022 percent) is captured by green plants and used to make the glucose essential to life.

Most of the incoming solar radiation not reflected away is degraded (or wasted) into longer-wavelength heat (in accordance with the second law of thermodynamics) and flows into space. The actual amount of energy that returns to space is affected by the presence of molecules of water, methane, carbon dioxide, and ozone, and by various forms of particulate matter in the atmosphere. Many of these barriers are created by man-made activities and might affect global climate patterns by disrupting the rate at which incoming solar energy flows through the biosphere and returns to space. We'll discuss the possible effects of human activities on climate later.

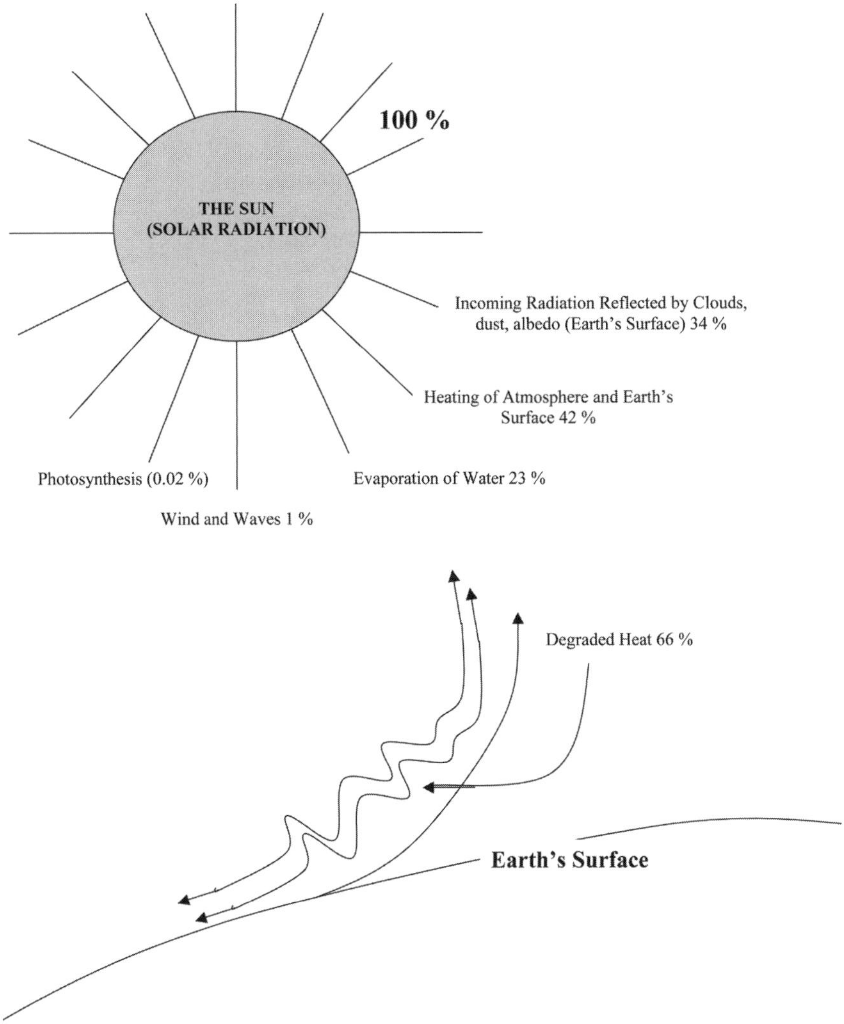

FIGURE 2.11
Flow of energy to and from the earth.

Energy Flow in the Ecosystem

For an ecosystem to exist and to maintain itself, it must have energy. All activities of living organisms involve work—the expending of energy—the degradation of a higher state of energy to a lower state. The flow of energy through an ecosystem is governed by the two laws mentioned earlier: The First and Second Laws of Thermodynamics.

Remembering that the first law, sometimes called the *conservation law*, states that energy may not be created nor destroyed, and that the second law states that no energy transformation is 100 percent efficient, sets the stage for a discussion of energy flow in the ecosystem. Hand in hand with the second law (some energy is always lost, dissipated as heat) is another critically important concept—*entropy*. Used as a measure of the non-availability of energy to a system, entropy increases with an increase in heat dissipation. Because of entropy, input of energy into any system is higher than the output or work done; the resultant efficiency is less than 100 percent.

Environmental scientists and technicians are primarily concerned with the interaction of energy and materials in the ecosystem. Earlier we discussed biogeochemical nutrient cycles and pointed out that the flow of energy drives these cycles. Energy does not cycle as nutrients do in biogeochemical cycles. For example, when food passes from one organism to another, energy contained in the food is reduced step by step until all the energy in the system is dissipated as heat. This process has been referred to as a *unidirectional flow* of energy through the system, with no possibility for recycling of energy. When water or nutrients are recycled, energy is required. The energy expended in the recycling is not recyclable. As Odum (1975) points out, this is a "fact not understood by those who think that artificial recycling of man's resources is somehow an instant and free solution to shortages" (p. 61).

Did You Know?

When an organism loses heat, it represents one-way flow of energy out of the ecosystem. Plants only absorb a small part of energy from the sun. Plants store half of the energy and lose the other half. The energy plants lose is metabolic heat. Energy from a primary source will flow in one direction through two different types of food chains. In a grazing food chain, the energy will flow from plants (producers) to herbivores, and then through some carnivores. In detritus-based food chains, energy will flow from plants through detrivores and decomposers. In terms of the weight (or biomass) of animals in many ecosystems, more of their body mass can be traced back to detritus than to living producers. Most of the time the two food webs will intersect one another. For example, in the Chesapeake Bay bass fish of the grazing food web will eat a crab of the detrital food web (Spellman, 2007).

The principal source of energy for any ecosystem is sunlight. *Producers* (green plants—flowers, trees, ferns, mosses, and algae), through the process of *photosynthesis*, transform the sun's energy into carbohydrates, which are consumed by animals. This transfer of energy, as stated earlier, is unidirectional—from producers to consumers. Often the transfer of energy to different organisms is called a *food chain*. Figure 2.12 shows a simple aquatic food chain.

All organisms, alive and dead, are potential sources of food for other organisms. All organisms that share the same general type of food in a food chain are said to be at the same *trophic level* (feeding level). Since green plants use sunlight to produce food for animals, they are called producers of the first tropic level. The herbivores eat plants directly, and are called the second trophic level, or the *primary consumers*. The carnivores are flesh-eating consumers; they include several trophic levels from the third on up. At each transfer,

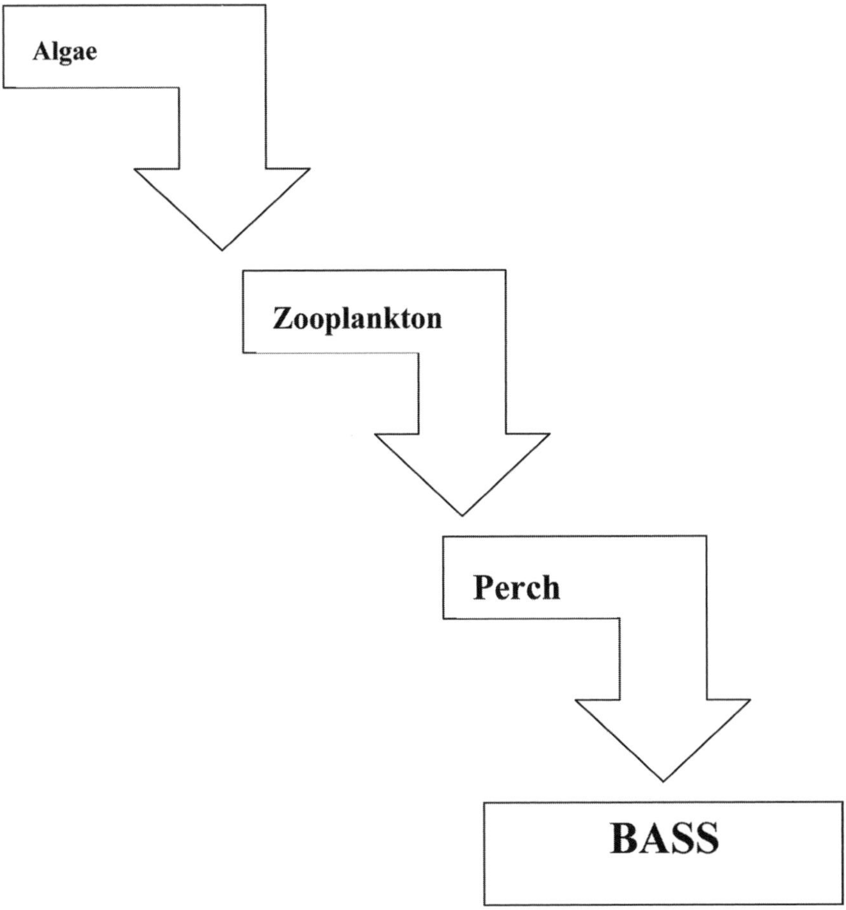

FIGURE 2.12
Aquatic food chain.

a large amount of energy (about 80 to 90 percent) is lost as heat and wastes. Nature normally limits food chains to four or five links. Note, however, that in aquatic food chains, the links are commonly longer than they are on land, because several predatory fish may be feeding on the plant consumers. Even so, the built-in inefficiency of the energy transfer process prevents development of extremely long food chains.

Only a few simple food chains are found in nature, and most are interlocked. This interlocking of food chains forms a *food web*—a map that shows what eats what. An organism in a food web may occupy one or more trophic levels. Food chains and webs help to explain how energy moves through the ecosystem.

Another important trophic level of the food web is comprised of *decomposers*. The decomposers feed on dead plants or animals and play an important role in recycling nutrients in the ecosystem. Healthy ecosystems produce no wastes. All organisms, alive or dead, are potential sources of food (and energy) for other organisms.

Ecological Pyramids

As we proceed in the food chain from the producer to the final consumer, it becomes clear that a particular community in nature often consists of several small organisms associated with a smaller and smaller number of larger organisms. A grassy field, for example, has a larger number of grass and other small plants, a smaller number of herbivores like rabbits, and an even smaller number of carnivores like foxes. The practical significance of this is that we must have several more producers than consumers.

This pound-for-pound relationship, where it takes more producers than consumers, can be demonstrated graphically by building an ecological pyramid. In an ecological pyramid, the number of organisms at various trophic levels in a food chain is represented by separate levels or bars placed one above the other with a base formed by producers and the apex formed by the final consumer. The pyramid shape is formed due to a great amount of energy loss at each trophic level. The same is true if numbers are substituted by the corresponding biomass or energy. Ecologists generally use three types of ecological pyramids: *pyramids of number*, *biomass*, and *energy*. Obviously, there will be differences among them. Some generalizations:

1. Energy pyramids must always be larger at the base than at the top (because of the Second Law of Thermodynamics, and has to do with dissipation of energy as it moves from one trophic level to another).

2. Likewise, biomass pyramids (in which biomass is used as an indicator of production) are usually pyramid-shaped. This is particularly true of terrestrial systems and aquatic ones dominated by large plants (marshes), in which consumption by heterotroph is low and organic matter accumulates with time. It is important to point out, however, biomass pyramids can sometimes be inverted. This is especially common in aquatic ecosystems, in which the primary producers are microscopic planktonic organisms that multiply very rapidly, have very short life spans, and experience heavy grazing by herbivores. At any single point in time, the amount of biomass in primary producers is less than that in larger, long-lived animals that consume primary producers.
3. Numbers pyramids can have various shapes (and not be pyramids at all, actually) depending on the sizes of the organisms that make up the trophic levels. In forests, the primary producers are large trees and the herbivore level usually consists of insects, so the base of the pyramid is smaller than the herbivore level above it. In grasslands, the number of primary producers (grasses) is much larger than that of the herbivores above (large grazing animals) (Spellman, 2001).

To get a better idea of how an ecological pyramid looks and how it provides information, we need to look at an example. The example to be used here is the energy pyramid. According to Odum (1983), the energy pyramid is a fitting example because among the "three types of ecological pyramids, the energy pyramid gives by far the best overall picture of the functional nature of communities" (p. 154).

In an experiment conducted in Silver Springs, Florida, Odum measured the energy for each trophic level in terms of kilocalories. A kilocalorie is the amount of energy needed to raise 1 cubic centimeter of water 1 degree centigrade. When an energy pyramid is constructed to show Odum's findings, it takes on the typical upright form (as it must because of the Second Law of Thermodynamics) as shown in figure 2.13.

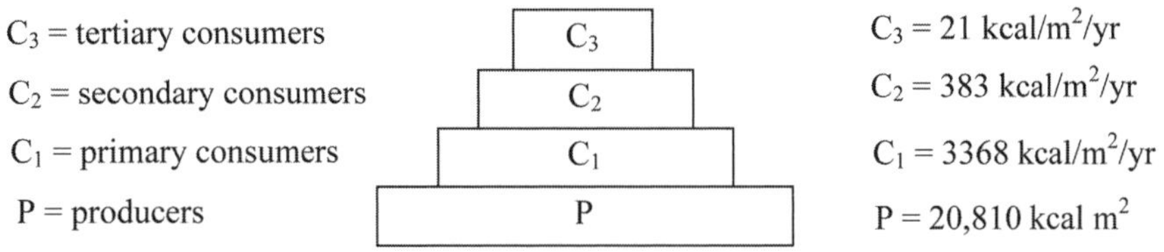

FIGURE 2.13
Energy flow pyramid. Adapted from Odum, 1971, *Fundamentals of Ecology*, p. 80.

In summary, as reflected in figure 2.13 and according to the second law of thermodynamics, no energy transformation process is 100 percent efficient. This fact is demonstrated, for example, when a horse eats hay. The horse cannot obtain, for his own body, 100 percent of the energy available in the hay. For this reason, the energy productivity of the producers must be greater than the energy production of the primary consumers. When human beings are substituted for the horse, it is interesting to note that according to the Second Law of Thermodynamics, only a small population could be supported. But this is not the case. Humans also feed on plant matter, which allows a larger population. Therefore, if meat supplies become scarce, we must eat more plant matter. This is the situation we see today in countries where meat is scarce. Consider this: if we all ate soybeans, there would be at least enough food for ten times as many of us as compared to a world in which we all eat beef (or pork, fish, chicken, etc.). There is another way of looking at this: every time we eat meat, we are taking food out of the mouths of nine other people, who could be fed with the plant material that was fed to the animal we are eating (EBE, 1999). It's not quite that simple, of course, but you get the general idea.

Productivity

(This section is from F. R. Spellman [2007]. *Ecology for Non-Ecologists.* Lanham, MD: Government Institutes Press.) As mentioned previously, the flow of energy through an ecosystem starts with the fixation of sunlight by plants through photosynthesis. In evaluating an ecosystem, the measurement of photosynthesis is important. Ecosystems may be classified into highly productive or less productive. Therefore, the study of ecosystems must involve some measure of the productivity of that ecosystem.

Smith (1974) defines production (or more specifically primary production, because it is the basic form of energy storage in an ecosystem) as being "the energy accumulated by plants." Stated differently, primary production is the rate at which the ecosystem's primary producers capture and store a given amount of energy, in a specified time interval. In even simpler terms, primary productivity is a measure of the rate at which photosynthesis occurs; that is, the rate of generation of biomass in an ecosystem via photosynthesis. Odum (1971) lists four successive steps in the production process, as follows:

Gross primary productivity—the total rate of photosynthesis in an ecosystem during a specified interval at a given trophic level.

Net primary productivity—the rate of energy storage in plant tissues in excess of the rate of aerobic respiration by primary producers.

Net community productivity—the rate of storage of organic matter not used.

Secondary productivity—the rate of energy storage at consumer levels.

When attempting to comprehend the significance of the term *productivity* as it relates to ecosystems, it is wise to consider an example. Consider the productivity of an agricultural ecosystem such as a wheat field. Often its productivity is expressed as the number of bushels produced per acre. This is an example of the harvest method for measuring productivity. For a natural ecosystem, several one-square-meter plots are marked off, and the entire area is harvested and weighed to give an estimate of productivity as grams of biomass per square meter per given time interval. From this method, a measure of net primary production (net yield) can be measured.

Productivity, both in the natural and cultured ecosystem, may vary considerably, not only between types of ecosystems, but also within the same ecosystem. Several factors influence year-to-year productivity within an ecosystem. Such factors as temperature, availability of nutrients, fire, animal grazing, and human cultivation activities are directly or indirectly related to the productivity of a particular ecosystem.

The following description of an aquatic ecosystem is used as an example of productivity. Productivity can be measured in several different ways in the aquatic ecosystem. For example, the production of oxygen may be used to determine productivity. Oxygen content may be measured in several ways. One way is to measure it in the water every few hours for a period of twenty-four hours. During daylight, when photosynthesis is occurring, the oxygen concentration should rise. At night the oxygen level should drop. The oxygen level can be measured by using a simple x-y graph. The oxygen level can be plotted on the y-axis with time plotted on the x-axis, as shown in figure 2.14.

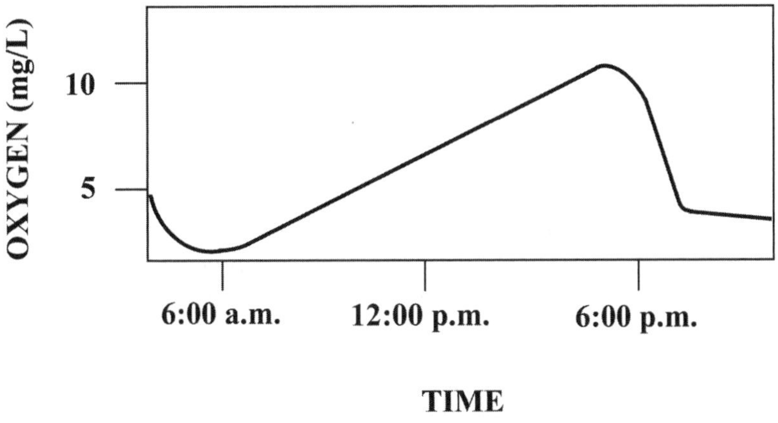

FIGURE 2.14
The diurnal oxygen curve for an aquatic ecosystem.

Another method of measuring oxygen production in aquatic ecosystems is to use light and dark bottles. Biochemical oxygen demand (BOD) bottles (300 ml) are filled with water to a particular height. One of the bottles is tested for the initial dissolved oxygen (DO), then the other two bottles (one clear, one dark) are suspended in the water at the depth they were taken from. After a twelve-hour period, the bottles are collected and the DO values for each bottle are recorded. Once the oxygen production is known, the productivity in terms of grams/m/day can be calculated.

Table 2.1 shows representative values for the net productivity of a variety of ecosystems—both natural and managed. Keep in mind that these values are only approximations derived from Odom's (1971; 1983) work and are subject to marked fluctuations because of variations in temperature, fertility, and availability of water.

In the aquatic (and any other) ecosystem, pollution can have a profound impact upon the system's productivity. For example, certain kinds of pollution may increase the turbidity of the water. This increase in turbidity causes a decrease in energy delivered by photosynthesis to the ecosystem. Accordingly, this turbidity and its aggregate effects decrease net community productivity on a large scale (Laws, 1993).

TABLE 2.1
Estimated Net Productivity of Certain Ecosystems

Ecosystem	*Kilocalories/m²/year*
Temperate deciduous forest	5,000
Tropical rain forest	15,000
Tall-grass prairie	2,000
Desert	500
Coastal marsh	12,000
Ocean close to shore	2,500
Open ocean	800
Clear (oligotrophic) lake	800
Lake in advanced state of eutrophication	2,400
Silver Springs, Florida	8,800
Field of alfalfa (Lucerne)	15,000
Corn (maize) field, U.S.	4,500
Rice paddies, Japan	5,500
Lawn, Washington, D.C.	6,800
Sugar cane, Hawaii	25,000

Productivity: The Bottom Line

The ecological trends paint a clear picture. Wherever we look, ecological productivity is limping behind human consumption. Since 1984, the global fish harvest has been dropping, and so has the per capita yield of grain crops (Brown, 1994). Moreover, stratospheric ozone is being depleted, the release of greenhouse gases has changed the atmospheric chemistry and might lead to climate change; erosion and desertification is reducing nature's biological productivity; irrigation water tables are falling; contamination of soil and water is jeopardizing the quality of food; other natural resources are being consumed faster than they can regenerate; and biological diversity is being lost—to reiterate only a small part of a long list. These trends indicate a decline in the quantity and productivity of nature's assets (Wachernagel, 1997).

Population Ecology

Population ecology owes its beginning to the contributions of Thomas Malthus, an English clergyman, who in 1798 published his *Essay on the Principle of Population*. Malthus introduced the concept that at some point in time an expanding population must exceed supply of prerequisite natural resources—the "Struggle for Existence Concept." Malthus's theories profoundly influenced Charles Darwin's *On the Origin of Species* (1859)—for example, the "Survival of the Fittest Concept." Let's begin with the basics. The following is a definition of the word *population* by *Webster's Third New International Dictionary*:

- The total number or amount of things, especially within a given area.
- The organisms inhabiting a particular area or biotype.
- A group of interbreeding biotypes that represents the level of organization at which speciation begins.

The following is a definition of the word *population* by an ecologist (Abedon, 2007):

- A population in an ecological sense is a group of organisms, of the same species, which roughly occupy the same geographical area at the same time.
- Individual members of the same population can either interact directly, or may interact with the dispersing progeny of the other members of the same population (e.g., pollen).

- Population members interact with a similar environment and experience similar environmental limitations.

Population: Simply Defined

A population is a set of individuals of the same species living in a given place at a given time.

A population system, or life-system ("population system" is definitely a better term, however), is a population with its effective environment (Clark et al., 1967; Berryman, 1981; Sharov, 1992).

Major Components of a Population System

1. Population itself: Organisms in the population can be subdivided into groups according to their age, stage, sex, and other characteristics.
2. Resources: These include food, shelters, nesting places, space, etc.
3. Enemies: These include predators, parasites, pathogens, etc.
4. Environment: This includes air (water, soil) temperature, composition, variability of these characteristics in time and space (Sharov, 1997).

Population Ecology Defined (Sharov, 1996): Population ecology is the branch of ecology that studies the structure and dynamics of populations. Population ecology relative to other ecological disciplines is shown in figure 2.15.

The term "population" is interpreted differently in various sciences. For example, in human demography a population is a set of humans in a given area. In genetics a population is a group of interbreeding individuals of the same species, which is isolated from other groups. In population ecology a population is a group of individuals of the same species inhabiting the same area.

Important Point: The main axiom of population ecology is that organisms in a population are ecologically equivalent. Ecological equivalency means:

1. Organisms undergo the same life-cycle.
2. Organisms in a particular stage of the life-cycle are involved in the same set of ecological processes.
3. The rates of these processes (or the probabilities of ecological events) are basically the same if organisms are put into the same environment (however, some individual variation may be allowed) (Sharov, 1996).

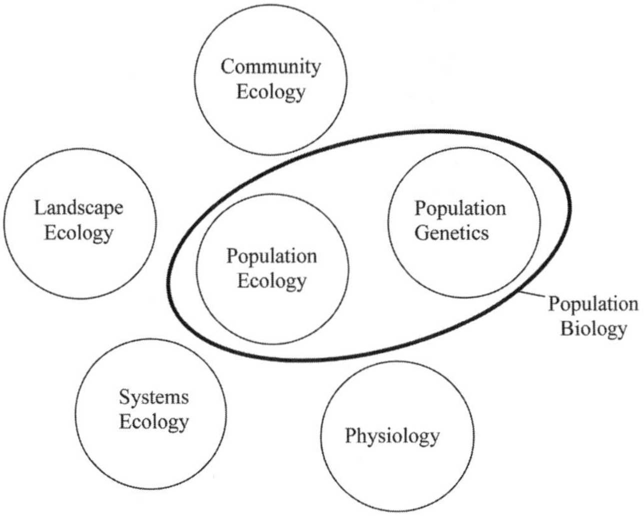

Population ecology – the branch of ecology that studies the structure and dynamics of populations.

Physiology –study of individual characteristics and individual processes. Used as a basis for prediction of processes at the population level.

Community ecology – study of the structure and dynamics of animal and plant communities. Population ecology provides modeling tools that can be used for predicting community structure and dynamics.

Population genetics – the study of gene frequencies and microevolution in populations. Selective advantages depend on the success of organisms in their survival, reproduction and competition. These processes are studied in population ecology. Population ecology and population genetics are often considered together and called 'population biology'. Evolutionary ecology is one of the major topics in population biology.

Systems ecology – a relatively new ecological discipline which studies interaction of human population with environment. One of the major concepts are optimization of ecosystem exploitation and sustainable ecosystem management.

Landscape ecology – another relatively new area in ecology. It studies regional large-scale ecosystems with the aid of computer-based geographic information systems. Population dynamics can be studied at the landscape level, and this is the link between landscape and population ecology

FIGURE 2.15
Population ecology relative to other ecological disciplines. Adapted from Alexi Sharov, 1966. Dept. of Entomology, Virginia Tech, VA, USA, p. 1.

Did You Know?

Recent USGS studies have shown that non-native annual grasses in the genera *Bromus* and *Schismus* now dominate most plant communities in the Mojave Desert. Unlike most native annual plants, these grasses grow in many different situations and can create continuous fuel beds across the landscape by filling in the plant-free space that once separated and protected native perennials from fire.

The following is a list of properties of populations (Abedon, 2007):

1. Population size (size)—depends on how the population is defined.
2. Population density (density)—the number of individual organisms per unit area; the relation between the number of individuals of a population and the area or volume they occupy.
3. Patterns of dispersion (dispersion)—individual members of populations may be distributed over a geographical area in a number of different ways, including: clumped, uniform, and random distribution.
4. Demographics (demographics)—a population's vital statistics, including education, parental status, work environment, geographic location, religious beliefs, marital status, and income, as well as race, gender, ethnicity, age, sexual orientation, and physical ability.
5. Population growth (growth)—simply, population growth occurs when there are no limitations on growth within the environment. When this occurs, two situations develop: (1) the population displays its intrinsic rate of increase (i.e., the rate of growth of a population when that population is growing under ideal conditions and without limits); and (2) the population experiences exponential growth (i.e., exponential growth means that a population's size at a given time is equal to the population's size at an earlier time, times some greater-than-one number).
6. Limits on population growth (limits)—exponential growth cannot go on forever; sooner or later any population will run into limits in their environment.

Did You Know?

The population growth rate (PGR) can be positive or negative because it is the percent variation between the number of individuals in a population at two different times.

Important Point: Note that all of these properties are not those of individual organisms but instead are properties that exist only if one considers more than one organism at any given time, or over a period of time.

Laws of Population Ecology

(The information in this section is based on and adapted from Haemig's [2006] *Laws of Population Ecology*.) According to Haemig (2006), the discovery of laws in ecology has lagged behind many of the other sciences (e.g., chemistry, physics, etc.) because ecology is a much younger science.

However, as Colyvan & Ginzburg (2003) point out, misunderstandings and unrealistic expectations of what laws are have also hindered the search, as have mistaken beliefs that ecology is just too complex a science to have laws. Nevertheless, over the years, researchers have been able to identify some of the laws that exist in ecology.

Ginzburg (1986) points out that while much remains to be learned, it now appears that laws of ecology resemble laws of physics. Colyvan & Ginzburg (2003) and Ginzburg and Colyvan (2004) point out that laws of ecology describe idealized situations, have many exceptions, and need not be explanatory or predictive. The laws of population ecology are listed and described below.

Malthusian Law—says that when birth and death rates are constant, a population will grow (or decline) at an exponential rate.

Allee's Law—says that there is a positive relationship between individual fitness and either the numbers or density of conspecifics (conspecifics are other individuals of the same species).

Verhulst's Law—deals with one factor: intra-specific competition (i.e., competition between members of the same species). Because the organisms limiting the population are also members of the population, this law is also called "population self-limitation" (Turchin, 2001).

Lotka-Volterra's Law—says that "when populations are involved in negative feedback with other species, or even components of their environments," oscillatory (cyclical) dynamics are likely to be seen (Berryman, 2002, 2003).

Liebig's Law—says that of all the biotic or abiotic factors that control a given population, one has to be limiting (i.e., active, controlling the dynamics) (Berryman, 1993, 2003). Time delays produced by this limiting factor are usually one or two generations long (Berryman, 1999). Krebs (2001) defines a factor as "limiting if a change in the factor produces a change in average or equilibrium density."

Fenchel's Law—says that species with larger body sizes generally have lower rates of population growth—the maximum rate of reproduction decreases with body size at a power of approximately ¼ the body mass (Fenchel, 1974). Fenchel's Law is expressed by the following equation: $r = aW^{-1/4}$ where:

r = the intrinsic rate of nature increase of the population
a = constant (has 3 different values)
W = average body weight (mass) of the organism

Calder's Law—says that species with larger body sizes generally have longer population cycles—the length of the population cycle increases with increasing body size at a power of approximately ¼ the body mass (Calder, 1983), so $t = aW^{1/4}$ where:

t = average time of the population cycle
a = a constant
W = average body weight (mass) of the organism

Damuth's Law—says that species with larger body sizes generally have lower average population densities—the average density of a population decreases with body size at a power of approximately ¾ the body mass (Damuth, 1981, 1987, 1991). Damuth's Law is expressed by the equation $d = aW^{-3/4}$ where:

d = the average density of the population
a = a constant
W = average body weight (mass) of the organism

Generation-Time Law—says that species with larger body sizes usually have longer generation-times—that the generation-time increases with increasing body size at a power of approximately ¼ the body mass (Bonner, 1965). Note: the body mass used in this law is the body mass of the organism at the time of reproduction. The Generation-Time Law is expressed by the equation $g = aW^{1/4}$ where:

g = average generation-time of the population
a = a constant
W = average body weight (mass) of the organism

Ginzburg's Law—says that the length of a population cycle (oscillation) is the result of the maternal effect and inertial populating growth. According to this law, the period lengths in the cycles of a population must be either two generations long or six or more generations long (Ginzburg & Colyvan, 2004).

Applied Population Ecology

In attempting to explain any concept, it is always best to do so with an example in mind—an illustrative example. In the following, a stream ecosystem is the illustrative example used to help explain population ecology.

If environmental scientists wanted to study the organisms in a slow-moving stream or stream pond, they would have two options. They could study each fish, aquatic plant, crustacean, insect, and macroinvertebrate one by one. In that case, they would be studying individuals. It would be easier to do this if the subject were trout, but it would be difficult to separate and study each aquatic plant.

The second option would be to study all of the trout, all of the insects of each specific kind, all of a certain aquatic plant type in the stream or pond at the time of the study. When stream ecologists study a group of the same kind of individuals in a given location at a given time, they are investigating a population. "Alternately, a population may be defined as a cluster of individuals with a high probability of mating with each other compared to their probability of mating with a member of some other population" (Pianka, 1988). When attempting to determine the population of a particular species, it is important to remember that time is a factor. Whether it is at various times during the day, during the different seasons, or from year to year, time is important because populations change.

When measuring populations, the level of species or density must be determined. Density (D) can be calculated by counting the number of individuals in the population (N) and dividing this number by the total units of space (S) the counted population occupies. Thus, the formula for calculating density becomes:

$$D = N/S \tag{2.3}$$

When studying aquatic populations, the occupied space (S) is determined by using length, width, and depth measurements. The volumetric space is then measured in cubic units.

Population density may change dramatically. For example, if a dam is closed off in a river midway through spawning season, with no provision allowed for fish movement upstream (a fish ladder), it would drastically decrease the density of spawning salmon upstream. Along with the swift and sometimes unpredictable consequences of change, it can be difficult to draw exact boundaries between various populations. Pianka (1988) makes this point in his comparison of European starlings that were introduced into Australia with starlings that were introduced into North America. He points out that these starlings are no longer exchanging genes with each other; thus, they are separate and distinct populations.

The population density, or level of a species, depends on natality, mortality, immigration, and emigration. Changes in population density are the result of both births and deaths. The birth rate of a population is called natality and the death rate mortality. In aquatic populations, two factors besides natality and mortality can affect density. For example, in a run of returning salmon to their spawning grounds, the density could vary as more salmon migrated in or as others left the run for their own spawning grounds. The arrival of new salmon to a population from other places is termed immigration (ingress). The departure of salmon from a population is called emigration (egress). Thus, natality and immigration increase population density, whereas mortality and emigration decrease it. The net increase in population is the difference between these two sets of factors.

Population regulation is the control of the size of a population. Population is limited by various factors. There are basically two different types of population-limiting factors—classified according to the types of factors that control the size of the population. The population limiting factors are (1) density-dependent control and (2) density-independent control (Winstead, 2007).

Density-Dependent Factors

These are factors where the effect of the factor on the size of the population depends upon the original density or size of the population. Density-dependent factors include:

- Density-dependent limits on population growth are ones that stem from intraspecific competition.
- Typically, the organisms best suited to compete with another organism are those from the same species.
- Thus, the actions of conspecifics (again, an organism belonging to the same species) can very precisely serve to limit the environment (e.g., eat preferred food, obtain preferred shelter, etc.).
- Actions that serve to limit the environment for conspecifics (e.g., eating, excreting wastes, using up non-food resources, taking up space, defending territories) are those that determine carrying capacity (K).
- They are referred to as *density dependent* because the greater the density of the *population*, the greater their effects.
- Density-dependent factors may exert their effect by reducing birth rates, increasing death rates, extending generation times, or by forcing the migration of conspecifics to new regions (Abedon, 2007).
- "The impact of disease on a population can be density dependent if the transmission rate of the disease depends on a certain level of crowding the population."
- "A death rate that rises as population density rise is said to be density dependent, as in a birth rate that falls with rising density. Density-dependent rates are an example of negative feedback. In contrast, a birth rate or death rate that does change with population density is said to be

density independent. . . . Negative feedback prevents unlimited population growth."

- Predation can also be density dependent since predators often can switch prey preferences to match whatever prey organisms are more plentiful in a given environment.
- "Many predators, for example, exhibit switching behavior: They begin to concentrate on a particularly common species of prey when it becomes energetically efficient to do so" (Campbell & Reece, 2004).

Migration Basics (NPS, 2012)

The National Park Service (NPS, 2012) points out that migration is a fascinating aspect of animal ecology. Migration inspires us whether we are studying salmon migrating thousands of miles back to their spawning grounds, huge flocks of sandhill cranes migrating across the northern skies, or caribou crossing rivers in the fall.

Migration has captured the interest of humans for centuries. Ancient civilizations devised many myths to explain the periodic appearance and disappearance of vast numbers of animals. For instance, people once thought that tiny birds called swallows buried themselves in the mud at the bottom of lakes to get through the winter. Instead, scientists found out that swallows fly all the way from Europe to Africa and back in one year. Perhaps the truth was harder to believe than the myth.

What is migration and why do it?

Animals that live in habitats that are difficult to survive in year-round, must evolve a way to cope with the difficult time of year. A strategy used by many mammals and other species is hibernation. Migration is another option for animals that can move across long distances. They survive by leaving the area for part of the year, or part of their life, and moving to habitats that are more hospitable.

The most common reason to migrate is to take advantage of food, shelter, and water that vary with seasons, or life stage. The availability of food and water can change throughout the year. For instance, the lack of insects and leaves in the winter means there is less food to eat. Some environments have a rainy and a dry season that are very different. Temperatures change between the seasons: some areas get very cold or very hot, which can be hard on some species. Sometimes it is not about getting food but about staying safe. Deep snow may make animals easier to catch by predators, or animals may go to special breeding grounds to keep their young safe when they are especially vulnerable.

There are many types of migration. These terms are used to describe attributes of migration such as timing, direction, the reason for migration, and how many of the species migrate. More than one term can be used to describe one species migration pattern. Some common types of migration are:

Seasonal migration—is migration that corresponds with the change in seasons. Most migrations fall within this category. Many altitudinal, longitudinal, latitudinal, and reproductive migrations take place when the seasons change.

Latitudinal migration—is the movement of animals north and south. The geese flying south for the winter is one of the most recognizable examples of latitudinal migration. By moving north and south, animals are changing their climate. In the northern hemisphere, the winters are colder as you move north and warm as you move south. On the other hand, summers in the north can be rich in food, especially in the far north where summers are short, but the days are very long.

Altitudinal migration—is the movement of animals up and down major land features such as mountains. While food may be plentiful in alpine meadows in summer, the winters will be colder and have more snow as you move higher up. Many animals take advantage of the summers, and then move to lower, more moderate elevations during the winter.

Reproductive migration—is the movement of animals to bear young. The areas may be safer for the young because of fewer predators or more shelter from predators. In other cases, the area is safer because the animal requires a different type of habitat when it is young than when it is older.

Nomadic migration—is the movement of animals between not known areas; it looks to us more like wandering. Grazing animals will move across larger expanses as the grass gets eaten and they travel to greener pastures.

Removal migration—is the migration of animals that don't come back. This can take place when resources such as food, water, or shelter are no longer available to animals where they are. The environment can have changed, through fire, flooding, invasive plant species, human development, or other causes, and the animals need to leave to survive. Another cause of removal migration is when the resources haven't changed, but the population gets too big. There are too many animals, and many of them leave to find food, water, and shelter elsewhere. Removal migration is what brought many immigrants to America in the 1800s.

Complete migration—is when virtually all members of the species leave their breeding range during the non-breeding season. Many North American birds are complete migrants. Most complete migrants breed in the northern temperate and arctic areas (such as Alaska) of North America, Europe, and Asia. Complete migrants travel incredible distances, sometimes more than 15,000 miles (25,000 kilometers) per year. The wintering areas for most complete North American migrants are South and Central America, the Caribbean basin, and the southernmost United States.

Partial migration—is the most common type of migration. Partial migrant means that some, but not all, members of a species move away from their breeding grounds during the nonbreeding season. There is an overlap between breeding and nonbreeding ranges of the species. Species like red-tailed hawks, herring gulls, and golden eagles are partial migrants over much of their North American range.

Irruptive migration—are not seasonally or geographically predictable. Such migration may occur one year, but not again for many years. The distances and numbers of individuals and involved area also are less predictable than with complete or partial migrants. In some years, irruptions can be over long distances and involve many individuals, or they can be short and involve only a few.

Migration Examples

- Humpback whales of the Pacific Ocean head south in the fall to give birth to their young in subtropical waters off of Hawaii, and then in late spring head north to spend the summer in the cold waters off of Alaska that are rich with food.
- Salmon are reproductive migrants that start their lives in freshwater streams, move to the ocean for their adult lives, then return to their home stream to lay eggs.
- Dall sheep are seasonal, altitudinal migrants that spend summers near the top of mountain ranges and then winter at lower elevations, where there is less snow and food is easier to find.
- Arctic terns are complete migrants that spend all year in summer by alternating subpolar regions in the northern and southern hemispheres.
- Golden eagles of Denali National Park and Preserve in Alaska spend the summer in the north where there is plenty of food, and head south for the winter when there is less food in the north and the temperatures drop far below zero. While all of the golden eagles of Denali do migrate, golden eagles are considered partial migrants because those that live far enough south do not migrate.
- Sea turtles return from ocean waters to the coast to lay eggs in the sand—where they hatch and head to the open ocean until it is their turn to lay eggs. They are another example of reproductive migrants.
- Locusts change when they get too crowded and become more active and social, creating large groups of insects that move across the land in search

of new places with plenty of food (and fewer locusts). This adaptation to overcrowding is called removal migration.
- Great gray owls are an irruptive migrant, migrating southward only occasionally and in numbers that vary greatly. Northern finches and crossbills are also irruptive migrants.

Migration Cues

How do animals know when to migrate? That depends on the type of migration. For many types of migration, it is the change of seasons that spurs animals on. As summer becomes fall, days become shorter and that can trigger animals to prepare for migration. Closer to the equator, the days don't change in length and one theory is that animals become restless after too many days with a constant length.

Other migrations are initiated by seasonal conditions. Food availability can be a motivator for some longitudinal and altitudinal migrators. For example, as plant foods in upper elevations become hidden under snow, animals move down toward the valleys, and then in the spring as the plants come out again, animals move back into the upper areas following the plants as they appear. Nomadic animals move to the next feeding ground as they run out of food where they are. As ponds dry with seasonal changes, animals will move to find available water supplies, and then return during the seasonal rains.

In some species, migration happens when there are just too many animals too close together. The overcrowding causes many of the individuals to leave in hopes of finding another habitat with less competition. Or migration takes place when there isn't enough food, not because of the changing seasons, but because the food where they are has been eaten. The animals start moving in search of new food.

Navigation

How do animals know how to get where they are going? That depends on the animal and where it is going. There is strong evidence that genetics plays a large role in migratory behavior and that animals inherit migratory routes from their parents genetically.

Animals use a variety of different information and senses to navigate. Researchers believe that most animals use a combination of navigation cues, depending on where they are and what the conditions are. In shorter migrations, animals do not need complicated navigation abilities. They can simply follow the food to the water, or head downhill to the valleys in winter and back up toward the ridges in summer.

Researchers have learned a lot about animal migration by studying animal movements. Starlings, for instance, orient themselves using the sun, compensating for how the sun moves across the sky throughout the day. Mallard ducks can find north using the stars of the night sky. Animals as diverse as migratory birds, salamanders, salmon, or hamsters use the geomagnetic field for orientation. Studies of loggerhead turtles revealed that hatchlings have the ability to sense the direction and strength of the earth's magnetic field, which they use for navigating along the turtles' regular migration route. Scientists have discovered a collection of nerve cells in the brains of subterranean Zambian mole rats that enable the animal to process magnetic information used in navigation.

Animals can also use mental maps. Just like people they become familiar with an area and navigate their way using land features like mountain ranges, coastlines, rivers, and even, in the case of dolphins, the shape of the sea floor.

Smell can be a powerful tool for many animals. Many land animals can create mental maps based on the smell rather than just the sight of major land features. Salmon use smell to find the exact stream where they were born. Fish can use water currents that circulate around the oceans, or they can swim against the current or with the current in streams and rivers.

Adaptations for Migration

Migratory animals that travel long distances have special adaptations to help them get there. The most obvious are birds. They have wings that allow them to fly long distances, their bodies are especially light (they have hollow bones) so they can stay high in the air, and they don't have unnecessary weight to carry around. Geese fly in formation, in the shape of a *V*, which decreases the wind drag on all the birds along both sides.

Birds add on extra fat stores to give them enough energy for long flights north and south, because they do not eat during the migration. Similarly, whales stock up well on food in the northern seas before heading south for the winter, because they don't eat on the way. Land animals must rely on their legs and feet to get them where they need to go.

Density-Independent Factors

These are "where the effect of the factor on the size of the population is independent and does not depend upon the original density or size of the population. The effect of weather is an example of a density-independent factor. A severe storm and flood coming through an area can just as easily wipe out a large population as a small one. Another example would be a harmful pollutant put into the environment (e.g., a stream). The probability of that harmful substance at some concentration killing an individual would not change depending on the size of the population. For example, populations of small mammals are often regulated more by this type of regulation" (Winstead, 2007).

- Density-independent effects on population sizes (or structures) occur to the same extent regardless of population size.
- These can be things like sudden changes in the weather.
- "Over the long term, many populations remain fairly stable in size and are presumably close to a carrying capacity that is determined by density-dependent factors. Superimposed on this general stability, however, are short-term fluctuations due to density-independent factors" (Campbell & Reece, 2004).

Distribution or Dispersion

Each organism occupies only those areas that can provide for its requirements, resulting in an irregular distribution or dispersion. How a particular population is distributed within a given area has considerable influence on density. As shown in figure 2.16, organisms in nature may be distributed (dispersed) in three ways, as a result of complex interactions among ecological variables.

In a *random* distribution, there is an equal probability of an organism occupying any point in space, and "each individual is independent of the others" (Smith 1974). In other words, the position of each individual is not determined or influenced by the other members of the population.

In a *regular* or *uniform* distribution, in turn, organisms are spaced more evenly; they are not distributed by chance. Animals compete with each other and effectively defend a specific territory, excluding other individuals of the

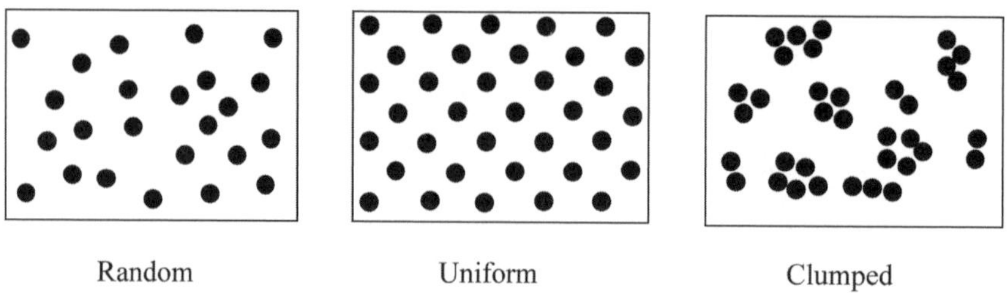

FIGURE 2.16
Basic patterns of distribution. Adapted from Odum, 1971, *Fundamentals of Ecology*, p. 205.

same species. In regular or uniform distribution, the competition between individuals can be quite severe and antagonistic to the point where spacing generated is quite even (Odum, 1983). This is seen in forest areas where trees are uniformly distributed so that each receives adequate water and light.

The most common distribution is the *contagious* or *clumped* distribution where organisms are found in groups, clumped together; this may reflect the heterogeneity of the habitat. Smith (1974) points out that contagious or clumped distribution "produces aggregations, the result of response by plants and animals to habitat differences."

Organisms that exhibit a contagious or clumped distribution may develop social hierarchies in order to live together more effectively. Animals within the same species have evolved many symbolic aggressive displays that carry meanings that are not only mutually understood but also prevent injury or death within the same species. For example, in some mountainous regions, dominant male bighorn sheep force the juvenile and subordinate males out of the territory during breeding season (Hickman et al., 1990). In this way, the dominant male gains control over the females and need not compete with other males.

As mentioned, distribution patterns are the result of complex interactions among ecological variables. For example, consider a study conducted by Hubbell and Johnson (1977) of five tropical bee colonies (the bees live in colonies in suitable trees) in the tropical dry forests of Costa Rica. The researchers set out to examine the relationship between aggressiveness and patterns of colony distribution.

The researchers mapped locations of suitable nest trees. They found that the number of suitable trees was greater than number of colonies—thus nest sites were not a limiting factor. Distribution of suitable trees was random. The researchers next mapped locations of bee colonies. They found that colonies' sites for one species were dispersed randomly. Members of this species do not exhibit aggression toward one another. The colonies were sometimes quite close to one another.

On the other hand, colony sites for the other four species were dispersed in a regular fashion. Members of all four species were aggressive to members of other colonies of the same species. They also mark their colony sites with pheromones. They also engage in ritualized battles for colony sites with conspecifics from other colonies.

Population Growth

The size of animal populations is constantly changing due to natality, mortality, emigration, and immigration. As mentioned, the population size will increase if the natality and immigration rates are high. On the other hand, it will decrease if the mortality and emigration rates are high. Each population has an upper limit on size, often called the *carrying capacity* (K). Carrying capacity can be defined as the "optimum number of species' individuals that can survive in a specific area over time" (Enger, Kormelink, Smith, and Smith, 1989). Stated differently, the carrying capacity is the maximum number of species that can be supported in a bioregion. A pond may be able to support only a dozen frogs depending on the food resource for the frogs in the pond. If there were thirty frogs in the same pond, at least half of them would probably die because the pond environment wouldn't have enough food for them to live. Carrying capacity, symbolized as *K*, is based on the quantity of food supplies, the physical space available, the degree of predation, and several other environmental factors.

Did You Know?

How do you count black and grizzly bear populations in the wild? USGS has found an apparent answer to this question by launching a 2009 grizzly bear research project in the Northern Continental Divide Ecosystem of northwestern Montana. This work uses hair collection and DNA analysis. This method estimates population growth by collecting hair at natural bear rubs along trails, roads, and fence and power lines. Short pieces of barbed wire were attached to the rubbed surface to facilitate hair collection at most sites; however, barbless wire was used on trees bumped by pack animals. No lures or attractants were used to attract bears to these sites. Using this method, USGS was able to count 258 individual grizzlies that had deposited hair (USGS, 2010).

The carrying capacity is of two types: ultimate and environmental. Ultimate carrying capacity is the theoretical maximum density; that is, it is the maximum number of individuals of a species in a place that can support itself without rendering the place uninhabitable. The environmental carrying capacity is the actual maximum population density that a species maintains in an area. Ultimate carrying capacity is always higher than environmental.

The population growth for a certain species may exhibit several types of growth. Smith (1974) points out that "the rate at which the population grows can be expressed as a graph of the numbers in the population against time." Figure 2.17 shows one type of growth curve.

The J-shaped curve shown in figure 2.17 shows a rapid increase in size or exponential growth. Eventually, the population reaches an upper limit where exponential growth stops. The exponential growth rate is usually exhibited by

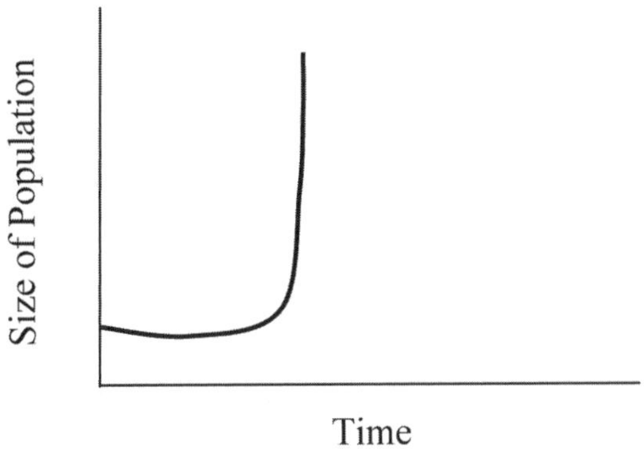

FIGURE 2.17
J-shaped growth curve.

organisms that are introduced into a new habitat, by organisms with a short life span such as insects, and by annual plants. A classic example of exponential growth by an introduced species is the reindeer transported to Saint Paul Island in the Pribolofs off Alaska in 1911. A total of twenty-five reindeer were released on the island and by 1938 there were over 2,000 animals on the small island. As time went by, however, the reindeer overgrazed their food supply and the population decreased rapidly. By 1950 only eight reindeer could be found (Pianka, 1988).

Did You Know?

Biotic potential is the capability of growth of a given population under hypothetical optimum conditions, that is, in an environment without limiting factors to such growth. Under such conditions the population tends to grow indefinitely. Biotic Potential is a fundamental species characteristic, defined by Chapman (1931) as "a sort of algebraic sum of the number of young produced at each reproduction, number of reproductions over a period of time, sex ratio of the species, and their general ability to survive under given physical conditions." Chapman relates a "vital index": Vital Index = (number of births/number of deaths) x 100.

Another example of exponential growth is demonstrated by the Lilly Pond Parable. The "parable" (not really a parable) is an excellent example providing insight into long-term carrying capacity and population growth.

A Lily Pond Parable

Question 1: If a pond lily doubles every day and it takes 30 days to completely cover a pond, on what day will the pond be ¼ covered?

Question 2: When will it be half covered?

Question 3: Does the size of the pond make a difference?

Question 4: What kind of environmental, social, and economic developments can be expected as the 30th day approaches?

Question 5: What will begin to happen at one minute past the 30th day?

Question 6: At what point (what day) would preventive action become necessary to prevent unpleasant events?

Answer 1: Day 28. Growth will be barely visible until the final few days. (On the 25th day, the lilies cover 1/32nd of the pond; on the 21st day, the lilies cover 1/512th of the pond).

Answer 2: The 29th day.

Answer 3: No. The doubling time is still the same. Even if you could magically double the size of the pond on day 30, it would still hold only one day's worth of growth!

Answer 4: The pond will become visibly more crowded each day, and this crowding will begin to exhaust the resources of the pond.

Answer 5: The pond will be completely covered. Even through the lilies will be reproducing, there will be no more room for additional lilies, and the excess population will die off. In fact, since the resources of the pond have been exhausted, a significant proportion of the original population may die off, as well.

Answer 6: It depends on how long it takes to implement the action and how full you want the lily pond to be. If it takes two days to complete a project to reduce lily reproductive rates, that action must be started on day 28, when the pond is only 25 percent full—and that will still produce a completely full pond. Of course, if the action is started earlier, the results will be much more dramatic.

Doubling Time

Population growing at a constant rate will have a constant doubling time . . . the time it takes for the population to double in size. Population growing at a constant rate can be modeled with exponential growth equation:

$$dN/dt = rN$$

The integral of the equation is:

$$N_t = N_0 e^{rt}$$

How long will it take for the population to double growing at a constant rate ("r")?

$$.69/r = T$$

The Rule of 70

The Rule of 70 is useful for financial as well as demographic analysis. It states that to find the doubling time of a quantity growing at a given annual percentage rate, divide the percent into 70 to obtain the approximate number of years required to double. For example, at a 10 percent annual growth rate, doubling time is 70/10 = 7 years.

Similarly, to get the annual growth rate, divide 70 by the doubling time. For example, 70/14 years doubling time = 5, or a 5 percent annual growth rate. Table 2.2 shows some common doubling times. Another type of growth curve is shown in figure 2.18. This logistic or S-shaped (sigmoidal) curve is used for populations of larger organisms having a longer lifespan. This type of curve has been successfully used by ecologists and biologists to model populations of several different types of organisms, including water fleas, pond snails, and sheep, to name only a few (Masters, 1991). The curve suggests an early exponential growth phase, while conditions for growth are optimal. As the number of individuals increases, the limits of the environment, or environmental resistance, begin to decrease the number of individuals, and the population size levels off near the carrying capacity, shown as K in figure 2.18. Usually there is some oscillation around K before the population reaches a stable size as indicated on the curve.

TABLE 2.2

Growth Rate (% Per Year)	*Doubling Time in Years*
0.1	700
0.5	140
1	70
2	35
3	23
4	18
5	14
6	12
7	12
10	7

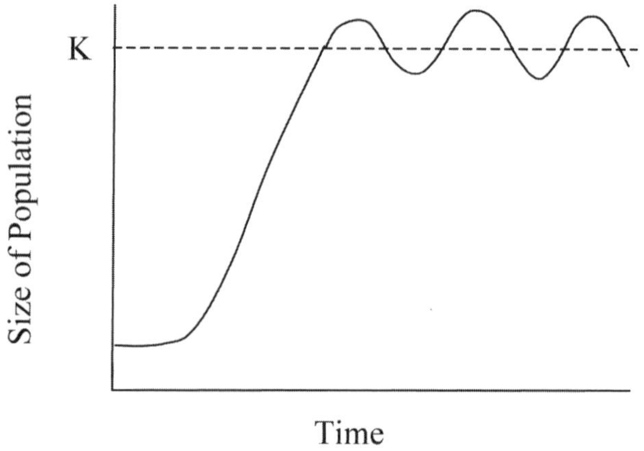

FIGURE 2.18
S-shaped (sigmoidal) growth curve.

Did You Know?

Environmental resistance is the action of limiting abiotic and biotic factors that disallow the growth of a population as it would grow to its biotic potential. Environmental resistance factors are those aspects of an environment that constrain the growth of a population and establish the maximum number of individuals that can be sustained. Such factors include the availability of essential resources (e.g., food, water), predation, disease, the accumulation of toxic metabolic wastes, and, in some species, behavior changes due to stress caused by overcrowding (Allaby, 2010).

As pointed out in the Rule of 70 example, the S-shaped curve in figure 2.18 is derived from the following differential equation:

$$Dn/dt = Rn(1 - N/K)$$

where N is population size, R is a growth rate, and K is the carrying capacity of the environment. The factor (1 - N/K) is the environmental resistance. As population grows, the resistance to further population growth continuously increases.

It is interesting to note that the S-shaped curve can also be used to find the maximum rate that organisms can be removed without reducing the population size. This concept in population biology is called the maximum sustainable yield value of an ecosystem. For example, imagine fishing steelhead fish from a stream. If the stream is at its carrying capacity, theoretically, there will be no population growth, so that any steelheads removed will reduce the population. Thus, the maximum sustainable yield will correspond to a population size less than the carrying capacity. If population growth is logistic or S-shaped, the maximum sustainable yield will be obtained when the population is half the carrying capacity. The slope of the logistic curve is given by the following:

$$Dn/dt = Rn\ (1 - N/K)$$

Setting the derivative to zero gives:

$$d/dt\ (Dn/dt = r\ dn/dt - r/k\ (2N\ Dn/dt) = 0$$

yielding:

$$1 - 2N/K = 0$$
$$N = K/2$$

The logistic growth curve is said to be density conditioned. As the density of individuals increases, the growth rate of the population declines.

As stated previously, after reaching environmental carrying capacity, population normally oscillates around the fixed axis due to various factors that affect the size of population. These factors work against maintaining the level of population at the K level due to direct dependence on resource availability. The factors that affect the size of populations are known as population controlling factors. They are usually grouped into two classes, density dependent and density independent. Table 2.3 shows factors that affect population size.

Density-dependent factors are those that increase in importance as the size of the population increases. For example, as the size of a population grows, food and space may become limited. The population has reached the carrying capacity. When food and space become limited, growth is suppressed

TABLE 2.3
Factors Affecting Population Size

Density Independent	*Density Dependent*
Drought	Food
Fire	Pathogens
Heavy rain	Predators
Pesticides	Space
Human destruction of habitat	Psychological disorders and physiological disorders

by competition. Odum (1983) describes density-dependent factors as acting "like governors on an engine and for this reason are considered one of the chief agents in preventing overpopulation."

Density-independent factors are those that have the same effect on population regardless of size. Typical examples of density-independent factors are devastating forest fires, stream beds drying up, or the destruction of the organism's entire food supply by disease.

Thus, population growth is influenced by multiple factors. Some of these factors are generated within the population, others from without. Even so, usually no single factor can account fully for the curbing of growth in a given population. It should be noted, however, that humans are, by far, the most important factor; their activities can increase or exterminate whole populations.

Population Response to Stress

As mentioned earlier, population growth is influenced by multiple factors. When a population reaches its apex of growth (its carrying capacity), certain forces work to maintain population at a certain level. On the other hand, populations are exposed to small or moderate environmental stresses. These stresses work to affect the stability or persistence of the population. Ecologists have concluded that a major factor that affects population stability or persistence is species diversity.

Species diversity is a measure of the number of species and their relative abundance. There are several ways to measure species diversity. One way is to use the straight ratio, D = S/N. In this ratio, D = species diversity, N = number of individuals, and S = number of species. As an example, a community of 1000 individuals is counted; these individuals are found to belong to fifty different species. The species diversity would be 50/1000 or 0.050. This calculation does not take into account the distribution of individuals of each species. For this reason, the more common calculation of species diversity is called the Shannon-Weiner Index. The Shannon-Weiner Index measures diversity by the following:

$$H = -\sum_{i=1}^{s} (pi)(\log pi)$$

where:

H= the diversity index
s = the number of species

i = the species number
pi = proportion of individuals of the total sample belonging to the ith species.

The Shannon-Weiner Index is not universally accepted by ecologists as being the best way to measure species diversity, but it is an example of a method that is available.

Species diversity is related to several important ecological principles. For example, under normal conditions, high species diversity, with a large variety of different species, tends to spread risk. This is to say that ecosystems that are in a fairly constant or stable environment, such as a tropical rain forest, usually have higher species diversity. However, as Odum (1983) points out, "diversity tends to be reduced in stressed biotic communities."

If the stress on an ecosystem is small, the ecosystem can usually adapt quite easily. Moreover, even when severe stress occurs, ecosystems have a way of adapting. Severe environmental change to an ecosystem can result from such natural occurrences as fires, earthquakes, and floods, and from people-induced changes such as land clearing, surface mining, and pollution.

One of the most important applications of species diversity is in the evaluation of pollution. As stated previously, it has been determined that stress of any kind will reduce the species diversity of an ecosystem to a significant degree. In the case of domestic sewage, for example, the stress is caused by a lack of dissolved oxygen (DO) for aquatic organisms.

Ecological Succession

Ecosystems can and do change. For example, if a forest is devastated by a fire, it will grow back, eventually, because of ecological succession. Ecological succession is the observed process of change (a normal occurrence in nature) in the species structure of an ecological community over time; that is, a gradual and orderly replacement of plant and animal species takes place in a particular area over time. The result of succession is evident in many places. For example, succession can be seen in an abandoned pasture. It can be seen in any lake and any pond. Succession can even be seen where weeds and grasses grow in the cracks in a tarmac, roadway, or sidewalk. Additional specific examples of observable succession include:

1. Consider a red pine planting area where the growth of hardwood trees (including ash, poplar, and oak) occurs. The consequence of this hardwood tree growth is the increased shading and subsequent mortality of the sun-loving red pines by the shade-tolerant hardwood seedlings.

The shaded forest floor conditions generated by the pines prohibit the growth of sun-loving pine seedlings and allow the growth of the hardwoods. The consequence of the growth of the hardwoods is the decline and senescence of the pine forest.

2. Consider raspberry thickets growing in the sunlit forest sections beneath the gaps in the canopy generated by wind-thrown trees. Raspberry plants require sunlight to grow and thrive. Beneath the dense shade canopy, particularly of red pines but also dense stands of oak, there is not sufficient sunlight for the raspberries' survival. However, in any place in which there has been a tree fall the raspberry can proliferate into dense thickets. Within these raspberry thickets, by the way, are dense growths of hardwood seedlings. The raspberry plants generate a protected "nursery" for these seedlings and prevent a major browser of tree seedlings (the white tail deer) from eating and destroying the trees. By providing these trees a shaded haven in which to grow, the raspberry plants are setting up the future tree canopy that will extensively shade the future forest floor and consequently prevent the future growth of more raspberry plants!

Succession usually occurs in an orderly, predictable manner. It involves the entire system. The science of ecology has developed to such a point that ecologists are now able to predict several years in advance what will occur in a given ecosystem. For example, scientists know that if a burned-out forest region receives light, water, nutrients, and an influx or immigration of animals and seeds, it will eventually develop into another forest through a sequence of steps or stages.

Two types of ecological succession are recognized by ecologists: primary and secondary. The particular type that takes place depends on the condition at a particular site at the beginning of the process.

Secondary succession is the most common type of succession. Secondary succession occurs in an area where the natural vegetation has been removed or destroyed but the soil is not destroyed. For example, succession that occurs in abandoned farm fields, known as old field succession, illustrates secondary succession. An example of secondary succession can be seen in the Piedmont region of North Carolina. Early settlers of the area cleared away the native oak-hickory forests and cultivated the land. In the ensuing years, the soil became depleted of nutrients, reducing the soil's fertility. As a result, farming ceased in the region a few generations later, and the fields were abandoned. Some 150 to 200 years after abandonment, the climax oak-hickory forest was restored.

The Process of Succession

Five steps are involved in the process of succession:

1. *Inertia*—the tendency of an ecosystem to maintain its overall structure.
2. *Disturbance*—an event that will instigate the process of succession.
3. *Primary succession*—when a community starts from bare rock.
4. *Secondary succession*—when succession starts from an area where humans once farmed.
5. *Tolerance*—when late succession plants are not disturbed by early succession plants.

Discussion Questions

1. How can we best conserve our energy resources?
2. Describe several advantages for an animal that can occupy more than one trophic level.
3. Suggest ways in which the transportation systems can be modified that will result in environmental improvement compared to present practices.
4. Why is finding a solution to an environmental conflict so complex? Explain your answer.
5. What is the quality of life?
6. What is the "good life"?
7. Do you believe the society you live in is on an unsustainable path? Explain.
8. Are humans a part of or separate from nature?
9. Do you think technology can solve our environmental problems?
10. Why are decomposers so important to our ecosystem? Explain.

References and Recommended Reading

Abedon, S.T., 2007. *Population Ecology*. Accessed 02/27/07 at abdeon.1@osu.edu.

Able, K.P., 1999. *Gathering of Angels, Migrating Birds and Their Ecology*. London: Comstock Books.

Allaby, M., 2010. *A Dictionary of Zoology*. New York: Oxford University Press.

Allaby, A., and M. Allaby, 1991. *The Concise Dictionary of Earth Sciences*. Oxford: Oxford University Press.

Allee, W.C., 1932. *Animal aggregations: a study in general sociology*. USA: University of Chicago.

Arms, K., *Environmental Science*, 2nd ed., 1994. HBJ College and School Division, Saddle Brook, NJ.
Associated Press, 1997, in *The Virginian-Pilot* (Norfolk, VA), "Does warming feed El Niño?" p. A-15, December 7.
Associated Press, 1998, in *The Lancaster New Era* (Lancaster, PA), "Ozone hole over Antarctica at record size," September 28.
Associated Press, 1998, in *The Lancaster New Era* (Lancaster, PA), "Tougher air pollution standards too costly, Midwestern states say," September 25.
Baden, J., and R.C. Stroup (eds.), 1981. *Bureaucracy vs. Environment.* Ann Arbor: University of Michigan.
Baker, R.R., 1978. *The Evolutional Ecology of Mammal Migration.* New York: Holmes & Meier Publishers, Inc.
Berryman, A.A., 1981. *Population Systems: A general introduction.* New York: Plenum Press.
Berryman, A.A., 1993. Food web connectance and feedback dominance, or does everything really depend on everything else? *Oikos* 68: 13–185.
Berryman, A.A., 1999. *Principles of population dynamics and their application.* UK: Cheltenham.
Berryman, A.A., 2002. *Population cycles: the case for trophic interactions.* New York: Oxford University.
Berryman, A.A., 2003. On principles, laws and theory in population ecology. *Oikos* 103: 695–701.
Bolin, B., and R.B. Cook, 1983. *The major biogeochemical cycles and their interactions.* New York: Wiley.
Bonner, J.T., 1965. *Size and Cycle.* USA: Princeton University.
Botkin, D.B., 1995. *Environmental Science: Earth as a Living Planet.* New York: Wiley.
Box, G., and N. Draper, 1987. *Empirical model building and response surfaces.* New York: Wiley, 74.
Brown, L.R., 1994. "Facing food insecurity." In *State of the World.* Brown, L.R., et al. (eds.). New York: W.W. Norton.
Calder, W.A., 1983. An allometric approach to population cycles of mammals. *Journal of Theoretical Biology* 100: 275–282.
Calder, W.A., 1996. *Size, function and life history.* Mineola, NY: Dover Publications.
Campbell, N.A., and J.B. Reece, 2004. *Biology*, 7th ed. Benjamin Cummings.
Carbonify.com, 2009. *Global warming—a hoax?* Accessed 11/07/09 at http://www.carbonify. com/articles/global-warming-hoax.htm.
Carpi, A., and A.E. Egger, 2009. *The Scientific Method.* Accessed 09/03/09 at http://www.visionlearing.com/linraary/module_view.php?print=1 &mid=45&mcid=.
Chapman, R.N., 1931. *Animal Ecology.* New York: McGraw-Hill.
Clarke, T., and S. Clegg (eds.), 2000. *Changing Paradigms.* London: Harper Collins.
Clark, L.R., P.W. Gerier, R.D. Hughes, and R.F. Harris, 1967. *The ecology of insect populations.* London: Methuen.
Cobb, R.W., and C.D. Elder, 1983. *Participation in American Politics,* 2nd ed. Baltimore: John Hopkins.
Colinvaux, P., 1986. *Ecology.* New York: Wiley.

Colyvan, M., and L.R. Ginzburg, 2003. Laws of nature and laws of ecology. *Oikos* 101: 649-653.

Cramer, H., 1963. *Mathematical methods of statistics*. Princeton, NJ: Princeton U. Press.

Damuth, J., 1981. Population density and body size in mammals. *Nature* 290: 699-700.

Damuth, J., 1987. Interspecific allometry of population density in mammals and other animals: the independence of body mass and population energy-use. *Biological Journal of the Linnean Society* 31: 193–246.

Damuth, J., 1991. Of size and abundance. *Nature* 351: 268–269.

Dasmann, R.F., 1984. *Environmental Conservation*. New York: Wiley.

Davis, M.L., and D.A. Cornwell, 1991. *Introduction to Environmental Engineering*. New York: McGraw-Hill, Inc.

Dawid, P.P., 1983. Inference, statistical: I. In Kotz, S., and N.L. Johnson. *Encyclopedia of statistical science*. New York: Wiley, 89–105.

Diamond, J., 1996. *Guns, Germs, and Steel: The Fates of Human Societies*. New York: W.W. Norton.

Dolan, E.F., 1991. *Our Poisoned Sky*. New York: Cobblehill Books.

Downing, P.B., 1984. *Environmental Economics and Policy*. Boston: Little, Brown.

Easterbrook, G., 1995. *A Moment on the Earth: The Coming Age of Environmental Optimism*. Bergenfield, NJ: Viking.

EBE, 1999. Environmental Biology-Ecosystems at www.marietta.edu.biol.102/ecosystem.html.

Ehrlich, P.R., A H. Ehrlich, and J.P. Holdren, 1977. *Ecoscience: Population, Resources, and Environment*. San Francisco: W.H. Freeman.

Enger, E., J.R. Kormelink, B.F. Smith, and R.J. Smith, 1989. *Environmental science: The study of interrelationships*. Dubuque, IA: William C. Brown.

EPA, 2005. Basic Air Pollution Meteorology. Accessed 01/15/08 at www.epa.gov/apti.

EPA, 2007. National Ambient Air Quality Standards (NAAQS). Accessed 01/12/08 at www.epa.gov/air/criteria/html.

EPA, 2009. Regulatory Atmospheric Modeling Accessed 03/02/09 at http://www.epa.gov./scram001/).

Euclid, 1956. *The Elements*. New York: Dover.

Fenchel, T., 1974. Intrinsic rate of natural increase: the relationship with body size. *Oecologia* 14: 317–326.

Field, B.C., 1996. *Environmental Economics: An Introduction*, 2nd ed. New York: McGraw-Hill.

Ford, A., 2009. *Modeling the Environment*, 2nd ed. Washington, D.C.: Island Press.

Franck, I., and D. Brownstone, 1992. *The Green Encyclopedia*. New York: Prentice Hall.

Frazer, D.A.S., 1983. Inference, statistical: II. In Kotz, S. and N.L. Johnson. *Encyclopedia of statistical science*. New York: Wiley, 105–14.

Freese, F., 1967. Elementary statistical methods for foresters. *USDA Agric. Hand.* 317. Washington, D.C.: U.S. Department of Agriculture.

Ginzburg, L.R., 1986. The theory of population dynamics: Back to first principles. *Journal of Theoretical Biology* 122: 385–399.

Ginzburg, L.R., and M. Colyvan, 2004. *Ecological Orbits: How planets move and populations grow.* New York: Oxford University Press.

Ginzburg, L.R., and C.X.J. Jensen, 2004. Rules of thumb for judging ecological theories. *Trends in Ecology and Evolution* 19: 121–126.

Haemig, P.D., 2006. Laws of Population Ecology. *ECOLOGY.INFO* #23.

Hansen, J.E., et al., 1986. Climate Sensitivity to Increasing Greenhouse Gases. In *Greenhouse Effect and Sea Level Rise: A Challenge for this Generation*, Barth M.C., and J.G. Titus, eds. New York: Van Nostrand Reinhold.

Hansen, J.E., et al., 1989. Greenhouse Effect of Chlorofluorocarbons and Other Trace Gases, *Journal of Geophysical Research* 94 November, 16: 417–416, 421.

Hegerl, G.C., F.W. Zwiers, P. Braconnot, N.P. Gillett, Y. Luo, J.A. Marengo Orsini, N. Nicholls, J.E. Penner, et al., 2007. Understand and Attributing Climate Change—Spatial and Temporal Patterns of the Response to Different Forcings and their Uncertainties. In Solomon, D. Quin, M. Manning, et al., *Climate Change 2007: The Physical Science Basis. Contribution of Working Group I to the Fourth Assessment Report of the Intergovernmental Panel on Climate Change*, Intergovernmental Panel on Climate Change, Cambridge, UK: Cambridge University Press.

Henry, J.G., and G.W. Heinke, 1995. *Environmental Science and Engineering*, 2nd ed., New York: Prentice Hall.

Hestenes, D., 2011. Modeling Methodology for Physics Teachers. Accessed 04/11/11 at http://Modeling.la.asu.edu/modeling/ModMeth.html.

Hickman, C.P., L.S. Roberts, and F.M. Hickman, 1990. *Biology of Animals.* St Louis: Time Mirror/Mosby College Publishing.

Hubbell, S.P., and L.K. Johnson, 1977. Competition and next spacing in a tropical stingless bee community. *Ecology.* 58: 949–963.

Jackson, A. R., and J.M. Jackson, 1996. *Environmental Science: The Natural Environment and Human Impact.* New York: Longman.

Kerlinger, P., 1995. *How Birds Migrate.* Mechanicsville, PA: Stackpole Books.

Kish, L., 1967. *Survey sampling*, 2nd ed. New York: Wiley.

Koren, H., and M. Bisesi, 2003. *Handbook of Environmental Health*, Vol. 1, 4th ed. Boca Raton, FL: CRC Press.

Kormondy, E.J., 1984. *Concepts of Ecology*, 3rd ed. Englewood Cliffs, NJ: Prentice-Hall.

Krebs, R.E., 2001. Scientific laws, principles and theories. Westport, CT: Greenwood Press.

Kuhn, T.S., 1996. *The Structure of Scientific Revolutions*, 3rd ed. Chicago and London: Univ. of Chicago Press.

Ladurie, E.L., 1971. *Times of Feast, Times of Famine: A History of Climate Since the Year 1000.* New York: Doubleday.

Lave, L.B., 1981. *The Strategy of Social Regulations: Decision Frameworks for Policy.* Washington, D.C.: Brookings.

Laws, E.A., 1993. *Environmental Science: An Introductory Text.* New York: Wiley.

Leopold, A., 1948. *A Sand County Almanac, And Sketches Here and There.* New York: Oxford University Press.

Leopold, A., 1970. *A Sand County Almanac.* New York: Ballentine Books.

Liebig, J., 1840. *Chemistry and its application to agriculture and physiology.* London: Taylor and Walton.
Malthus, T.R., 1798. *An Essay on the Principle of Population.* London: J. Johnson.
Manahan, S.E., 1997. *Environmental Science and Technology*, Boca Raton, FL: Lewis.
Masters, G.M., 1991. *Introduction to Environmental Engineering and Science.* Englewood Cliffs, NJ: Prentice Hall.
McHibben, B., 1995. *Hope, Human and Wild: True Stories of Living Lightly on the Earth.* Boston: Little Brown & Company.
Merriam-Webster, 2009. Online Dictionary. *Science.* Accessed 09/03/09 at http://www.m-w.com/dictionary/science.
Miller, G.T., 1997. *Environmental Science: Working with the Earth,* 5th ed. Belmont, CA: Wadsworth.
Miller, G.T., 2004. *Environmental Science*, 10th ed. Australia: Brookscole.
Miller, G.T., 1988. *Environmental Science: An Introduction.* Belmont, CA: Wadsworth.
Miller, T., Jr., and Brewer, R., 2008. *Living in the Environment.* Belmont, CA: Brooks Cole.
Molina, M.J., and F.S. Rowland, 1974. Stratospheric sink for chlorofluoromethanes: Chlorine atom-catalyzed destruction of ozone. *Nature* 249: 810–812.
Moran, J.M., M.D. Morgan, and H.H. Wiersma, 1986. *Introduction to Environmental Science.* New York: W.H. Freeman.
Nature, 2002. Contrails reduce daily temperature range, Vol. 418. Accessed 03/02/09 at www.nature.com/nature.
NCES, 2011. National Center for Education Statistics, U.S. Dept of Education-Environmental Science: Facts and Discussion Forum. Accessed at http://nces.ed.gov.
Odum, E.P., 1971. *Fundamentals of Ecology.* Philadelphia: Saunders College Publishing.
Odum, E.P., 1975. *Ecology: The Link between the Natural and the Social Sciences.* New York: Hold, Rinehart, and Winston, Inc.
Odum, E.P., 1983. *Basic Ecology.* Philadelphia: Saunders College Publishing.
Ophuls, W., *Ecology and the Politics of Scarcity.* New York: W.H. Freeman, 1977.
Pepper, I.L., C.P. Gerba, and M.L. Brusseau, 1996. *Pollution Science.* San Diego: Academic Press Textbooks.
Pianka, E.R., 1988. *Evolutionary Ecology.* New York: Harper Collins.
Popper, K., 1959. *The Logic of Scientific Discovery,* 2nd ed. New York: Routledge.
Porteous, A., 1992. *Dictionary of Environmental Science and Technology.* New York: John Wiley.
Price, P.W., 1984. *Insect Ecology.* New York: Wiley.
Ramade, F., 1984. *Ecology of Natural Resources.* New York: Wiley.
Ramanathan, V., 2006. Atmospheric brown clouds: Health, climate and agriculture impacts. *Pontifical Academy of Sciences Scripta Varia,* 106: 47–60.
Sagan, C., 1996. *The Demon-Haunted World: Science as a Candle in the Dark.* NY: Ballantine.
Sharov, A., 1992. Life-system approach: A system paradigm in Population ecology. *Oikos* 63: 485–494.

Sharov, A., 1996. *What Is Population Ecology*? Blacksburg, VA: Department of Entomology, Virginia Tech University.
Sharov, A., 1997. *Population Ecology*. Accessed 02/28/07 at http://www.gypsymoth.ent.ut.edu/Sharov/population/welcome.
Shiver, B.D., and B.E. Borders, 1996. Systematic sampling with multiple random starts. *For. Sci.* 6: 42–50.
Smith, T.M.F., 1994. Sample surveys: 1975-1990; an age of reconciliation? *International Statistical Review*. 62: 5–34.
Smith, R.L., 1974. *Ecology and Field Biology*. New York: Harper & Row.
Spellman, F.R., 1996. *Stream Ecology and Self-Purification: An Introduction for Wastewater and Water Specialists*. Lancaster, PA: Technomic.
Spellman, F.R., and N. Whiting, 2006. *Environmental Science and Technology: Concepts and Applications*. Boca Raton, FL: CRC Press.
Spellman, F.R., 2007. *Ecology for Non-Ecologists*. Lanham, MD: Government Institutes Press.
Stanhill, G., and S. Moreshet, 2004. Global radiation climate changes in Israel. *Climatic Change* 22: 121–138.
Taylor, J.D., 2005. *Gunning the Eastern Uplands*. Lancaster, PA: Bonasa Press.
Taylor, J.D., 2002. *The Wild Ones: A Quest for North America's Forest and Prairie Grouse*. Lancaster, PA: Bonasa Press.
Time, 1998. Global Warming: It's Here . . . And Almost Certain to Get Worse. August 24.
Tomera, A.N., 1989. *Understanding Basic Ecological Concepts*. Portland, ME: J. Weston Walch.
Tower, E., 1995. *Environmental and Natural and Natural Resource Economics*. New York: Eno River Press.
Townsend, C.R., J.L. Harper, and M. Begon, 2000. *Essentials of Ecology*. Blackwell Science.
Trefil, J., 2008. *Why Science?* New York: Teachers College Press.
Turchin, P., 2001. Does population ecology have general laws? *Oikos* 94: 17–26.
Turchin, P., 2003. *Complex Population Dynamics: A Theoretical/Empirical Synthesis*. USA: Princeton University Press.
Upton, G, and I. Cook, 2008. *Oxford Dictionary of Statistics*. New York: Oxford University Press.
USA Today, 1997, Global Warming: Politics and economics further complicate the issue. A-1, 2, December 1.
USA Today, 2009. Your eyes aren't deceiving you: Sikes are dimmer. Accessed 03/13/09 at http://www.usadtodya.com/weaterh/environment.
USDA, 2004. *Statistical Techniques for Sampling and Monitoring Natural Resources*. Schreuder, H.T., R. Ernst, H. Ramirez-Maldonado. Washington, D.C.: U.S. Department of Agriculture.
USDA, 2011. *Pond Fact Sheet No. 17*. Washington, D.C.: U.S. Department of Agriculture.
USDA, FS 1995. *Forested Wetlands*. NA-PR-01-95. Washington, D.C.: U.S. Dept. of Agriculture.

USEPA, 2011. *Sustainability*. Accessed 04/11/11 at http://epa.gov/sustainability/basicinfo.htm.

USGS, 2009. *Hydroelectric Power Water Use*. Accessed 11/05/09 at http://ga.water.usgs.bof/edu/suhy.html.

USFS, 2009. *U.S. Forest Resource Facts and Historical Trends*. Accessed 03/26/11 at http://fia.fs.fed.us.

USGS, 1999. *Hawaiian Volcano Observatory*. Accessed 03/01/07 at http://hvo.wr.usgs.gov /volcano-watch/1999.

Verhulst, P.F., 1838. Notice sur la loi que la population suit dans son accrossement. *Corr. Math. Phys.* 10: 113–121.

Volterra, V., 1926. Variazioni e fluttuazioni del numero d'indivudui in specie animali conviventi. *Mem. R. Accad. Naz. die Lincei Ser. VI* 2.

Wachernagel, M., 1997. Framing the Sustainability Crisis: Getting from Concerns to Action. Accessed 02/26/07 at http://www.sdri,ubc.ca/publications/wacherna.html.

Wadsworth, H.M. 1990. *Handbook of Statistical Methods for Engineers and Scientists*. New York: McGraw-Hill.

Walker, M., 1963. *The Nature of Scientific Thought*. Englewood Cliffs, N.J., Prentice-Hall, Spectrum Books.

Wanielista, M.P., Y.A. Yousef, J.S. Taylor, and C.D. Cooper, C.D., 1984. *Engineering and the Environment*. Monterey, CA: Brooks/Cole Engineering Division.

WCED, 1987. World commission on environment and development. *Our Common Future*. New York: Oxford University Press.

Wessells, N.K., and J.L. Hopson, 1988. *Biology*. New York: Random House.

Winstead, R.L., 2007. *Population Regulation*. Accessed 02/28/07 at http://nsm1.nsm.iup.edu/.

WMO, 2009. *Manual of Codes*. Accessed 03/02/09 at http://www.wmo.ch/pages/prog/www/WMOCodes/Manual/WMO306_vol-1-2-PartB.pdf.

Wood, R., et al., 2007. Climate Models and their Evaluation. In Solomon, S., D. Qin, M. Manning, Z. Chen, M. Marquis, K.B. Averyt, M. Tignor, and H.L. Miller, *Climate Change 2007: The Physical Science Basins*. Cambridge, UK: Cambridge University Press.

Yassi, A.T., Kjellstri, T. deKok, and T. Guidotti. *Basic Environmental Health*. New York: Oxford University Press.

Zurer, P.S., 1988. Studies on Ozone Destruction Expand Beyond Antarctic, *C & E News*, 18–25 May.

3
Toxicology

> We are rightly appalled by the genetic effects of radiation; how then, can we be indifferent to the same effects in chemicals we disseminate widely in our environment?
>
> —Rachel Carson

Funeral Homes: A Toxicological Concern

FUNERAL HOMES PERFORM SERVICES that involve health and safety concerns for their employees. One of the main health concerns at funeral homes is exposure to formaldehyde—formaldehyde is used in embalming. Formaldehyde is listed as a human carcinogen in the Twelfth Report on Carcinogens published by the National Toxicology Program because it causes cancer of the throat, nose, and blood. It is also a volatile organic compound (VOC). Proper ventilation is the most effective way to control exposure to formaldehyde in embalming rooms. Personal and respiratory protective equipment is also required to reduce exposure to formaldehyde. Note that formaldehyde as it is used in funeral homes poses no danger to people attending funerals. Potential risks from formaldehyde are to funeral home employees only.

Funeral homes also use glutaraldehyde, methanol ethanol, and phenol as preservatives and disinfectants. They use solvents such as methyl alcohol, glycerol, perchloroethylene (PERC), trichloroethylene (TCE), and other chlorinated compounds. Moisturizing agents include glycerine, ethylene glycol, and propylene glycol. Sodium phosphate and sodium citrate are used to stabilize embalming solutions.

Funeral employees can also be exposed to bloodborne pathogens that can cause disease. At times, employees must wear personal and respiratory protective equipment when they work with potential hazards, such as bloodborne pathogens, preservatives, and disinfectants. Respiratory protective equipment can prevent exposure to respiratory hazards such as flu viruses, bacteria, and organ vapors.

Other concerns at funeral homes include the need for proper ventilation and disposal of medical waste, solid waste, and wastewater. Wastewater and hazardous waste disposal practices must comply with state and federal regulations.

* * *

As defined by CDC (2009), *toxicology* is "the study of how natural or man-made poisons cause undesirable effects in living organisms." Another definition is that it is "the science that deals with the effects, antidotes, and detection of poisons" (Random House, 2009). For our purposes we prefer the following simplified definition, one we have used and taught for years: *Toxicology is the study of poisons . . . or the study of the harmful effects of chemicals.*

Why should we care about toxicology? Isn't it enough to know that we don't want pesticides on our food or chemicals in our water? Yes; it is important to know about the pesticides on our food or the chemicals in our water, but, as the discussion that follows confirms, the challenges in the field are enormous. Challenges? Absolutely! Consider the emphasis placed on the toxicological challenges that confront us now and in the future and the central role of toxicology in identifying the potential hazards of numerous chemicals in use in the United States. According to the US HHS (2004):

> More than 80,000 chemicals are registered for use in the United States [more than 70,000 are in common use]. Each year, an estimated 2,000 new ones are introduced for use in such everyday items as foods, personal care products, prescription drugs, household cleaners, and lawn care products. We do not know the effects of many of these chemicals on our health, yet we may be exposed to them while manufacturing, distributing, using, and disposing of them or when they become pollutants in our air, water, or soil. Relatively few chemicals are thought to pose a significant risk to human health. However, safeguarding public health depends on identifying both what the effects of these chemicals are and at what levels of exposure they may become hazardous to humans—that is understanding toxicology. (p. 4)

In this chapter we define terms used in environmental toxicology and provide an overview of the discipline in order to establish a basic foundation for the environmental health applications covered in the remainder of the text.

The Chemical World We Live In

Chemicals are everywhere in our environment. All matter on our planet consists of chemicals. We are made up of a few thousand different types of chemicals, some of which are considered toxic. The vast majority of these chemicals are natural; in fact, the most potent chemicals on the planet are those occurring naturally in plants and animals.

Did You Know?

Arsenic is a naturally occurring element widely distributed in the earth's crust. In the environment, arsenic is combined with oxygen, chlorine, and sulfur to form inorganic arsenic compounds. Arsenic in animals and plants combine with carbon and hydrogen to form organic arsenic compounds.

Natural chemicals are sometimes presented by the media as being "safe" relative to manufactured chemicals. As a result, they may also be considered "safe" by much of the public.

For example, "organic" produce and livestock are becoming more popular. But does this mean pesticides are unsafe? Actually, manufactured pesticides used on crops and animals are heavily regulated and rarely contain enough chemicals to be harmful at typically encountered levels. This book is designed, formatted, and presented in such as manner so as to enable you to make your own conclusions about health hazards from chemicals in the environment, and to be a more informed practitioner of environmental health, professional, and consumer. Also, lack of knowledge about a topic often leads to unnecessary fear of the unknown. Moreover, ignorance can lead to disregard of the potential catastrophic consequences of our actions. Consider the following discussion from a historical perspective of the latter concept.

For decades the Roman Empire was the crown jewel of the world. The expanse of land controlled by the Romans grew at an amazing rate during this period, and their rule led to the development of irrigation systems, roads, public sporting events, and a government that, at least in part, helped improve the lot of its citizens. With the success of the empire and the expansion of other societies came the trappings of wealth. However, this wealth was retained in the ruling class. Great feasts were held daily in which the ruling class, especially the emperors, ate tremendous meals and imbibed wines from all over the world. Some of the poorer quality wines were augmented with flavor enhancers before the aristocracy consumed them. This wine was served in the best goblets, made by master craftsmen and collected from around Greece and other European countries. The meals were served in the best bronze and copper pots. Emperors such as Nero, Claudius, and Caligula were perhaps those who took this feasting to its greatest extent.

Meanwhile, the common folk were unable to afford the great wine and lovely goblets, and instead were only able to buy cheap wine flasks and poor quality wine. They were also unable to afford the bronze and copper pots for cooking, and were only able to buy the cheaper metallic pots.

Did You Know?

Inorganic arsenic compounds are mainly used to preserve wood. Copper chromate arsenate (CCA) is used to make "pressure-treated" lumber. CCA is no long used in the United States for residential uses; it is still used in industrial applications. Organic arsenic compounds are used as pesticides, primarily on cotton fields and orchards.

After several years, the rulers began to behave strangely. Claudius started forgetting things and slurring his speech. He also began to slobber and walk with a staggering gait. Many of the decisions he made adversely impacted the empire, and there was little basis for these decisions. Eventually, he was replaced as emperor because he could no longer function as a ruler. In another famous story, the mad emperor Nero fiddled while his city burned. He became insane while ruling, and his actions had an adverse impact on the empire. Caligula became sexually depraved and suffered a mental breakdown. The ruling class seemed plagued by neurological diseases. Eventually, the repeated succession of apparently incompetent rulers began the decline of the Roman Empire.

For hundreds of years the actions of these rulers were not understood. It was a mystery why similar diseases did not appear with nearly the same frequency in the common class. Why was it that the aristocracy had such a high rate of neurological disorders?

Historians interested in Roman architecture and society soon discovered some facts about Roman and Greek manufacturing that at first seemed unrelated to these events. The Romans and Greeks would typically coat bronze and copper cooking pots and goblets with lead to prevent copper or other metals from being dissolved into food or drink. Adding lead compounds prior to being served to the aristocracy enhanced the flavor of poor wines.

Writers of the Roman Empire era noted that the excessive use of lead-treated wines led to "paralytic hands" and "harm to the nerves." Numbness, paralysis, seizures, insomnia, stomach distress, and constipation are other symptoms of lead toxicity.

Now we know that lead is a neurological poison when even relatively low amounts are consumed. A theory was advanced that the neurological problems of the emperors may have been caused, at least in part, by lead toxicity. To test the theory, researchers prepared a liter of wine according to an ancient Roman recipe and extracted 237 milligrams of lead. Based on the known excessive habits of these emperors, it was estimated using risk assessment techniques that the average Roman aristocrat consumed 250 milligrams of lead daily. Some aristocrats, notably the emperors discussed above, may have consumed a gram or more of lead daily. Today, Americans that live in cities are known to have the highest lead exposures from a variety of sources, including leaded water pipes, lead-based paint, and food.

Even with the concern about lead exposure by these people, the average city-dwelling American consumes only about 30 to 50 micrograms of lead daily. By this comparison, the Roman aristocracy consumed on average eight times what a highly exposed city-dwelling American now consumes. Some emperors may have consumed twenty times more than city-dwelling Americans.

Despite the more obvious neurological effects, the more important effects of lead may have contributed to the decline of the Roman Empire through contributing to a declining birth rate and shorter lifespan among the ruling class.

This example illustrates that general lack of knowledge about toxicology led to everyday practices that may have contributed to shaping our history. Similarly, toxicology has an impact on our lives today. The foods we eat and the additives put into them are routinely tested for chemicals. Levels of pesticides are randomly measured in food. Our drinking water is purified to eliminate harmful chemicals. Air pollution is monitored, and many efforts are under way to improve our global air quality. All of these routine actions in today's world originated when the effects of various substances on our health and the health of wildlife and the environment became evident.

Did You Know?

When arsenic enters the environment:

- It cannot be destroyed in the environment. It can only change form.
- Rain and snow remove arsenic dust particles from the air.
- Many common arsenic compounds can dissolve in water.

What about the chemicals we are exposed to daily? Are the effects of our daily exposure to these substances on our health and the health of the wildlife and the environment so evident? The best way to answer these questions is by providing another example, an example we call Sick Water—for reasons that will become obvious to the reader.

Sick Water

The term *sick water* was coined by the United Nations in a 2010 press release addressing the need to recognize that it is time to arrest the global tide of sick water. The gist of the UN's report pointed out that transforming waste from a major health and environmental hazard into a clean, safe, and economically attractive resource is emerging as a key challenge in the twenty-first century. As practitioners of environmental health, we certainly support the UN's view on this important topic.

However, when we discuss sick water, in the context of this text and in many others we have authored on the topic we go a few steps further than the UN in describing the real essence and tragic implications of potable water that makes people or animals sick or worse.

Water that is sick is actually a filthy medium, spent water, wastewater—a cocktail of fertilizer runoff and sewage disposal alongside animal, industrial, agricultural, and other wastes. In addition to these listed wastes of concern, other wastes are beginning to garner widespread attention; they certainly have earned our attention in our research on the potential problems related to these so-called other wastes.

What are these other wastes? Any waste or product we dispose of in our waters, that we flush down the toilet, pour down the sink or bathtub, pour down the drain of a worksite deep sink. Consider the following example of "pollutants" we discharge to our wastewater treatment plants or septic tanks—wastes we don't often consider as waste products, but that in reality are waste products.

Each morning a family of four wakes up and prepares for the day: the two parents to go to work and the two teenagers to go to school. Fortunately, this family has three upstairs bathrooms to accommodate everyone's need to prepare for the day; via the morning natural waste disposal, shower and soap usage, cosmetic application, hair treatments, vitamins, sunscreen, fragrances, and prescribed medications end up down the various drains. In addition, the overnight deposit of cat and dog waste is routinely picked up and flushed down the toilet. Let's examine a short inventory of what this family of four has disposed of or has applied to themselves as they prepare for their day outside the home.

- Toilet-flushed animal wastes
- Prescription and over-the-counter therapeutic drugs
- Veterinary drugs
- Fragrances
- Soap

- Shampoo, conditioner, and other hair treatment products
- Body lotion, deodorant, and body powder
- Cosmetics
- Sunscreen products
- Diagnostic agents
- Nutraceuticals (e.g., vitamins, medical foods, functional foods, etc.)

Even though these bioactive substances have been around for decades, today we group all of them (the exception being animal wastes), substances and/or products, under the title of pharmaceuticals and personal care products called "PPCPs" (EPA, 2012).

We pointed to the human activities of the family of four in contributing PPCPs to the environment, but other sources of PPCPs should also be recognized. For example, residues from pharmaceutical manufacturing; residues from hospitals, clinics, doctors, veterinary offices, or urgent care facilities; illicit drug disposal (i.e., which occurs when the police knock on the door and the frightened user flushes the illicit drugs down the toilet [along with $100 bills, weapons, dealers' phone numbers, etc.] and into the wastewater stream); veterinary drug use, especially antibiotics and steroids; and agribusiness are all contributors of PPCPs in the environment.

In our examination of the personal deposit of PPCPs to the environment and to the local wastewater supply, let's return to that family of four. After having applied or ingested the various substances mentioned earlier, the four individuals involved unwittingly add traces (or more than traces) of these products, PPCPs, to the environment through excretion (the elimination of waste material from the body) and when bathing, and then possibly through disposal of any unwanted medications to the sewers or trash. How many of us have found old medical prescriptions in the family medicine cabinet and decided they were no longer needed? How many of us have grabbed up such unwanted medications and disposed of them with a single toilet flush? Many of these medications (for example, antibiotics) are not normally found in the environment.

Earlier we stated that wastewater is a cocktail of fertilizer runoff and sewage disposal with additions of animal, industrial, agricultural, and other wastes. When we factor in PPCPs to this cocktail, we can state analogously that we are simply adding mix to the mix.

The questions about our mixed waste cocktail are obvious: Does the disposal of antibiotics or other medications into the local wastewater treatment system cause problems for anyone or anything else downstream? When we ingest locally treated water, are we also ingesting flushed-down-the-toilet or flushed-down-the-drain antibiotics, other medications, illicit drugs, animal excretions, cosmetics, vitamins, various personal or household cleaning products, sunscreen products, diagnostic agents, crankcase oil, grease, oil, fats, and veterinary drugs and hormones anytime we drink a glass of tap water?

The jury is still out on these questions. Simply, we do not know what we do not know about the fate of PPCPs or their impact on the environment once they enter our wastewater treatment systems, the water cycle, and eventually our drinking water supply systems. Even though some PPCPs are easily broken down and processed by the human body or degraded quickly in the environment, we have recognized the disposal of certain wastes as problematic for some time—in fact, since the time of the mythical hero Hercules (arguably the world's first environmental engineer), when he performed his fifth labor by cleaning up King Augeas's stables. Hercules, faced literally with a mountain of horse and cattle waste piled high in the stable area, had to devise some method to dispose of the waste; he diverted a couple of rivers to the stable interior. All the animal waste was simply deposited into the river: Out of sight, out of mind. The waste just followed the laws of gravity; it flowed downstream, becoming someone else's problem. Hercules understood the principal point in pollution control technology, one pertinent to this very day; that is, *dilution is the solution to pollution.*

As applied to today, the fly in the ointment in the Pollution Solution is that the dilution approach is today's modern PPCPs. Although Hercules was able to dispose of animal waste into a running water system where eventually the water's self-purification process cleaned the stream, he didn't have to deal with today's personal pharmaceuticals and the hormones that are given to many types of livestock to enhance health and growth.

Studies show that pharmaceuticals are present in our nation's waterbodies. Further research suggests that certain drugs may cause ecological harm. The EPA and other research agencies are committed to investigating this topic and developing strategies to help protect the health of both the environment and the public. To date, scientists have found no hard evidence of adverse human health effects from PPCPs in the environment. Others might argue that even if PPCPs were present today or in ancient (and mythical) times, the amount in local water systems and other supply areas would represent only a small fraction (ppt—parts per trillion, 10^{-12}) of the total volume of water. Critics would be quick to point out that when we are speaking of parts per trillion (ppt) we are speaking of a proportion equivalent to one-twentieth of a drop of water diluted into an Olympic-size swimming pool. One student in our environmental health classes stated that he did not think the water should be termed "sick water"; he wondered, if the water contained so many medications, how could it be sick? Instead it might be termed "well water—making anyone who drinks it well, cured, no longer sick, etc."

We point out that the term "sick water" can be applied to not only PPCP-contaminated water but also to any filthy, dirty, contaminated, vomit-contaminated, polluted, pathogen-filled drinking water source. The fact is, dirty or sick water means that, worldwide, more people now die from contaminated and polluted water than from all forms of violence, including wars (UN, 2010[1]). The United Nations also points out that dirty or sick water is also a key factor in the rise of de-oxygenated dead zones that have been emerging in seas and oceans across the globe.

Did You Know?

We may be exposed to arsenic by:

- Ingesting small amounts present in your food and water or breathing air containing arsenic.
- Breathing sawdust or smoke from wood treated with arsenic.
- Living in areas with unusually high natural levels of arsenic in rock.
- Working in a job that involves arsenic production or use, such as copper or lead smelting, wood treating, or pesticide application.

Toxicology and Risk

The study of the effects of chemicals on our health and the health of wildlife combines toxicology with risk assessment. Risk assessment evaluates the relative safety of chemicals from an exposure approach, considering that both contact with a chemical and the inherent toxicity of the chemical are needed to have an effect. We can't eliminate the toxicity of a chemical, but we can limit our exposure to it.

This is the general approach behind the development of most pesticides. Many chemicals are designed to target only insects, leaving the crops essentially free of pesticides. Even though the pesticides are toxic, there is very little risk from eating the crops because almost no chemical remains in the food.

Did You Know?

Ingesting large levels of arsenic can result in death. Ingesting or breathing low levels of inorganic arsenic for a long time can cause a darkening of the skin and the appearance of small "corns" or "warts" on the palms, soles, and torso.

Toxicological Terms

To gain an understanding of toxicology and how the discipline interrelates with environmental health, it is necessary to understand the key terms used in the practice. In light of this need, we have provided the following questions with answers and selected definitions from the Agency for Toxic Substances & Disease Registry (ATSDR) of the Centers for Disease Control and Prevention (CDC, 2009).

What is toxic? This term relates to poisonous or deadly effects on the body by inhalation (breathing), ingestion (eating), or absorption, or by direct contact with a chemical.

What is a toxicant? A toxicant is any chemical that can injure or kill humans, animals, or plants; a poison. The term "toxicant" is used when talking about toxic substances that are produced by or are a by-product of human activities. For example, dioxin (2,3-7,8-tetrachlorodibenzo-p-dioxidn [TCDD]), produced as a by-product of certain chlorinated chemicals, is a toxicant. On the other hand, arsenic, a toxic metal, may occur as a natural contaminant of groundwater or may contaminate groundwater as a by-product of industrial activities. If the second case is true, such toxic substances are referred to as toxicants, rather than toxins.

What is a toxin? The term "toxin" usually is used when talking about toxic substances produced naturally. A toxin is any poisonous substance of microbial (bacteria or other tiny plants or animals), vegetable, or synthetic chemical origin that reacts with specific cellular components to kill cells, alter growth or development, or kill the organisms.

What is a toxic symptom? This term includes any feeling or sign indicating the presence of a poison in the system.

What are toxic effects? This term refers to the health effects that occur due to exposure to a toxic substance; also known as a poisonous effect on the body.

What is toxicity? This can be defined as the degree to which a chemical inherently causes adverse effects. The concept of toxicity is not clear-cut. Cyanide, for example, is more toxic than table salt because toxic effects can result from much lower amounts of cyanide. However, as Stelljes (2000) points out, we are not usually exposed to cyanide, but we are exposed to table salt daily. Which chemical is more dangerous to us?

What is selective toxicity? "Selective toxicity" means that a chemical will produce injury to one kind of living matter without harming another form of life, even though the two may exist close together.

How does toxicity develop? Before toxicity can develop, a substance must come into contact with a body surface such as skin, eye, or mucosa of the digestive or respiratory tract. The dose of the chemical, or the amount one comes into contact with, is important when discussing how "toxic" a substance can be.

What is the dose? The dose is the actual amount of a chemical that enters the body. The dose received may be due to either acute (short-term) or chronic (long-term) exposure. An acute exposure occurs over a very short period of time, usually twenty-four hours. Chronic exposures occur over long periods of time such as weeks, months, or years. The amount of exposure and the type of toxin will determine the toxic effect.

What is dose-response? Dose-response is a relationship between exposure and health effect that can be established by measuring the response relative to an increasing dose. This relationship is important in determining the toxicity of a particular substance. It relies on the concept that a dose, or a time of exposure (to chemical, drug, or toxic substance), will cause an effect (response) on the exposed organism. Usually, the larger or more intense the dose, the greater the response, or the effect. This is the meaning behind the statement "the dose makes the poison."

It is important to point out that the dose-response relationship, which, again, defines the potency of a toxin, is typically the primary thrust of basic toxicology training. This is the case because the dose-response relationship is the most fundamental and pervasive concept in toxicology. To understand the potential hazard of a specific toxin (chemical), toxicologists must know both the type of effect it produces and the amount, or dose, required to produce that effect.

The relationship of dose to response can be illustrated as a graph called a *dose-response curve*. There are two types of dose-response curves: one that describes the graded responses of an *individual* to varying doses of the chemical and one that describes the distribution of response to different doses in a *population* of individuals. The dose is represented on the x-axis. The response is represented on the y-axis (figure 3.1).

An important aspect of dose-response relationships is the concept of *threshold* (figure 3.1). For most types of toxic responses, there is a dose, called a threshold, below which there are no adverse effects from exposure to the chemical. This is important because it identifies the level of exposure to a toxin at which there is no effect.

Threshold and non-threshold types of basic dose-response relations are illustrated in figure 3.1. As shown in the figure, any dose can cause an effect in a non-threshold relationship (Chemical B). As the dose increases, the relative effect will also increase. For a threshold relationship, which is the more common of the two across all chemicals, effects only occur above a certain dose (Chemical A). Above the threshold level, there is little difference between the

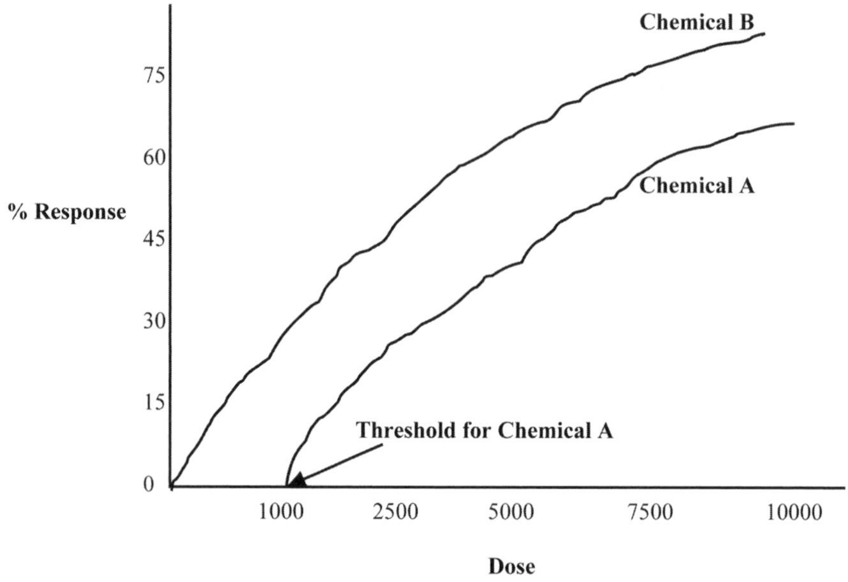

FIGURE 3.1
Dose-response curves for threshold and non-threshold reponses.

two relationships. At high doses, the lines are essentially parallel, indicating that the response changes at the same rate for both chemicals. The rate at which effects increase with the dose defines the potency of a chemical. This is the slope of the line in Figure 3.1. We can use this principle to compare the potencies of different chemical by comparing their slopes (Stelljes, 2000).

What is the threshold dose? As mentioned, given the idea of a dose-response, there should be a dose or exposure level below which the harmful or adverse effects of a substance are not seen in a population. That dose is referred to as the "threshold dose." This dose is also referred to as the *no observed adverse effect level (NOAEL)* or the *no effect level (NEL)*. These terms are often used by toxicologists when discussing the relationship between exposure and dose. However, for substances causing cancer (carcinogens), no safe level of exposure exists, since any exposure could result in cancer.

What is meant by "individual susceptibility"? This term describes the differences in types of responses to hazardous substances, between people. Each person is unique, and because of that, there may be great differences in the response to exposure. Exposure in one person may have no effect, while a second person may become seriously ill, and a third may develop cancer.

What is a "sensitive sub-population"? A sensitive sub-population describes those persons who are more at risk from illness due to exposure to hazardous substances than the average, healthy person. These persons usually include the very young, the chronically ill, and the very old. It may also include pregnant women and women of childbearing age. Depending on the type of contaminant, other factors (e.g., age, weight, lifestyle, gender) could be used to describe the population.

What is LD50/LC50? A common measure of the acute toxicity is the lethal dose *(LD50)* or lethal concentration *(LC50)* that causes death (resulting from a single or limited exposure) in 50 percent of the treated animals, known as the population. LD50 is generally expressed as the dose, milligrams (mg) of chemical per kilogram (kg) of body weight. LC50 is often expressed as mg of chemical per volume (with results expressed in terms of an air concentration, e.g., ppm) the organism is exposed to. Chemicals are considered highly toxic when the LD50/LC50 is small and practically non-toxic when the figure is large.

What are routes of entry? A toxin is not toxic to a living system unless there is exposure. Chemical exposure is defined as contact with a chemical by an organism. For humans, there are three primary routes of chemical exposure, in order of importance:

- Through breathing (i.e., inhalation)
- By touching (i.e., dermal contact)
- By eating (i.e., ingestion)

Exposures to toxic chemicals can be in the form of liquids, gases, mists, fumes, dusts, and vapors. Generally, toxic agents are classified in terms of their target organs, use, source, and effects.

Other Pertinent Toxicological Definitions

Absorption—passage of a chemical across a membrane and into the body.

Acceptable risk—a risk that is so low that no significant potential for toxicity exists, or a risk society considers is outweighed by benefits.

Acute—a single or short-term exposure period.

Alkaloid—adverse group of structurally related chemicals naturally produced by plants; many of these chemicals have high toxicity.

Ames Assay—a popular laboratory *in vitro* test for mutagenicity using bacteria.

Anesthetic—a toxic depressant effect on the central nervous system.

Bioassay—a toxicity study in which specific toxic effects from chemical exposure are measured in the laboratory using living organisms.

Carcinogen—a cancer-causing substance.

Chronic—an exposure period encompassing the majority of the lifespan for a laboratory animal species, or covering at least 10 percent of a human's lifespan.

Dermal contact—exposure to a chemical through the skin.

Exposure—contact with a chemical by a living organism.

Hazard—degree of likelihood of non-cancer adverse effects occurring from chemical exposure.

Insecticide—a pesticide that targets insects.

Mutagen—a change in normal DNA structure.

Pesticide—a chemical used to control pests.

Potency—the relative degree of toxic effects caused by a chemical at a specific dose.

Risk—the probability of an adverse effect resulting from an activity or from chemical exposure under specific conditions.

Sensitivity—the intrinsic degree of an individual's susceptibility to a specific toxic effect.

Target organ—the primary organ where a chemical causes non-cancer toxic effects.

Teratogen—a chemical causing a mutation in the DNA of a developing offspring.

Threshold dose—a dose below which no adverse effects will occur.

Xenobiotic—a chemical foreign to a living organism.

Classification of Toxic Agents

(Material in this section is from ATSDR 2009 Toxicology Curriculum—Module 1—Lecture Notes. Accessed at http://www.atsdr.cdc.gov/training/

TABLE 3.1
Scale of Relative Toxicity (Adapted from Stelljes, 2000)

Category	*Concentration*	*Amount for Average Adult*	*Example*
Extremely toxic	<1 mg/kg	taste	botulinum
Highly toxic	1–50 mg/kg cyanide	7 drops to a teaspoon	nicotine
Moderately toxic	50–500 mg/kg	teaspoon to an ounce	DDT
Slightly toxic	500–5,000 mg/kg	ounce to a pint	salt
Practically nontoxic	5,000–15,000 mg/kg	pint to a quart	ethanol
Relatively harmless	>15,000 mg/kg	>1 quart	water

Source: Adapted from Stelljes, 2000.

toxmanal/modules/1/lecturenotes.html; Klaassen, 1996.) Toxic substances are classified into the following:

Heavy metals—Metals differ from other toxic substances in that they are neither created nor destroyed by humans. Their use by humans plays an important role in determining their potential for health effects. Their effect on health could occur through at least two mechanisms: first, by increasing the presence of heavy metals in air, water, soil, and food; and second, by changing the structure of the chemical. For example, chromium III can be converted to or from chromium VI, the more toxic form of the metal.

Solvents and vapors—Nearly everyone is exposed to solvents. Occupational exposures can range from the use of "white out" by administrative personnel to the use of chemicals by technicians in a nail salon. When a solvent evaporates, the vapors may also pose a threat to the exposed population.

Radiation and radioactive materials—Radiation is the release and propagation of energy in space or through a material medium in the form of wave, the transfer of heat or light by waves of energy, or the stream of particles from a nuclear reactor (Koren, 1996).

Dioxins/furans—Dioxin (or TCDD) was originally discovered as a contaminant in the herbicide Agent Orange. Dioxin is also a by-product of chlorine processing in paper producing industries.

Pesticides—The EPA defines *pesticide* as any substance or mixture of substances intended to prevent, destroy, repel, or mitigate any pest. Pesticides may also be described as any physical, chemical, or biological agent that will kill an undesirable plant or animal pest (Klaassen, 1996).

Plant toxins—Different portions of a plant may contain different concentrations of chemicals. Some chemicals made by plants can be lethal. For example, taxon, used in chemotherapy to kill cancer cells, is produced by a species of the yew plant (Klaassen, 1996).

Animal toxins—These toxins can result from venomous or poisonous animal releases. Venomous animals are usually defined as those that are capable of producing a poison in a highly developed gland or group of cells, and can deliver that toxin through biting or stinging. Poisonous animals are generally regarded as those tissues, either in part or in their whole, that are toxic (Klaassen, 1996).

Subcategories of toxic substance classifications—All of these substances may also be further classified according to their:

- Effect on target organs (liver, kidney, hematopoietic [body's blood-producing system])
- Use (pesticide, solvent, food additive)
- Source of the agent (animal and plant toxins)

- Effects (cancer mutation, liver injury)
- Physical state (gas, dust, liquid)
- Labeling requirements (explosive, flammable, oxidizer)
- Chemistry (aromatic amine, halogenated hydrocarbon)
- Poisoning potential (extremely toxic, very toxic, slightly toxic)

Significant Chemical and Biological Toxins and Effects

From the environmental health practitioner's point of view, toxicology is concerned with exposure to and the harmful effects of chemicals in all environmental settings. In the definitions that follow, we also include key biological agents along with certain chemical agents because of the nature of our present circumstances in regards to worldwide terrorism. Typically, under normal circumstances (whatever they may be?), environmental health specialists do not concern themselves with smallpox, anthrax, botulinum, etc. The problem is that these are not "normal" times. Another one of the environmental health specialist's principal tasks is to "anticipate" hazards in the workplace. The possibility of and the potential for deliberate exposure to both chemical and biological toxins in today's workplace (e.g., anthrax in the post office), though caused by cowardly, uncivilized, and beastly acts of terrorism, is very real. Moreover, exposure to toxins from bloodborne pathogens, foodborne disease, etc., can occur anywhere, including in the workplace. In a footnote, concerning the foodborne disease issue, note that the landmark Delaney Clause, adopted in 1958 in the United States as part of the Food Additives Amendment, required that any food additive be found "safe" before the FDA approves it for use in food. This means that no chemical can be used as a food additive if there is a known potential for it to cause cancer. The industrial hygienist must be aware of eating habits in the workplace because of the possibility of inadvertent mixing of workplace chemicals (contaminants) with the worker's brown-bag lunch.

The bottom line: at present, we think it is prudent for today's environmental specialists to "anticipate" (and expect) worker exposure to workplace contaminants—no matter the source.

Did You Know?

There are several hundred thousand chemicals either naturally produced or manufactured that have some use for humans. We have adequate human toxicology information for less than 100 of these, and adequate animal toxicology information for less than 1,000. Therefore, we do not have adequate information for almost all chemicals (Stelljes, 2000).

Toxins

Anthrax—Anthrax is an acute infectious disease caused by a spore-forming bacterium called *Bacillus anthracis*. It is generally acquired following contact with anthrax-infected animals or anthrax-contaminated animal products. It can also be acquired following a deliberate terrorist act.

Asbestos—Asbestos is the name given to a group of six different fibrous minerals (amosite, chrysotile, crocidolite, and the fibrous varieties of tremolite, actinolite, and anthophyllite) that occur naturally in the environment. Asbestos minerals have separable long fibers that are strong, flexible enough to be spun and woven, and heat resistant. Because of these characteristics, asbestos has been used for a wide range of manufactured goods, mostly in building materials (roofing shingles, ceiling and floor tiles, paper products, and asbestos cement products), friction products (automobile clutch, brake, and transmission parts), heat-resistant fabrics, packaging, gaskets, and coatings. Some vermiculate or talc products may contain asbestos. Exposure to asbestos fibers can cause mesothelioma or asbestosis.

Avian flu—Avian influenza is a highly contagious disease of birds that is currently an epidemic among poultry in Asia. Despite the uncertainties, poultry experts agree that immediate culling of infected and exposed birds is the first line of defense for both the protection of human health and the reduction of further losses in the agricultural sector.

Bloodborne pathogens and needle stick prevention—OSHA estimates that 5.6 million workers in the health care industry and related occupations are at risk of occupational exposure to bloodborne pathogens, including human immunodeficiency virus (HIV), hepatitis B virus (HBV), hepatitis C virus (HCV), and others.

Botulism—Cases of botulism are usually associated with consumption of preserved foods. However, botulinum toxins are currently among the most common compounds explored by terrorists for use as biological weapons.

Carbon monoxide—Carbon monoxide is a vapor that can pass across the alveoli into the lungs through inhalation. Carbon monoxide causes carboxyhemoglobin formation (CO binds strongly with hemoglobin), replacing oxygen in red blood cells, leading to asphyxiation.

Cotton dusts—Cotton dust is often present in the air during cotton handling and processing. Cotton dust may contain many substances, including ground-up plant matter, fiber, bacteria, fungi, soil, pesticides, non-cotton matter, and other contaminants that may have accumulated during growing, harvesting, and subsequent processing or storage periods. The OSHA Cotton Dust Standard 1910.1043 specifically lists the operations that are covered; operations not specifically listed are not covered by the standard. Covered operations include yarn manufacturing, textile waste houses, slashing and weaving operations, waste recycling, and garneting. Occupational exposure to cotton dusts can cause byssinosis (tightness in chest, chronic bronchitis).

Cyanide—Cyanide is a rapidly acting, potentially deadly chemical that can exist in various forms. Cyanide can be a colorless gas, such as hydrogen cyanide (HCN) or cyanogens chloride (CNCl), or a crystal form such as sodium cyanide (NaCN) or potassium cyanide (KCN). Cyanide sometimes is described as having a "bitter almond" smell, but it does not always give off an odor and not everyone can detect this odor. Cyanide is released from natural substances in some foods and in certain plants such as cassava. Cyanide is contained in cigarette smoke and the combustion products of synthetic materials such as plastics. (Combustion products are substances given off when things burn.) In manufacturing, cyanide is used to make paper, textiles, and plastics. It is present in the chemicals used to develop photographs. Cyanide salts are used in metallurgy for electroplating, metal cleaning, and removing gold from its ore. Cyanide gas is used to exterminate pests and vermin in ships and buildings. If accidentally ingested, chemicals found in acetonitrile-base products that are used to remove artificial nails can produce cyanide.

Foodborne disease—Foodborne illnesses are caused by viruses, bacteria, parasites, toxins, metals, and prions (microscopic protein particles). Symptoms range from mild gastroenteritis to life-threatening neurologic, hepatic, and renal syndromes.

Hantavirus—Hantaviruses are transmitted to humans from the dried droppings, urine, or saliva of mice and rats. Animal laboratory workers and persons working in infested buildings are at increased risk to this disease.

Isocyanates (MIC)—Methyl isocyanate (MIC) is used to produce carbarnate pesticides. Methyl isocyanate is extremely toxic to humans from acute (short-term) exposure. In Bhopal, India, accidental acute inhalation exposure to methyl isocyanate resulted in the deaths of several thousand people and adverse health effects in greater than 170,000 survivors. Pulmonary edema was the probable cause of death in most cases, with many deaths resulting from secondary respiratory infections.

Legionnaires' disease—Legionnaires' disease is a bacterial disease commonly associated with water-based aerosols. It is often the result of poorly maintained air-conditioning cooling towers and portable water systems.

Mercuric nitrate—"Mad Hatter's" downfall; attacks the central nervous system (CNS).

Methyl alcohol—Methanol; wood alcohol; Columbian spirits; carbinol—all cause eye, skin, and upper respiratory irritation; headache; drowsiness and dizziness; nausea and vomiting; dilation of the pupils, visual disturbance, and blindness; excessive sweating and dermatitis; ingestion (acute) abdominal; shortness of breath; cold, clammy extremities; blurring of vision and hyperemia of the optic disc.

Methylene chloride—Employees exposed to methylene chloride are at increased risk of developing cancer; adverse effects on the heart, central nervous system, and liver; and skin or eye irritation. Exposure may occur through inhalation, by absorption through the skin, or through contact with the skin. Methylene chlorine is a solvent that is used in many different types of work activities, such as paint stripping, polyurethane foam manufacturing, cleaning, and degreasing.

Molds and fungi—Molds and fungi produce and release millions of spores small enough to be air-, water-, or insect-borne that may have negative effects on human health, including allergic reactions, asthma, and other respiratory problems.

Organochlorine insecticides—One of the many chlorinated insecticides, for example, DDT, dieldrin, chlordane, BHC, Lindane, etc.—neurotoxins.

Organophosphate (OP) insecticides—Chlorpyrifos, dimethoate, malathion, and trichlorfon are organophosphate (OP) insecticides. These insecticides interfere with nerve-impulse transmission, blocking the action of cholinesterase enzymes essential to proper nerve function. Symptoms of OP poisoning include headache, sweating, nausea and vomiting, diarrhea, loss of coordination, and, in extreme cases, death.

Paradichlorobenzene—Paradichlorobenzene is a mild respiratory irritant and hepatotoxic and is used in tobacco growing as a plant bed treatment for disease control. It is also used as a fumigant for clothes moths in fabric, and for ant control. It is used on apricots, cherries, nectarines, peaches, and plums for insect control. It is also used as a fumigant and repellent in combination with other materials to control squirrels, moles, gophers, and rats and to repel cats and dogs.

PCBs—PCBs belong to a family of organic compounds known as chlorinated hydrocarbons. Most PCBs were sold for use as dielectric fluids (insulating liquids) in electrical transformers and capacitors. When released into the environment, PCBs do not easily break apart and form new chemical arrangements (i.e., they are not readily biodegradable). Instead they persist for many years, bioaccumulate, and bioconcentrate in organisms. Exposure to PCBs in humans can cause chloracne (a painful, disfiguring skin ailment), liver damage, nausea, dizziness, eye irritation, and bronchitis.

Plague—The World Health Organization reports 1,000 to 3,000 cases of plague every year. A bioterrorist release of plague could result in a rapid spread of the pneumonic form of the disease, which could have devastating consequences.

Ricin—Ricin is one of the most toxic and easily produced plant toxins. It has been used in the past as a bioterrorist weapon and remains a serious threat.

Severe Acute Respiratory Syndrome (SARS)—SARS is an emerging, sometimes fatal respiratory illness. According to the Centers for Disease Control and Prevention (CDC), the most recent human cases of SARS were reported in China in April 2004, and there is currently no known transmission anywhere in the world.

Silica—Silicosis (fibrosis).

Smallpox—Smallpox is a highly contagious disease unique to humans. It is estimated that no more than 20 percent of the population has any immunity from previous vaccinations.

Thalidomide—Thalidomide is probably one of the most well-known teratogens. Teratogens are agents that cause offspring to be born with abnormalities (e.g., heart malformation, cleft palate, undeveloped or underdeveloped limbs). Teratogens cause their damage when the fertilized embryo is first forming an organ. At that time the teratogen interferes with the proper development of that organ. By contrast, mutagens cause birth defects by altering sperm or egg cell DNA before the egg is fertilized.

Tri-ortho-cresyl phosphate (TOCP) (Jamaica ginger)—According to the American Botanical Council (1995), early in the year 1930, a strange new paralytic illness was affecting relatively large numbers of individuals. Victims of the disease would typically notice numbness in the legs, followed by weakness and eventual paralysis with "foot drop." In most cases, this was followed within about a week by a similar process in the arms, resulting in many cases in "wrist drop." The disease was rarely fatal, but recovery was very slow and in many cases the damage to the nervous system left the patient with permanent disabilities. It did not take long after the first appearance of the illness to link it to the consumption of fluid extract of Jamaica ginger, commonly referred to as "Jake" by many who used the product. During investigation of the incidents, chemists soon discovered the presence of a cresol compound, a substance they had never before encountered, in adulterated fluid extracts of ginger. Later, it was certain that the compound was tri-ortho-cresyl phosphate (TOCP) and that this substance was present to the extent of about 2 percent in samples allegedly associated with paralysis.

Tularemia—Tularemia is also known as "rabbit fever" or "deer fly fever." Tularemia is a zoonotic disease and is extremely infectious. Relatively few bacteria are required to cause the disease, which is why it is an attractive weapon for use in bioterrorism.

Vinyl chloride—Most vinyl chloride is used to make polyvinyl chloride (PVC) plastic and vinyl products. Acute (short-term) exposure to high levels of vinyl chloride in the air has resulted in central nervous system effects (CNS), such as dizziness, drowsiness, and headaches in humans. Chronic (long-term) exposure to vinyl chloride through inhalation and oral exposure in humans has resulted in liver damage. Cancer is a major concern from exposure to vinyl chloride via inhalation, as vinyl chloride exposure has been shown to increase the risk of a rare form of liver cancer in humans. USEPA has classified vinyl chloride as a Group A, human carcinogen.

Viral hemorrhagic fevers (VHFs)—Viral hemorrhagic fevers (VHFs) refer to a group of illnesses that are caused by several distinct families of viruses. In general, the term "viral hemorrhagic fever" is used to describe a severe multisystem syndrome. Characteristically, the overall vascular system is damaged, and the body's ability to regulate itself is impaired. These symptoms are often accompanied by hemorrhage (bleeding); however, the bleeding is itself rarely life-threatening. While some types of hemorrhagic fever viruses can cause relatively mild illnesses, many of these viruses cause severe, life-threatening disease.

Did You Know?

Breathing high levels of inorganic arsenic can give you a sore throat or irritated lungs. Ingesting very high levels of arsenic can result in death. Exposure to lower levels can cause nausea and vomiting, decreased production of red and white blood cells, abnormal heart rhythm, damage to blood vessels, and a sensation of "pins and needles" in the hands and feet.

Factors Affecting Toxicity

The amount of a toxin that reaches the target tissue is dependent upon four factors. These four in combination govern the degree of toxicity, if any, from chemical exposure. These four factors are:

- Absorption
- Distribution
- Metabolism
- Excretion

Absorption is defined as the passage of a chemical across a membrane into the body. There are four major factors that affect absorption and subsequent distribution, metabolism, and excretion: 1) size of the molecule, 2) lipid solubility, 3) electrical charge, and 4) cell membrane carrier molecules (Kent, 1998). Until a chemical is absorbed, toxic effects are only rarely observed and then only at points of contact with the body (e.g., acid burns on the skin) (Stelljes, 2000).

Once a chemical is absorbed into the body, it is *distributed* to certain organs via the blood stream (circulatory system). The rate of distribution to each organ is related to the blood flow through the organ, the ease with which the chemical crosses the local capillary wall and the cell membrane, and the affinity of components of the organ for the toxin (Lu and Kacew, 2002).

Metabolism is the sum of all physical and chemical changes that take place in an organism; it includes the breakdown of substances, the formation of new substances, and changes in the energy content of cells. Metabolism can either increase or decrease the toxicity, but it typically increases the water solubility of a chemical, which leads to increased excretion (Stelljes, 2000).

Excretion is defined as elimination from the body, either as urine, feces, or through sweat or tears. The rate at which excretion of toxic substances occurs is important in determining the toxicity of a substance. The faster a substance is eliminated from the body, the more unlikely a biological effect will be (Kent, 1998). Other factors affecting toxicity include:

- Rate of entry and route of exposure; that is, how fast the toxic dose is delivered and by what means.
- Age, which can affect the capacity to repair damage.
- Previous exposure, which can lead to tolerance or increased sensitivity, or make no difference.
- State of health, medications, physical condition, and lifestyle, which can affect the toxic response. Preexisting disease can result in increased sensitivity.
- Host factors, including genetic predisposition and the sex of the exposed individual.

Classification of Toxic Materials

Toxic materials can be classified according to physical properties such as:

Gas: a form of matter that is neither solid nor liquid. In its normal state at room temperature and pressure, it can expand indefinitely to completely fill a container. A gas can be changed to its liquid or solid state under the right temperature and pressure conditions.

Vapor: the gaseous phase of a material that is ordinarily a solid or a liquid at room temperature and pressure. Vapors may diffuse. Evaporation is the process by which a liquid is changed into the vapor state and mixed with the surrounding air. Solvents with low boiling points will volatize readily.

Aerosol: liquid droplets or solid particles dispersed in air that are of fine enough size (less than 100 micrometers) to remain dispersed for a period of time. The toxic potential of an aerosol is only partially described by its concentration in mg/m^3. For a proper assessment of the toxic hazard, the size of the aerosol's particles is important. Particles between 5 and 10 micrometers (μm) will only deposit in the upper respiratory tract. Those between 1 and 5μm will deposit in the lower respiratory tract. Very small particles (<0.5μm) are generally not deposited.

Dust: solid particles suspended in air produced by some physical process such as crushing, grinding, abrading, or blasting. Dusts may be an inhalation, fire, or dust-explosion hazard.

Mist: liquid droplets suspended in air produced by some physical process such as spraying, splashing, boiling, or condensation of vapor.

Fume: an airborne dispersion of minute solid particles arising from the heating of a solid (such as molten metal). Gases and vapors are not fumes, although the terms are often mistakenly used interchangeably.

Smoke: minute airborne particles, either liquid or solid (but usually carbon or soot), generated as a result of incomplete combustion of an organic material.

Target Systems/Organs Commonly Affected by Toxins

Once a toxin enters the body, it is distributed by the circulatory system, where it may be absorbed by and accumulated in a variety of tissues. The composition of the tissue and the physio-chemical properties of the toxicant will determine where toxic substances will concentrate and exert their effects (Kent, 1998).

Central nervous system (CNS): The effects of toxic substances on neurons in the CNS may be placed into two categories: (1) those that affect the neuron structure, and 2) those that affect the neurotransmission between the presynaptic terminal and postsynaptic membrane. In each situation normal neuron functioning is disrupted. Damage to the structure of the neuron can be caused by a variety of toxic substances. For example, toxins in alcohol, carbon disulfide, and organomercury compounds cause damage in the CNS.

Hematoxic (blood) effects: Toxic substances may have a direct effect on mature blood cells; further, they may affect the bone marrow, where the blood cells are produced.

Immune system: Immunotoxins affect the cellular component of the immune system by inhibiting the production of leukocytes in the bone marrow, or by inhibiting their proliferation in response to an antigen.

Cardiotoxic effects: Toxic substances affect the heart in several ways. The toxicant may have a direct effect on the cardiac tissues by affecting the cell membrane integrity or cellular metabolism, such as enzyme synthesis, or ATP production.

Pulmonary (lungs) effects: Toxic substances that are inhaled can be divided into two categories: particulates and non-particulates. These toxic substances have various effects on different regions of the respiratory tract, including irritation, sensitization, scar formation, cancer, pneumonia, and emphysema.

Hepatotoxic (liver) effects: The liver is one of the primary organs of the body involved in the detoxification of harmful substances that enter the bloodstream. When some toxic substances are retained within the liver cells they cause intracellular damage. They interfere with normal protein synthesis and enzyme functions. This interference, as well as damage to structural components of the cell, results in death of the liver cells.

Nephrotoxic (kidneys) effects: Because the kidneys filter unwanted toxins out of our bloodstream and excrete them in urine, they are the targets of unwanted toxins. Further, the kidneys have a high amount of blood flowing through them. This facilitates the accumulation of toxic substances.

Reproductive system effects: Some toxins impact the ability to reproduce. Chemicals may act on both sexes, or many only affect one sex. A fumigant, DBCP, for example, impacts the reproductive ability of only one sex. Formaldehyde is an example of a toxin that impacts the reproductive ability of females only.

Carcinogens, Mutagens, and Teratogens

A *carcinogen* is an agent that may produce cancer (uncontrolled cell growth), either by itself or in conjunction with another substance. A *suspect carcinogen* is an agent that is suspected of being a carcinogen based on chemical structure, animal research studies, or mutagenicity studies.

The IARC—International Agency for Research on Cancer—classifies carcinogens in the following manner:

1: Carcinogenic to humans with sufficient human evidence.
2A: Probably carcinogenic to humans with some human evidence.
2B: Probably carcinogenic to humans with no human evidence.
3: Sufficient evidence of carcinogenicity in experimental animals.

The NIOSH (National Institute for Occupational Safety and Health) classifies carcinogens as either carcinogenic or non-carcinogenic with no further categorization. The NTP (National Toxicology Program) classifies carcinogens as either a (carcinogenic with human evidence) or b (carcinogenic with limited human evidence but sufficient animal evidence). The ACGIH (American Conference of Governmental Industrial Hygienists) classifies carcinogens in its TLV's (Threshold Limit Values) as:

A1. Confirmed Human Carcinogen: The agent is carcinogenic to humans based on the weight of evidence from epidemiologic studies.

A2. Suspected Human Carcinogen: Human data are accepted as adequate in quality but are conflicting or insufficient to classify the agent as a confirmed human carcinogen; OR, the agent is carcinogenic in experimental animals at doses, by routes of exposure, at sites, of histologic types, or by mechanisms considered relevant to worker exposure. The A2 is used primarily when there is limited evidence of carcinogenicity in humans and sufficient evidence of carcinogenicity in experimental animals with relevance to humans.

A3. Confirmed Animal Carcinogen with Unknown Relevance to Humans: The agent is carcinogenic in experimental animals at a relatively high dose, by routes of administration, at sites, or histologic types, or by mechanisms that may not be relevant to the worker. Available epidemiological studies do not confirm an increased risk of cancer in exposed humans. Available evidence does not suggest that the agent is likely to cause cancer in humans except under uncommon or unlikely routes or levels or exposure.

A4. Not Classifiable as a Human Carcinogen: Agents that cause concern that they could be carcinogenic for humans but that cannot be assessed conclusively because of a lack of data.

A5. Not Suspected as a Human Carcinogen: The agent is not suspected to be a human carcinogen on the basis of properly conducted epidemiologic studies in humans.

A *teratogen* (literal translation is "monster making") is a substance that can cause physical defects in a developing embryo. A *mutagen* is a material that induces genetic changes (mutations) in the DNA. Mutations may or may not relate to cancer.

Did You Know?

One of the most important concepts in toxicology can be summarized by the following simple relationship: Toxic Effects = Potency × Exposure. A toxic effect is an adverse response in an organism caused by a chemical. Potency generally describes the rate at which a chemical causes effects. More potent chemicals have higher rates than those that are less potent. For example, salt has low potency because there is a wide range of amounts over which the degree of toxic effects, if any, changes very slowly. On the one hand, one drop of strychnine, a natural chemical from the seeds of *Nux vomica* that attacks our nervous system, might kill us. For this chemical, very little change in amount is needed to have toxic effects range from nothing to death. This is a very potent chemical. The potency of chemicals varies widely, but is never zero. Why then are we not constantly at risk from chemicals? Because in order for toxic effects to occur, we must be exposed to these chemicals. Exposure can be controlled so that it can be zero. If exposure is zero, there are no toxic effects because the product of potency times exposure is zero.

Risk Assessment

So far, we have discussed various aspects of toxicology. This discipline identifies the manner in which chemicals exert toxicity, and the potency of chemicals on various species. The majority of toxicology studies are conducted under controlled conditions in the lab. This is necessary to establish cause-and-effect relationships and to develop dose-response information on specific chemicals. However, humans are not typically exposed to concentrations tested in these lab studies. We learned about the uncertainty in trying to extrapolate toxicity information to humans or other species. In spite of this uncertainty, we are ultimately concerned with the potential impact of chemicals released into the environment. This issue concerns all of us because of the myriad ways we might interact with these chemicals. They can be present in our water, air, soil, or food. Estimating the likelihood to toxicity from exposure to chemicals in the environment is the focus of the discipline of risk assessment.

Risk has been defined as the expected frequency of undesirable effects arising from exposure to a pollutant. It may be expressed in absolute terms as the risk due to exposure to a specific pollutant. It may also be expressed as a relative risk, which is the ratio of the risk among the exposed population to that among the unexposed. The term was first adopted by the International Commission on Radiological Protection (ICRP) in evaluating the health hazards related to ionizing radiation. The use of this term stems from the realization that often a clear-cut "safe" or "unsafe" decision cannot be made.

There is risk associated with every aspect of our lives. As long as we live, there is a finite risk that harm could come from everyday activities. For example, we could be involved in a car accident, or get struck by lightning, or get skin cancer from ultraviolet light exposure. These risks might differ from one another, but each of them could happen. However, most of us don't avoid the sun because of the possibility of getting skin cancer. Similarly, we don't avoid driving just because we could get in an accident. Even if we decided to walk everywhere, there is a risk of tripping and injuring ourselves, or getting hit by a car while walking. The same is true for chemicals. There is a risk of toxicity from exposure to any amount of a chemical. This risk might be so large that we can be fairly certain toxicity would result (e.g., carbon monoxide poisoning from running a car in an enclosed space for a long period of time), or so small that it is essentially unmeasurable (e.g., one molecule of toluene in a reservoir). Part of the risk assessment discipline includes identifying what risks are so small that we can ignore them.

Risk assessment impacts many different areas and is the most relevant area of toxicology for the average person to understand. For example, in certain parts of the country, regulatory agencies announce "space the air" days because the degree of air pollution is expected to be higher than normal, most often due to weather patterns.

Discussion Questions

1. There are occasional accidental releases of chemicals into the air from industrial plants (e.g., oil refineries). Under what circumstances would we be concerned for our health when releases occur? What about industrial accidents (e.g., oil spills, train accidents, and spillage)? How much chemical needs to be released in such an accident to impact health?
2. Should we filter our water, or drink only bottled water? How safe is bottled water, which is usually processed?
3. What is the difference between toxins and toxicants? Give an example of each.
4. To what extent do you agree with the statement that "all substances are poisons"?
5. How does an agency know when to announce a "space the air" day, and what does it mean to those breathing the air?

Note

1. UNEP/GRID-Avendal and UN-HABITAT. Corcoran, E. (ed.) et al., 2010. *Sick Water.*

References and Recommended Reading

CDC, 2009. *Introduction to Toxicology*. Accessed 02/18/12 at http://www.atsdr.cdc.gov/training.toxmanual/modules/1/lecturenotes.html.

EPA, 2012. Origins and fate of PPCPs in the Environment. Accessed 02/09/12 at http://epa.gov/nerlesd1/chemistry/pharma/.

Klaassen, C., 1996. *Casarett and Doull's Toxicology: The basic science of poisons*, 5th ed. New York: MacMillan Publishing Company.

Kent, C., 1998. *Basics of Toxicology*. New York: Wiley.

Koren, H., 1996. *Illustrated dictionary of environmental health and occupational safety*. New York: CRC Lewis.

Lu, F.C., and S. Kacew, 2002. *Lu's Basic Toxicology*, 4th ed. Boca Raton, FL: CRC.

Random House, 2009. *Dictionary of the English Language*. The Unabridged Edition. New York: Random House.

Stelljes, M.E., 2000. *Toxicology for Non-Toxicologists*. Rockville, MD: Government Institutes.

US HHS, 2004. Public Health Service, National Toxicology Program. Current direction and evolving strategies: good science for good decisions. Washington, D.C.: United States Department of Health and Human Services.

4

Epidemiology

> Not everything that counts can be counted, and not everything that can be counted counts.
>
> —Albert Einstein

> Epidemiology is the study (scientific, systematic, data-driven) of the distribution (frequency, pattern) and determinants (causes, risk factors) of health-related states and events (not just disease) in specified populations (patient is community, individuals viewed collectively), and the application of (since epidemiology is a discipline within public health) this study to the control of health problems.
>
> —US HHS, 2006

Epidemiology = *epi* (upon) + *demos* (people) + *logos* (study) (Last, 2001).

Abandoned Mines: An Environmental Health Concern

(This section is adapted from NIH *Abandoned Mines* Toxtown at http://toxtown.him.nih.gov/text_version/locations.phg?id=153.) Abandoned mine lands pose serious threats to human health, safety, and the environment. There are as many as 500,000 abandoned mines in the United States. They exist on private and federal lands. Many are near recreational and fishing areas.

Most abandoned coal mines are in the East, primarily in Kentucky, Pennsylvania, and West Virginia. Most abandoned ore and metal mines are in the West. Most abandoned uranium mines are in Arizona, Colorado, New Mexico, Utah, and Wyoming.

There are approximately 31,000 abandoned mines on public lands, including hardrock mines, such as metals and ore mines, coal mines, and uranium mines. Approximately 26,721 of these abandoned mines are in the Southwest.

Abandoned mine sites include dangerous vertical mine shafts and unstable horizontal openings. Old support structures may be rotten and cause cave-ins. There may be pockets of air in the mines with little or no oxygen.

Lethal concentrations of deadly gases can accumulate in underground mine passages, including carbon monoxide, hydrogen sulfide, and radon, and invisible and odorless radioactive gas.

Abandoned mines can contain or release carbon dioxide and methane, which are greenhouse gases that contribute to global warming and climate change. The greenhouse effect from mines is small. However, methane can form an explosive mixture when mixed with air, especially in enclosed mines.

Soil and water in abandoned mines may be contaminated with cyanide, lead, arsenic, mercury, and other toxic chemicals.

Unused or misfired explosives left behind in abandoned mines can become unstable and deadly. Loose material in piles or trash heaps can collapse on hikers. Mine sites may include stockpiled waste rock and waste piles, power lines, abandoned heavy equipment, fuel storage tanks, electric machinery, and radioactive materials.

Open pits may be filled with water that can be highly acidic or contaminated with chemicals. Water-filled quarries can hide hazards such as rock ledges and old machinery.

Abandoned mines may serve as a habitat for rattlesnakes, bears, mountain lions, and bats. Abandoned coal mines may leave behind coal waste that can contaminate water drainages and cause coal fires.

Abandoned uranium mines pose a threat of exposure to radiation. They may contain radioactive waste, arsenic, lead, and naturally occurring radioactive material, including radon. Uranium can contaminate groundwater, surface water, dust, and soil.

Abandoned uranium mines may contain fuels, solvents, degreasers, and other chemicals used with heavy equipment and rock blasting operations.

Did You Know?

From 1944 to 1986, uranium was extracted from Navajo lands in the Four Corners (Arizona, New Mexico, Colorado, and Utah) areas of the

Southwest. Uranium mining is no longer allowed on Navajo lands. When uranium mining ceased, mining companies abandoned mines without sealing tunnel openings, filling pits, or removing uranium tailings. Depending on how they are counted, at least 520 and as many as 1,032 of the abandoned sites remain throughout the Navajo nation today. Abandoned uranium mines and pits on Navajo lands are sometimes used for swimming, recreation, and livestock containment, exposing people and livestock to radiation. Open mine pits are sometimes used for illegal trash dumping. Abandoned uranium mines expose families living nearby to radioactive waste. The Northeast Church Rock Mine in New Mexico is the EPA's highest priority for cleaning up abandoned mines on the Navajo nation.

Abandoned uranium mines may contain uranium tailings, which are radioactive, sand-like materials left over from uranium milling. Uranium tailings contain radium, which stays radioactive for thousands of years. Radium decays to produce radon. Tailings can also contain selenium and thorium.

A Sherlock Holmes-Type at the Pump

(This section is adapted from F. R. Spellman, 2006. *Environmental Science and Technology*, 2nd ed. Rockville, MD: Government Institutes Press.) He wandered the foggy, filthy, garbage-strewn, corpse-ridden streets of 1854 London searching, making notes, always looking . . . seeking a murdering villain (no; not the Ripper, but a killer just as insidious and unfeeling)—and find the miscreant, he did. He took action; he removed the handle from a water pump. And, fortunately for untold thousands of lives, his was the correct action—the lifesaving action.

He was a detective—of sorts. No, not the real Sherlock Holmes—but absolutely as clever, as skillful, as knowledgeable, as intuitive—and definitely as driven. His real name: Dr. John Snow. His middle name? Common Sense. Snow's master criminal, his target? A mindless, conscienceless, brutal killer: cholera.

Let's take a closer look at this medical super sleuth and at his quarry, the deadly cholera—and at Dr. Snow's actions to contain the spread of cholera. More to the point, let's look at Dr. Snow's subsequent impact on water treatment (disinfection) of raw water used for potable and other purposes.

An unassuming—and creative—London obstetrician, Dr. John Snow (1813–1858) achieved prominence in the mid-nineteenth century for proving his theory (in his *On the Mode of Communication of Cholera*) that cholera is a contagious disease caused by a "poison" that reproduces in the human body and is found in the vomitus and stools of cholera patients. He theorized that the main (though not the only) means of transmission was water contaminated with this poison. His theory was not held in high regard at first, because a commonly held and popular counter-theory stated that diseases are transmitted by inhalation of vapors. Many theories of cholera's cause were expounded. In the beginning, Snow's argument did not cause a great stir; it was only one of many hopeful theories proposed during a time when cholera was causing great distress. Eventually, Snow was able to prove his theory. We describe how Snow accomplished this later, but for now, let's take a look again at Snow's target: cholera.

Cholera

According to the U.S. Centers for Disease Control (CDC) and as mentioned, cholera is an acute, diarrheal illness caused by infection of the intestine with the bacterium *Vibrio cholera*. The infection is often mild or without symptoms, but sometimes can be quite severe. Approximately one in twenty infected persons have severe disease symptoms such as profuse watery diarrhea, vomiting, and leg cramps. In these persons, rapid loss of body fluids leads to dehydration and shock. Without treatment, death can occur within hours.

Did You Know?

You don't need to be a rocket scientist to figure out just how deadly cholera was during the London cholera outbreak of 1854. Comparing the state of "medicine" at that time to ours is like comparing the speed potential of a horse and buggy to a state-of-the-art NASCAR race car today. Simply stated: cholera was the classic epidemic disease of the nineteenth century, as the plague had been for the fourteenth. Its defeat was a reflection both of common sense and of progress in medical knowledge—and of the enduring changes in European and American social thought.

How does a person contract cholera? Good question. Again, we refer to the CDC for our answer. A person may contract cholera (even today) by drinking water or eating food contaminated with the cholera bacterium. In an epidemic, the source of the contamination is usually the feces of an infected person. The disease can spread rapidly in areas with inadequate treatment of sewage and drinking water. Disaster areas often pose special risks. For example, the aftermath of Hurricane Katrina in New Orleans raised concerns for a potential cholera outbreak.

Cholera bacterium also lives in brackish river and coastal waters. Raw shellfish have been a source of cholera, with a few people in the United States having contracted it from eating shellfish from the Gulf of Mexico. The disease is not likely to spread directly from one person to another; therefore, casual contact with an infected person is not a risk for transmission of the disease.

Flashback to 1854 London

The information provided in the preceding section was updated and provided by the Centers for Disease Control (CDC) in 1996. Basically, for our purposes, the CDC confirms the fact that cholera is a waterborne disease. Today, we know quite a lot about cholera and its transmission, as well as how to prevent infection and how to treat it. But what did they know about cholera in the 1850s? Not much. However, one thing is certain: they knew cholera was a deadly killer. And that was just about all they knew—until Dr. Snow proved his theory. He believed that cholera is a contagious disease caused by a poison that reproduces in the human body and is found in the vomitus and stools of cholera victims. He also believed that the main means of transmission was contaminated water.

Dr. Snow's theory was correct, of course, as we know today. The question is, how did he prove his theory correct twenty years before the development of the germ theory? The answer to this provides us with an account of one of the all-time legendary quests for answers in epidemiological research—and an interesting story!

Dr. Snow proved his theory in 1854, during yet another severe cholera epidemic in London. Though ignorant of the concept of bacteria carried in water (the germ theory), Snow traced an outbreak of cholera to a water pump located at the intersection of Cambridge and Broad Street (London). How did he isolate this source to this particular pump? He began his investigation by determining in which area in London persons with cholera lived and worked. He then used this information to map the distribution of cases on what epidemiologists call a "spot map." His map indicated that the majority of the deaths occurred within 250 yards of that communal water pump. The water pump was used regularly by most of the area residents. Those who did not use the pump remained healthy. Suspecting the Broad Street pump as the plague's source, Snow had the water pump handle removed and thus ended the cholera epidemic.

Sounds like a rather simple solution, doesn't it? For us, it is simple, but remember in that era, not even aspirin had yet been formulated—to say nothing of other medical miracles we now take for granted—antibiotics, for example. Dr. John Snow, by the methodical process of elimination and linkage (Sherlock Holmes would have been impressed—and he was!), proved his point and his theory. Specifically, he painstakingly documented the cholera cases and correlated the comparative incidence of cholera among subscribers to the city's two water companies. He learned that one company drew water from the lower Thames River, while the other company obtained water from the upper Thames. Snow discovered that cholera was much more prevalent in customers of the water company that drew its water from the lower Thames, where the river had become contaminated with London sewage. Snow tracked and pinpointed the Broad Street pump's water source. You guessed it: the contaminated lower Thames, of course.

Dr. Snow the obstetrician became the first effective practitioner of scientific epidemiology. His creative use of logic, common sense (by removing the handle from the pump), and scientific information enabled him to solve a major medical mystery—to discern the means by which cholera was transmitted—and this earned him the title "the father of field epidemiology." Today Dr. John Snow is known as the father of modern epidemiology.

Pump Handle Removal—To Water Treatment (Disinfection)

Dr. John Snow's major contribution to the medical profession, to society, and to humanity in general can be summarized rather succinctly: he determined and proved that the deadly disease cholera is a waterborne disease. (Dr. John Snow's second medical accomplishment was to be the first person to administer anesthesia during childbirth.)

What does all of this have to do with water treatment (disinfection)? Actually, Dr. Snow's discovery—his stripping of a mystery to its barest bones—has quite a lot to do with water treatment. Combating any disease is rather difficult without a determination of how the disease is transmitted—how it travels from vector or carrier to receiver. Dr. Snow established this connection, and from his work and the work of others, progress was made in understanding and combating many different waterborne diseases.

Today, sanitation problems in developed countries (those with the luxury of adequate financial and technical resources) deal more with the consequences that arise from inadequate commercial food preparation, and the results of bacteria becoming resistant to disinfection techniques and antibiotics. We simply flush our toilets to rid ourselves of unwanted wastes, and we turn on our taps to take in high-quality drinking water supplies, from which we've all but eliminated cholera and epidemic diarrheal diseases. This is generally the case in most developed countries today—but it certainly wasn't true in Dr. Snow's time.

The progress in water treatment from that notable day in 1854 (when Snow made the "connection" [actually the "disconnection" of the handle from the pump] between deadly cholera and its means of transmission, its "communication") to the present reads like a chronology of discovery leading to our modern water treatment practices. This makes sense, of course, because with the passage of time, pivotal events and discoveries occur—events that have a profound effect on how we live today. Let's take a look at a few elements of the important chronological progression that evolved from the simple removal of a pump handle to the advanced water treatment (disinfection) methods we employ today to treat our water supplies.

After Snow's discovery (that cholera is a waterborne disease emanating primarily from human waste), events began to drive the water/wastewater treatment process. In 1859, four years after Snow's discovery, the British Parliament was actually suspended during the summer because the stench coming from the Thames was unbearable. According to one account, the river began to "seethe and ferment under a burning sun." As was the case in many cities at this time, storm sewers carried a combination of storm water, sewage, street debris, and other wastes to the nearest body of water. In the 1890s, Hamburg, Germany, suffered a cholera epidemic. Detailed studies by Koch tied the outbreak to the contaminated water supply. In response to the epidemic, Hamburg was among the first cities to use chlorine as part of a wastewater treatment regimen. About the same time, the town of Brewster, New York, became the first U.S. city to disinfect its treated wastewater. Chlorination of drinking water was used on a temporary basis in 1896, and its first known continuous use for water supply disinfection occurred in Lincoln, England, and Chicago in 1905. Jersey City, New Jersey, became one of the first routine users of chlorine in 1908.

Time marched on, and with it came an increased realization of the need to treat and disinfect both water supplies and wastewater. Between 1910 and 1915, technological improvements in gaseous and then the solution feed of elemental chlorine (Cl_2) made the process more practical and efficient. Disinfection of water supplies and chlorination of treated wastewater for odor control increased over the next several decades. In the United States, disinfection, in one form or another, is now being used by more than 15,000 out of approximately 16,000 Publicly Owned Treatment Works (POTWs). The significance of this number becomes apparent when you consider that fewer than twenty-five of the 600-plus POTWs in the United States in 1910 were using disinfectants.

Pump Handle Removal: Lessons Learned

There are a number of lessons to be learned from the actions of Dr. John Snow when he removed the pump handle from the pump in London and effectively stopped the cholera outbreak. For our purposes, we will discuss four of these most important lessons.

In the first place, Dr. Snow's experiments with finding the culprit that caused the 1854 cholera outbreak in London demonstrate the most important aspect of why science is important and why we need it. Simply put, science saves lives. If there is anything more important than this, we can't find it.

Second, we simply can't go through life ignorant of our surroundings and the laws that literally make the world go 'round. We have too many pressing problems, such as the swine flu (H1N1 virus); the prospects of global climate change and/or global warming or global freezing and pending sea level rise; worldwide economic problems, which many people feel are going to require science and innovation to point the world's economies in the right direction; growing concerns about environmental pollution problems affecting all four environmental mediums (air, water, soil, and biota); the pending energy crisis and the urgent need to jump-start science and technology to bring renewable energy sources online. All of these issues are important, and obviously, they certainly point to a need for science.

In light of all of this we can't point out strongly enough the importance of an understanding of science for members of a democratic society. We simply cannot be ignorant of science and the facts. Think about it. How are we to participate in a democracy if we don't understand basic science concepts? When the naysayer who warns us of the need for this or for that because if we take no action doomsday is surely near, and when these so-called experts justify their expertise simply because of their positions of authority or power, we need to be able to filter the information to determine the truth. Science is the filter—the ultimate micro-filter.

Third, from the above account of Dr. Snow's activities in attempting to find the cause of the 1854 cholera epidemic in London, it should be apparent to the reader that he followed a step-by-step procedure in finding the culprit. We can say he used his toolbox (the scientific method) to track down the dreadful killer. In this particular case, Dr. Snow used the six major tools in the scientist's toolbox. He performed the following tasks:

- Observed—Dr. Snow observed that several residents of London were becoming ill and many had succumbed to some unknown agent.
- Questioned—Dr. Snow asked, why? Why are some Londoners getting ill and dying?

- Formed a hypothesis—Dr. Snow believed sewage dumped into the river or into cesspools near the town could contaminate the water supply, leading to a rapid spread of disease. In 1883 a German physician, Robert Koch, took the search for the cause of cholera a step further when he isolated the bacterium *Vibrio cholerae*, the "poison" Snow contended caused cholera. Dr. Koch determined that cholera is not contagious from person-to-person contact, but is spread only through unsanitary food or water supplies.
- Conducted an experiment—Dr. Snow made a spot map of the downtown London area. The spots indicated locations where people had contracted and died of cholera. His spot map indicated that most of the fatalities occurred in proximity to the Broad Street pump. Thus, he had the handle to the Broad Street pump removed.
- Collected and recorded data—Dr. Snow's data collected after removing the pump handle from the Broad Street pump indicated immediate results in that no further occurrences of cholera occurred in the Broad Street area.
- Drew a conclusion—Dr. Snow concluded that London's 1854 cholera epidemic was caused by sewage in the water pumped by the Broad Street pump.

The fourth lesson learned from Dr. Snow's cholera experiment is that science is flexible, never bends the truth, and has many branches, from acoustics, the branch of science related to the study of the transmission of sound waves, to zoology, a branch of biology (another science) that is related to the study of the animal kingdom, including evolution, classification, distribution, structure, habits, and the embryology of animals.

Finally, although Dr. Snow's experiments were more in line with epidemiology and the protocols and procedures followed therein, the important lesson to be learned by the environmental scientist from Dr. Snow's research and investigations is that in science it is wise to be a generalist. The study of the cause and distribution of diseases in human population is a narrow field of expertise that could easily be enhanced by the possession of a repertoire of wide-ranging knowledge. Because of the complex problems facing us all—the finding of how to prevent deadly exposure to toxins, to preventing environmental damage, to living the so-called good life without the huge costs accrued by ignoring and abusing our surroundings—in short, we advocate for the importance of a generalist view of the environment and all the sciences involved.

So, how does the work of Dr. John Snow affect the model of environmental epidemiology? As pointed out by Monson (1990) and listed and described by

Moeller (2011), several factors make Snow's study a classic, flagship model of environmental epidemiology:

- Snow recognized an association between exposure and disease—that is, between the source of the drinking-water supply and the incidence of cholera.
- He formulated a hypothesis—that fecal contamination of drinking water was the specific agent of transmission of the disease.
- He collected information to substantiate his hypothesis—in subdistricts where the drinking was supplied by only one company, the association was stronger.
- He recognized that there could be an alternative explanation of the association—that social class or place of residence might influence transmission of the disease.
- He applied a method to minimize the effects of the alternative explanation—he compared cholera rates within a single district or neighborhood, rather than between neighborhoods, on the basis of their water supply.
- He effectively minimized the collection of biased or false information—since most residents were not aware of the name of the company that supplied their water, he applied a chemical test to make this determination in a positive manner (Goldsmith, 1986).

Today, in the practice of modern epidemiology, do we still prescribe to Dr. Snow's precepts, procedures, and methodologies? Yes. No. Somewhat. Sort of.

The reality: Experience has shown that Snow's general criteria, as listed above, have stood the test of time. In short, Snow's criteria are regarded as foundational and fundamental to the design of all types of epidemiologic studies. Even though on his work and that of those who followed centered on disease and laboratory studies with little attention to study design, his logical, commonsense approach to problem-solving is still used. He certainly deserves the title of the Father of Epidemiology.

Today, instead of using the disease-centered approach of Dr. Snow and his early followers, we use an exposure-centered approach based on a multitude of factors. In this chapter, we focus on this modern approach to epidemiology—never forgetting that it was a Sherlock Holmes-type at the Broad Street pump that got the epidemiological ball rolling.

Did You Know?

Students of journalism are taught that a good news story, whether it be about a bank robbery, dramatic rescue, or presidential candidate's speech, must include the 5 Ws: what, who, where, when, and why (sometimes cited as why/how). The 5Ws are the essential components of a news story because if any of the five are missing, the story is incomplete.

The same is true in characterizing epidemiologic events, whether it be an outbreak of norovirus among cruise ship passengers or the use of mammograms to detect early breast cancer. The difference is that epidemiologists tend to use synonyms for the 5 Ws: diagnosis or health event (what), person (who), place (where), time (when), and causes, risk factors, and modes of transmission (why/how).

Definition of Epidemiology

(This section is based on U.S. HHS, 2006. *Principles of Epidemiology in Public Health Practice*, 3rd ed. Atlanta, GA, Centers for Disease Control and Prevention, or CDC.) The word *epidemiology* comes from the Greek words *epi*, meaning on or upon; *demos*, meaning people; and *logos*, meaning the study of. In other words, the word *epidemiology* has its roots in the study of what befalls a population. Though a large number of varying definitions have been proposed, the following definition captures the underlying principles and public health spirit of epidemiology:

> Epidemiology is the *study* of the *distribution* and *determinants* of *health-related states or events* in *specified populations*, and the *application* of this study to the control of health problems (Last, 2001).

Italicized key terms in this definition reflect some of the important principles of epidemiology.

Study

Epidemiology is a scientific discipline with sound methods of scientific inquiry at its foundation. Epidemiology is data-driven and relies on a systematic and unbiased approach to the collection, analysis, and interpretation of data. Basic epidemiologic methods tend to rely on careful observation and use of valid comparison groups to assess whether what was observed, such as the number of cases of disease in a particular area during a particular time period or the frequency of exposure among persons with disease, differs from what might be expected.

However, epidemiology also draws on methods from other scientific fields, including biostatistics and informatics, with biologic, economic, social, and behavioral sciences.

In fact, epidemiology is often described as the basic science of public health, and for good reason. First, epidemiology is a quantitative discipline that relies on a working knowledge of probability, statistics, and sound research methods. Second, epidemiology is a method of causal reasoning based on developing and testing hypotheses grounded in such scientific fields as biology, behavioral sciences, physics, and ergonomics to explain health-related behaviors, states, and events. However, epidemiology is not just a research activity but an integral component of public health, providing the foundation or direction of practical and appropriate public health action based on this science and casual reasoning (Cates, 1982; see Figure 4.1).

Distribution

Epidemiology is concerned with the frequency and pattern of health events in a population:

Frequency refers not only to the number of health events, such as the number of cases of meningitis or diabetes in a population, but also to the relationship of that number to the size of the population. The resulting rate allows epidemiologists to compare disease occurrence across different populations.

Pattern refers to the occurrence of health-related events by time, place, and person. Time patterns may be annual, seasonal, weekly, daily, hourly, weekly versus weekend, or any other breakdown of these that may influence disease or injury occurrence. Place patterns include geographic variation, urban/rural

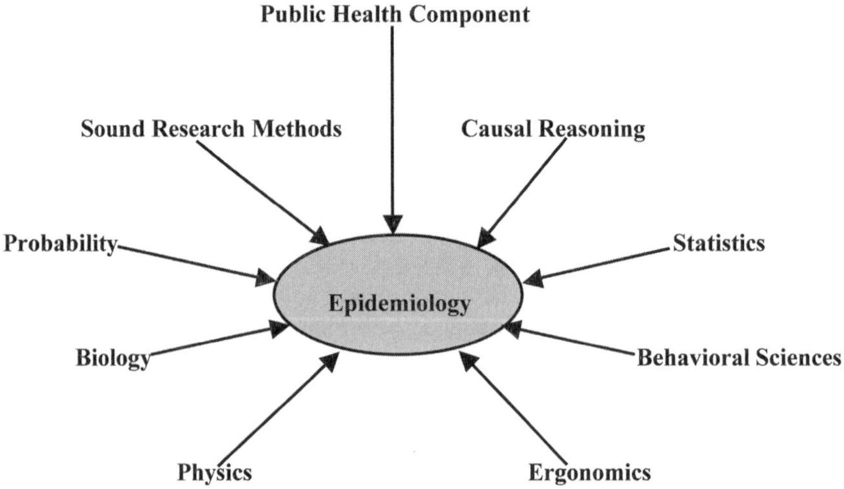

FIGURE 4.1
Primary elements making up epidemiology.

differences, and location of work sites or schools. Personal characteristics include demographic factors that may be related to risk of illness, injury, or disability such as age, sex, marital status, and socioeconomic status, as well as behaviors and environmental exposures.

Examples of distribution include the following:

- Marking on a map the residences of all children born with birth defects within two miles of a hazardous waste site
- Graphing the number of cases of congenital syphilis by year for the country
- Tabulating the frequency of clinical signs, symptoms, and laboratory findings among children with chicken pox in Cincinnati, Ohio

Did You Know?

Characterizing health events by time, place, and person are activities of *descriptive epidemiology*.

Determinants

Epidemiology is also used to search for determinants, which are the causes and other factors that influence the occurrence of disease and other health-related events. Epidemiologists assume that illness does not occur randomly in a population, but happens only when the right accumulation of risk factors or determinants exists in an individual. To search for these determinants, epidemiologists use analytic epidemiology or epidemiologic studies to provide the "why" and "how" of such events. They assess whether groups with different rates of disease differ in their demographic characteristics, genetic or immunological makeup, behaviors, environmental exposures, or other so-called potential risk factors. Ideally, the findings provide sufficient evidence to direct prompt and effective public health control and prevention measures.

Did You Know?

A determinant is any factor, whether event, characteristic, or other definable entity, that brings about a change in a health condition or other defined characteristic (Last, 2001).

Examples of determinants include the following:

- Comparing food histories between persons with Staphylococcus food poisoning and those without

- Comparing frequency of brain cancer among anatomists with frequency in general population

Health-Related States or Events

Epidemiology was originally focused exclusively on epidemics of communicable diseases (Greenwood, 1935), but was subsequently expanded to address endemic communicable diseases and non-communicable infectious diseases. By the middle of the twentieth century, additional epidemiological methods had been developed and applied to chronic diseases, injuries, birth defects, maternal-child heath, occupational health, and environmental health. Then epidemiologists began to look at behaviors related to health and well-being, such as amount of exercise and seat belt use. Now, with the recent explosion in molecular methods, epidemiologists can make important strides in examining genetic markers of disease risk. Indeed, the term *health-related states or events* may be seen as anything that affects the well-being of a population. Nonetheless, many epidemiologists still use the term *disease* as shorthand for the wide range of health-related states and events that are studied.

Specified Populations

Although epidemiologists and direct health-care providers (clinicians) are both concerned with occurrence and control of disease, they differ greatly in how they view "the patient." The clinician is concerned about the health of an individual; the epidemiologist is concerned about the collective health of the people in a community or population. In other words, the clinicians' "patient" is the individual; the epidemiologist's "patient" is the community. Therefore, the clinician and the epidemiologist have different responsibilities when faced with a person with illness. For example, when a patient with diarrheal disease presents, both are interested in establishing the correct diagnosis. However, while the clinician usually focuses on the threat and caring for the individual, the epidemiologist focuses on identifying the exposure of the source that caused the illness; the number of other persons who may have been similarly exposed; the potential for further spread in the community; and interventions to prevent additional cases or recurrences.

Application

Epidemiology is not just "the study of" health in a population; it also involves applying the knowledge gained by the studies to community-based practice. Like the practice of medicine, the practice of epidemiology is both a science and an art. To make the proper diagnosis and prescribe appropriate treatment for a patient, the clinician combines medical (scientific) knowledge with experience, clinical judgment, and understanding of the patient. Similarly, the epidemiologist uses the scientific methods of descriptive and analytic epidemiology as well as experience, epidemiologic judgment, and understanding of local conditions in "diagnosing" the health of a community and proposing appropriated, practical, and acceptable public health interventions to control and prevent disease in the community.

An example of application would be recommending that close contacts of a child recently reported with meningococcal meningitis receive Rifampin (a bactericidal antibiotic drug).

Cesspools: An Environmental Health Concern

A cesspool is an outdoor, underground drywell that receives waste and wastewater from homes, industries, or businesses. A cesspool may also be called a shallow disposal system or Class V well (used to inject non-hazardous fluids underground). After the waste enters the cesspool, it breaks down into liquids that seep into the ground through holes in the cesspool. The wastewater also leaves through the holes and enters the soil.

In a cesspool system, disease-causing pollutants from untreated waste can enter shallow groundwater and contaminate drinking water, streams, lakes, and eventually oceans. Failing cesspool systems can expose people to harmful bacteria and viruses. Odors can pose human health risks. If materials such as chemicals, oil, gas, pesticides, or paints enter a cesspool, they can harm the environment. These toxic liquids should never be poured down the drain.

Cesspool systems should be properly sited, constructed, and operated to protect the environment and human health and prevent contamination. State and local governments regulate cesspools for homes. In the past, many were built very close to shorelines and now that these cesspools are aging they may leak directly into ocean waters. Many are being replaced with modern septic systems. The federal government required the closing of all large-capacity cesspools by April 2005.

Epidemiology: Practical Uses

Epidemiology and the information generated by epidemiological methods have been used in many ways (Morris, 1957). Some common uses are described below.

Assessing the Community's Health

Public health officials responsible for policy development, implementation, and evaluation use epidemiologic information as a factual framework for decision-making. To assess the health of a population or community, relevant sources of data must be identified and analyzed by person, place, and time (descriptive epidemiology).

- What are the actual and potential health problems in the community?
- Where are they occurring?
- Which populations are at increased risk?
- Which problems have declined over time?
- Which ones are increasing or have the potential to increase?
- How do these patterns relate to the level and distribution of public health services available?

More detailed data may need to be collected and analyzed to determine whether health services are available, accessible, effective, and efficient. For example, public health officials used epidemiological data and methods to identify baselines, to set health goals for the nation in 2000 and 2010, and to monitor progress toward these goals (US HHS, 1991; 2000a; 2000b).

Making Individual Decisions

Many individuals may not realize that they use epidemiologic information to make daily decisions affecting their health. When persons decide to quit smoking, climb the stairs rather than wait for an elevator, eat a salad rather than a cheeseburger with fries for lunch, or use a condom, they may be influenced, consciously or unconsciously, by epidemiologists' assessments of risk. Since World War II, epidemiologists have provided information related to all those decisions. In the 1950s, epidemiologists reported the increased risk of lung cancer among smokers. In the 1970s, epidemiologists documented the role of exercise and proper diet in reducing the risk of heart disease. In the mid-1980s, epidemiologists identified the increased risk of HIV infection associated with certain sexual and drug-related behaviors. These and hundreds of other epidemiologic findings are directly relevant to the choices people make every day, choices that affect their health over a lifetime.

Completing the Clinical Picture

When investigating a disease outbreak, epidemiologists rely on health-care providers and laboratorians to establish the proper diagnosis of individual patients. But epidemiologists also contribute to physicians' understanding of the clinical picture and contribute to physicians' understanding of the clinical picture and natural history of disease. For example, later in 1989, a physician saw three patients with unexplained eosinophilia (an increase in the number of a specific type of white blood cell) and myalgias (severe muscle pains). Although the physician could not make a definitive diagnosis, he notified public health authorities. Edison et al. (1990) points out that within weeks, epidemiologists had identified enough other cases to characterize the spectrum and course of the illness that came to be known as eosinophilia-myalgia syndrome. More recently, epidemiologists, clinicians, and researchers around the world have collaborated to characterize SARS, a disease caused by a new type of coronavirus that emerged in China in later 2002 (Kamps and Hoffmann, 2003). Epidemiology has also been instrumental in characterizing many non-acute diseases, such as the number of conditions associated with cigarette smoking—from pulmonary and heart disease to lip, throat, and lung cancer.

Searching for Causes

Much epidemiologic research is devoted to searching for causal factors that influence one's risk of disease. Ideally, the goal is to identify a cause so that appropriate public health action might be taken. One can argue that epidemiology can never prove a causal relationship between an exposure and a disease, since much of epidemiology is based on ecologic reasoning. Nevertheless, epidemiology often provides enough information to support effective action. Examples date from the removal of the pump handle from the Broad Street pumping station following John Snow's investigation of cholera in the Golden Square area of London in 1854 (Snow, 1936), to withdrawal of a vaccine against rotavirus in 1999 after epidemiologists found that it increased the risk of intussusceptions (i.e., part of the intestine invaginating into another section of intestine), a potentially life-threatening condition (Murphy, et al., 2001). Just as often, epidemiology and laboratory science converge to provide the evidence needed to establish causation. For example, epidemiologists were about to identify a variety of risk factors during an outbreak of pneumonia among persons attending the American Legion Convention in Philadelphia in 1976, even though the Legionnaires' bacillus was not identified in the laboratory from the lung tissue of a person who had died from Legionnaires' disease until almost six months later (Fraser, 1977).

Did You Know?

A growing body of evidence suggests that numerous chemicals, both natural and man-made (including some plastic bottles and containers, liners of metal food cans, detergents, flame retardants, food, toys, cosmetics, and pesticides), may interfere with the endocrine system and produce adverse effects in laboratory animals, wildlife, and humans. Scientists often refer to these chemicals as "endocrine disruptors." Endocrine disruption is an important public health concern that is being addressed by various national health agencies.

Epidemiology Case Example: West Nile Virus Infection

In August 1999, epidemiologists learned of a cluster of cases of encephalitis caused by West Nile Virus infection among residents of Queens, New York. West Nile Virus infection, transmitted by mosquitoes, had never before been identified in North America. This information may be used for each of the following:

Assessing the community's health—Having identified a cluster of cases never before seen in the area, public health officials must seek additional information to assess the community's health. Is the cluster limited to people who have just returned from traveling where West Nile Virus infection is common, or was the infection acquired locally, indicating that the community is truly at risk? Officials could check whether hospitals have seen more patients than usual for encephalitis. If so, officials could document when the increase in cases began, where the patients live or work or travel, and personal characteristics such as age. Mosquito traps could be placed to catch mosquitoes and tested for presence of the West Nile Virus. If warranted, officials could conduct a serosurvey (test of blood serum from a group of individuals to determine seroprevalance) of the community to document the extent of infection. Results of these efforts would help officials assess the community's burden of disease and risk of infection.

Making decisions about individual patients—West Nile Virus infection is spread by mosquitoes. People who spend time outdoors, particularly at times such as dusk when mosquitoes may be most active, can make personal decisions to reduce their own risk or not. Knowing that the risk is present but may be small, an avid gardener might or might not decide to curtail the time spent gardening in the evening, or use insect repellent containing DEET, or wear long pants and long-sleeved shirts even though it is August, or empty the birdbath where mosquitoes breed.

Documenting the clinical picture of the illness—What proportion of people infected with West Nile Virus actually develops encephalitis? Do some infected people have milder symptoms or no symptoms at all? Investigators could conduct a serosurvey to assess infection, and ask about symptoms and illness. In addition, what becomes of the people who did develop encephalitis? What proportion survived? Did they recover completely or did some have continuing difficulties?

Searching for causes to prevent future outbreaks—Although the cause and mode of transmission were known (East Nile Virus and mosquitoes, respectively), public health officials asked many questions regarding how the virus was introduced (mosquito on an airplane? wayward bird? bioterrorism?), whether the virus had a reservoir in the area (e.g., birds), what types of mosquitoes could transmit the virus, what were the host risk factors for infection or encephalitis, etc.

Did You Know?

Endocrine disruptors are naturally occurring compounds or man-made substances that may mimic or interfere with the function of hormones in the body. Endocrine disruptors may turn on, shut off, or modify signals that hormones carry, which may affect the normal functions of tissues and organs. Many of these substances have been linked with developmental, reproductive, neural, immune, and other problems in wildlife and laboratory animals.

Core Epidemiologic Functions

Tyler and Last (1992) point out that in the mid-1980s five major tasks of epidemiology in public heath practice were indentified: public health surveillance, field investigation, analytic studies, evaluation, and linkages. A sixth task, policy development, was recently added. These tasks are described below.

Public Health Surveillance

Public health surveillance is the ongoing (as shown in figure 4.2), systematic collection, analysis, interpretation, and dissemination of health data to help guide public health decision-making and action. Surveillance is equivalent to monitoring the pulse of the community. The purpose of public health surveillance, which is sometimes called "information for action" (Orenstein and Bernier, 1990), is to portray the ongoing patterns of disease occurrence and disease potential so that investigation, control, and prevention measures can be applied efficiently and effectively. This is accomplished through the systematic collection and evaluation of morbidity and mortality reports and other relevant heath information, and the dissemination of these data and

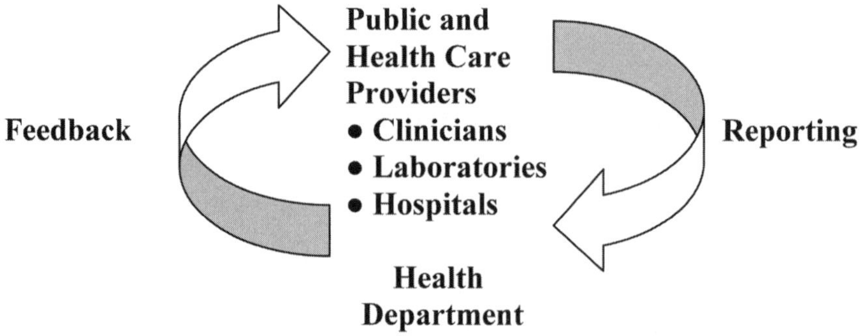

FIGURE 4.2
Surveillance cycle. From U.S. HHS, 2006, *Principles of Epidemiology in Public Health Practice*, 3rd ed., Atlanta, GA.

their interpretation to those involved in disease control and public health decision-making.

Morbidity and mortality reports are common sources of surveillance data for local and state health departments. These reports generally are submitted by health-care providers, infection control practitioners, or laboratories that are required to notify the health department of any patient with a reportable disease such as pertussis, meningococcal meningitis, or AIDS. Other sources of health-related data that are used for surveillance include reports from investigations of individual cases and disease clusters, public health program data such as immunization coverage in a community, disease registries, and health surveys.

Most often, surveillance relies on simple systems to collect a limited amount of information about each case. Although not every case of disease is reported, health officials regularly review the case reports they do receive and look for patterns among them. These practices have proven invaluable in detecting problems, evaluating programs, and guiding public health action.

While public health surveillance traditionally has focused on communicable diseases, surveillance systems now exist that target injuries, chronic disease, genetic and birth defects, occupational and potentially environmentally related diseases, and health behaviors. Since September 11, 2001, a variety of systems that rely on electronic reporting have been developed, including those that report daily emergency department visits, sales of over-the-counter medicines, and worker absenteeism (Wagner et al., 2001; CDC, 2004a). Because epidemiologists are likely to be called upon to design and use these and other new surveillance systems, an epidemiologist's core competencies must include design of data collection instruments, data management, descriptive methods and graphing, interpretation of data, and scientific writing and presentation.

Did You Know?

The endocrine system is one of the body's main communication networks and is responsible for controlling and coordinating numerous body functions. Hormones are first produced by the endocrine tissues, such as the ovaries, testes, adrenal, pituitary, thyroid, and pancreas, and then secreted into the blood to act as the body's chemical messengers, where they direct communication and coordination among other tissues throughout the body. For example, hormones work with the nervous system, reproductive system, kidneys, gut, liver, and fat to help maintain and control:

- Body energy levels
- Reproduction
- Growth and development
- Internal balance of body systems, or homeostasis
- Response to surroundings, stress, and injury

Endocrine-disrupting chemicals may interfere with the body's own hormone signals because of their structure and activity.

An example of public health surveillance is reviewing reports of test results for *Chlamydia trachomatis* (an obligate intracellular gram-negative human pathogen) for public health clinics.

Field Investigation

As noted above, surveillance provides information for action. One of the first actions that results from a surveillance case report or report of a cluster is investigation by the public health department. The investigation may be as limited as a phone call to the health-care provider to confirm or clarify the circumstances of the reported case, or it may involve a field investigation requiring the coordinated efforts of dozens of people to characterize the extent of an epidemic and to identify its cause.

The objectives of such investigations also vary. Investigations often lead to the identification of additional unreported or unrecognized ill persons who might otherwise continue to spread infection to others. For example, one of the hallmarks of investigations of persons with a sexually transmitted disease is the identification of sexual partners or contacts of patients. When interviewed, many of these contacts are found to be infected without knowing it, and are given treatment they did not realize they needed. Identification and treatment of these contacts prevents further spread.

For some diseases, investigations may identify a source or vehicle of infection that can be controlled or eliminated. For example, the investigation of a case of *Escherichia coli* 0157:H7 infection usually focuses on trying to identify the vehicle, often ground beef but sometimes something more unusual such as fruit juice. By identifying the vehicle, investigators may be able to determine how many other persons might have already been exposed and how many continue to be at risk. When a commercial product turns out to be the culprit, public announcements and recalling of the product may prevent many additional cases.

Occasionally, the objective of an investigation may simply be to learn more about the natural history, clinical spectrum, descriptive epidemiology, and risk factors of the disease before determining what disease intervention methods might be appropriate. Early investigations of the epidemic of SARS in 2003 were needed to establish a case definition based on the clinical presentation, and to characterize the populations at risk by time, place, and person. As more was learned about the epidemiology of the disease and communicability of the virus, appropriate recommendations regarding isolation and quarantine were issued (CDC, 2012).

Field investigations of the type described above are sometimes referred to as "shoe leather epidemiology," conjuring up images of dedicated, haggard epidemiologists beating the pavement in search of additional cases and clues regarding source and mode of transmission. This approach is commemorated in the symbol of the Epidemic Intelligence Service (EIS), CDC's training program for disease detectives—a shoe with a hole in the sole.

An example of a field investigation is interviewing persons infected with *Chlamydia trachomatis* to identify their sex partners.

Did You Know?

People may be exposed to endocrine disruptors through the food and beverages they consume, the medicine they take, the pesticides they apply, and the cosmetics they use. So, exposures may take place through the diet, air, skin, or water.

Analytic Studies

Surveillance and field investigations are usually sufficient to identify causes, modes of transmission, and appropriate control and prevention measures. But sometimes analytic studies employing more rigorous methods are needed. Often the methods are used in combination—with surveillance and field investigations providing clues or hypotheses about causes and modes of transmission, and analytic studies evaluating the credibility of those hypotheses.

Clusters or outbreaks of disease frequently are investigated initially with descriptive epidemiology. The descriptive approach involves the study of disease incidence and distribution by time, place, and person. It includes the calculation of rates and identification of parts of the population at higher risk than others. Occasionally, when the association between exposure and disease is quite strong, the investigation may stop when descriptive epidemiology is complete and control measures may be implemented immediately. John Snow's 1854 investigation of cholera is an example. More frequently, descriptive studies, like case investigations, generate hypotheses that can be tested with analytic studies.

The hallmark of an analytic epidemiologic study is the use of a valid comparison group. Epidemiologists must be skilled in all aspects of such studies, including design, conduct, analysis, interpretation, and communication of findings.

Design includes determining the appropriate research strategy and study design, writing justifications and protocols, calculating sample sizes, deciding on criteria for subject selection (e.g., developing case definitions), choosing an appropriate comparison group, and designing questionnaires.

Conduct involves securing appropriate clearances and approvals, adhering to appropriate ethical principles, abstracting records, tracking down and interviewing subjects, collecting and handling specimens, and managing the data.

Analysis begins with describing the characteristics of the subjects. It progresses to calculation of rates, creation of comparative tables (e.g., two-by-two tables), computation of measures of association (e.g., risk ratios or odds ratios), test of significance (e.g., chi-square test), confidence intervals, and the like. Many epidemiologic studies require more advanced analytic techniques such as stratified analysis, regression, and modeling.

Finally, **interpretation** involves putting the study findings into perspective, identifying the key take-home messages, and making sound recommendations. Doing so requires that the epidemiologist be knowledgeable about the subject matter and the strengths and weaknesses of the study.

An example of an analytic study would be comparing persons with symptomatic versus asymptomatic *Chlamydia* infection to indentify predictors.

Did You Know?

Some environmental endocrine-disrupting chemicals, such as the pesticide DDT, dioxins, and polychlorinated biphenyls (PCPs) used in electrical equipment, are highly persistent and slow to degrade in the environment, making them potentially hazardous over an extended period of time.

Evaluation

Epidemiologists, who are accustomed to using systematic and quantitative approaches, have come to play an important role in the evaluation of public health services and other activities. Evaluation is the process of determining, as systematically and objectively as possible, the relevance, effectiveness, efficiency, and impact of activities with respect to established goals (Beaglehole and Kjellstrom, 1993).

Effectiveness refers to the ability of a program to produce the intended results in the field; effectiveness differs from *efficacy*, which is the ability to produce results under ideal conditions.

Efficiency refers to the ability of the program to produce the intended results with a minimum expenditure of time and resources.

The evaluation itself may focus on plans (formative evaluation), operations (process evaluation), impact (summative evaluation), or outcomes—or any combination of these. Evaluation of an immunization program, for example, might assess the efficiency of the operations, the proportion of the target population immunized, and the apparent impact of the program on the incidence of vaccine-preventable disease. Similarly, evaluation of a surveillance system might address operations and attributes of the system, its ability to detect cases or outbreaks, and its usefulness (CDC, 2001a).

An example of an evaluation is conducting an analysis of patient flow at the public health clinic to determine waiting times for clinic patients.

Linkages

Epidemiologists working in public health settings rarely act in isolation. In fact, field epidemiology is often said to be a "team sport." During an investigation, an epidemiologist usually participates as either a member or the leader of a multidisciplinary team. Other team members may be laboratorians, sanitarians, infection control personnel, nurses or other clinical staff, and increasingly, computer information specialists. Many outbreaks cross geographical and jurisdictional lines, so co-investigators may be from local, state, or federal levels of government, academic institutions, clinical facilities, or the private sector. To promote current and future collaboration, the epidemiologists need to maintain relationships with staff of other agencies and institutions. Mechanisms for sustaining such linkages include official memoranda of understanding, sharing of published or online information for public health audiences and outside partners, and informational networking that takes place at professional meetings.

An example of linkage is meeting with directors of family planning clinics and college health clinics to discuss *Chlamydia* testing and reporting.

Did You Know?

Although there is limited evidence to prove that low-dose exposures are causing adverse human health effects, there is a large body of research in experimental animals and wildlife suggesting that endocrine disruptors may cause:

- Reductions in male fertility and a decline in the number of males born
- Abnormalities in male reproductive organs
- Female reproductive health issues, including fertility problems, early puberty, and early reproductive senescence
- Increases in mammary, ovarian, and prostate cancers
- Increases in immune and autoimmune diseases, and some neurodegenerative diseases

Policy Development

The definition of *epidemiology* ends with the following phrase: ". . . and the application of this study to the control of health problems." While some academic epidemiologists have stated that epidemiologists should stick to research and not get involved in policy development or even make recommendations (Rothman, 1993), public health epidemiologists do not have this luxury. Indeed, epidemiologists who understand a problem and the population in which it occurs are often in a uniquely qualified position to recommend appropriate interventions. As a result, epidemiologists working in public health regularly provide input, testimony, and recommendations regarding disease control strategies, reportable disease regulations, and health-care policy.

An example of policy development would be developing guidelines/criteria about which patients coming to the clinic should be screened (tested) for *Chlamydia* infection.

Did You Know?

There is data showing that exposure to BPA, as well as other endocrine-disrupting chemicals with estrogenic activity, may have effects on obesity and diabetes. This data, while preliminary and only tested in animals, indicates the potential for endocrine-disrupting agents to have effects on other endocrine systems not yet fully examined.

The Epidemiologic Approach

As with all scientific endeavors, the practice of epidemiology relies on a systematic approach. In very simple terms, the epidemiologist:

- **Counts** cases or health events, and describes them in terms of time, place, and person.
- **Divides** the number of cases by an appropriate denominator to calculate rates.
- **Compares** these rates over time or for different groups of people.

Before counting cases, however, the epidemiologist must decide what a case is. This is done by developing a case definition. Then, using this case definition, the epidemiologist finds and collects information about the case-patients. The epidemiologist then performs descriptive epidemiology by characterizing the cases collectively according to time, place, and person. To calculate the disease rate, the epidemiologist divides the number of cases by the size of the population. Finally, to determine whether this rate is greater than what one would normally expect, and if so to identify factors contributing to this increase, the epidemiologist compares the rate from this population to the rate in an appropriate comparison group, using analytic epidemiology techniques. These epidemiologic actions are described in more detail below. Subsequent tasks, such as reporting the results and recommending how they can be used for public health action, are just as important, but they are beyond the scope of this book.

Did You Know?

From animal studies, researchers have learned much about the mechanisms through which endocrine disruptors influence the endocrine system and alter hormonal functions.

Endocrine disruptors can:

- Mimic or partly mimic naturally occurring hormones in the body like estrogens (the female sex hormone), androgens (the male sex hormone), and thyroid hormones, potentially causing production overstimulation.
- Bind to a receptor within a cell and block the endogenous hormone from binding. The normal signal then fails to occur, and the body

fails to respond properly. Examples of chemicals that block or antagonize hormones are anti-estrogens and anti-androgens.
- Interfere or block the way natural hormones or their receptors are made or controlled, for example, by altering their metabolism in the liver.

Defining a Case

Before counting cases, the epidemiologist must decide what to count, that is, what to call a case. For that, the epidemiologist uses a case definition. A *case definition* is a set of standard criteria for classifying whether a person has a particular disease, syndrome, or other heath condition. Some case definitions, particularly those used for national surveillance, have been developed and adopted as national standards that ensure comparability. Use of an agreed-upon standard case definition ensures that every case is equivalent, regardless or when or where it occurred, or who identified it. Furthermore, the number of cases or rate of disease identified in one time or place can be compared with the number or rate from another time or place. For example, with a standard case definition, health officials could compare the number of cases of listeriosis that occurred in Forsyth County, North Carolina, in 2000 with the number that occurred there in 1999. Or they could compare the rate of listeriosis in Forsyth County in 2000 with the national rate in that same year. When everyone uses the same standard case definition and a difference is observed, the difference is likely to be real rather than a result variation in how cases are classified.

Did You Know?

The most common way to get listeriosis is by eating food contaminated with *Listeria*. Women who are infected during pregnancy can pass *Listeria* to their fetus or newborn baby.

To ensure that all health departments in the United States use the same case definitions for surveillance, the Council of State and Territorial Epidemiologists (CSTE), the CDC, and other interested parties have adopted standard case definitions for the notifiable infectious diseases (CDC, 1997a). These definitions are revised as needed. In 1999, to address the need for common definitions and methods for state-level chronic disease surveillance, CSTE, the Association of State and Territorial Chronic Disease Program Directors, and the CDC adopted standard definitions for seventy-three chronic disease indicators (CDC, 2004b).

Other case definitions, particularly those used in local outbreak investigations, are often tailored to the local situation. For example, a case definition developed for an outbreak of viral illness might require laboratory confirmation where such laboratory services are available, but likely would not if such services were not readily available.

Did You Know?

Usual symptoms of listeriosis include fever, muscle aches, diarrhea, and an upset stomach. Mild illnesses are generally not diagnosed. More serious infections cause severe headache, stiff neck, confusion, loss of balance, or convulsions. Although pregnant women themselves often have a mild, flu-like illness, for the unborn baby listeriosis can cause miscarriages, premature birth, or stillbirth. About 30 to 50 percent of newborns and 35 percent of nonpregnant adults with serious infection die from listeriosis. Note that a more detailed surveillance case definition of listeriosis is presented in the box below.

Components of a Case Definition for Outbreak Investigations

A case definition consists of clinical criteria and, sometimes, limitations on time, place, and person. The clinical criteria usually include confirmatory laboratory tests, if available, or combinations of symptoms (subjective complaints), signs (objective physical findings), and other findings. Case definitions used during outbreak investigations are more likely to specify limits on time, place, and/or person than those used for surveillance. Contrast the case definition used for the surveillance of listeriosis (see box below) with the case definition used during an investigation of a listeriosis outbreak in North Carolina in 2000 (CDC, 1997a; Boggs et al., 2000).

Both the national surveillance case definition and the outbreak case definition require a clinically compatible illness and laboratory confirmation of *Listeria monocytogenes* for a normally sterile site, but the outbreak case definition adds restrictions on time and place, reflecting the scope of the outbreak.

Listeriosis—Surveillance Case Definition (Source: CDC, 1997)

Clinical Description

Infection caused by *Listeria monocytogenes*, which may produce any of several clinical syndromes, including stillbirth, listeriosis of the newborn, meningitis, bacteriemia, or localized infections.

Laboratory Criteria for Diagnosis

Isolation of *Listeria monocytogenes* from a normally sterile site (e.g., blood or cerebrospinal fluid or, less commonly, joint, pleural, or pericardial fluid).

Case Classification

Confirmed: a clinically compatible case that is laboratory confirmed.

Listeriosis—Outbreak Investigation (Boggs et al., 2001)

Case Definition

Clinically compatible illness with *Listeria monocytogenes* isolated are:

- From a normally sterile site
- In a resident of Winston-Salem, North Carolina
- With onset between October 24, 2000, and January 4, 2001

Many case definitions, such as that shown for listeriosis, require laboratory confirmation. This is not always necessary, however; in fact, some diseases have no distinctive laboratory findings. Kawasaki syndrome, for example, is a childhood illness with fever and rash that has no known cause and no specifically distinctive laboratory findings. Notice that its case definition (see box below) is based on the presence of fever, at least four of five specified clinical findings, and the lack of a more reasonable explanation.

Kawasaki Syndrome—Case Definition (Source: CDC, 1990)

Clinical Description

A febrile (feverish) illness of greater than or equal to five days' duration, with at least four of the five following physical findings and no other more reasonable explanation for the observed clinical findings:

- Bilateral conjunctival infection
- Oral changes (erythema of lips or oropharynx, strawberry tongue, or fissuring of the lips)
- Peripheral extremity changes (edema, erythema, or generated) or periungual extremity changes (edema, erythema, or generalized or periungual desquamation)

- Rash
- Cervical lymphadenopathy (at least one lymph node greater than or equal to 1.5 cm in diameter)

Laboratory Criteria for Diagnosis

None

Case Classification

Confirmed: a case that meets the clinical case definition. (If fever disappears after intravenous gamma globulin therapy is started, fever may be less than five days' duration, and the clinical case definition may still be met.)

Criteria in Case Definitions

A case definition may have several sets of criteria, depending on how certain the diagnosis is. For example, during an investigation of a possible case or outbreak of measles, a person with a fever and rash might be classified as having a suspected, probable, or confirmed case of measles, depending on what evidence of measles is present (see below).

Measles (Rubeola)—1996 Case Definition (CDC 1997)

Clinical Description

An illness characterized by all the following:

- A generalized rash lasting greater than or equal to three days
- A temperature greater than or equal to 101.0°F (greater than or equal to 38.3°C)
- Cough, coryza, or conjunctivitis

Laboratory Criteria for Diagnosis

- Positive serologic test for measles immunoglobulin M antibody, or
- Significant rise in measles antibody level by any standard serologic assay
- Isolation of measles virus from a clinical specimen

Case Classification

Suspected: Any febrile illness accompanied by rash. *Probable*: A case that meets the clinical case definition, has noncontributory or no serologic or virologic testing, and is not epidemiologically linked to a confirmed case. *Confirmed*: A case that is laboratory confirmed or that meets the clinical case definition and is epidemiologically linked to a confirmed case. (A laboratory-confirmed case does not need to meet the clinical case definition.) Confirmed cases should be reported to National Notifiable Disease Surveillance System. An imported case has its source outside the country or state. Rash onset occurs within eighteen days after entering the jurisdiction, and illness cannot be linked to local transmission. Important cases should be classed as:

International. A case that is imported from another country.

Out-of-State. A case that is imported from another state in the United States. The possibility that a patient was exposed within his or her state of residence should be excluded; therefore, the patient either must have been out of state continuously for the entire period of possible exposure (at least seven to eighteen days before onset of rash) or have had one of the following types of exposure while out of state: a) face-to-face contact with a person who had either a probable or confirmed case, or b) attendance in the same institution as a person who had a case of measles (e.g., in a school, classroom, or day care center).

An indigenous case is defined as a case of measles that is not imported. Cases that are linked to imported cases should be classified as indigenous if the exposure to the imported case occurred in the reporting state. Any case that cannot be proved to be imported should be classified as indigenous.

A case might be classified as suspected or probable while waiting for the laboratory results to become available. Once the laboratory provides the report, the case can be reclassified as either confirmed or "not a case," depending on the laboratory results. In the midst of a large outbreak of a disease caused by a known agent, some cases may be permanently classified as suspected or probable because officials may feel that running laboratory tests on every patient with a consistent clinical picture and a history of exposure (e.g., chickenpox) is unnecessary and even wasteful. Case definitions should not rely on laboratory culture results alone, since organisms are sometimes present without causing disease.

Modifying Case Definitions

Case definitions can also change over time as more information is obtained. The first case definition of SARS, based on clinical symptoms and either contact with a case or travel to an area with SARS transmission, was published in CDC's Morbidity and Mortality Weekly Report (MMWR) on March 21, 2003 (see box below; CDC, 2003a). Two weeks later it was modified slightly. On March 29, after a novel coronavirus was determined to be the causative agent, an interim surveillance case definition was published that included laboratory criteria for evidence of infection with the SARS-associated coronavirus. By June, the case definition had changed several more times. In anticipation of a new wave of cases in 2004, a revised and much more complex case definition was published in December 2003 (CDC, 2003b).

CDC Preliminary Case Definition for Severe Acute Respiratory Syndrome (SARS)—March 21, 2003 (CDC, 2003a)

Suspected Case

Respiratory illness of unknown etiology with onset since February 1, 2003, and the following criteria:

- Documented temperature >100.4°F (>38°C)
- One or more symptoms with respiratory illness (e.g., cough, shortness of breath, difficulty breathing, or radiographic findings of pneumonia or acute respiratory distress syndrome)
- Close contact (defined as having cared for, having lived with, or having had direct contact with respiratory secretions and/or body fluids of a person suspected of having SARS) within ten days of onset of symptoms with a person under investigation for or suspected of having SARS or travel within ten days of onset of symptoms to an area with documented transmission of SARS as defined by the World Health Organization (WHO).

Epidemiology Case Example—SARS

In this example, consider the initial case definition for SARS presented above. In the following we demonstrate how the case definition might address the purposes listed below.

Diagnosing and caring for individual patients—the third criterion may be limiting because patients may not be aware of close contact.

Tracking the occurrence of disease—probably reasonable.

Doing research to identify the cause of the disease—criteria do not require sophisticated evaluation or testing, so they can be used anywhere in the world.

Deciding who should be quarantined (quarantine is the separation or restriction of movement of persons who are not ill but are believed to have been exposed to infection, to prevent further transmission)—too broad. Most people with cough and fever returned from Toronto, China, etc., are more likely to have upper respiratory infections than SARS.

Variation in Case Definitions

Case definitions may also vary according to the purpose for classifying the occurrences of a disease. For example, health officials need to know as soon as possible if anyone has symptoms of plague or anthrax so that they can begin planning what actions to take. For such rare but potentially severe communicable disease, for which it is important to identify every possible case, health officials use a sensitive case definition. A sensitive case definition is broad or "loose," in the hope of capturing most or all of the true cases. For example, the case definition for a suspected case of rubella (German measles) is "any generalized rash illness of acute onset" (CDC, 1997a). This definition is quite broad, and would include not only all cases of rubella, but also measles, chickenpox, and rashes due to other causes such as drug allergies. So while the advantage of a sensitive case definition is that it includes most or all of the true cases, the disadvantage is that it sometimes includes other illnesses, as well.

On the other hand, an investigator studying the causes of a disease outbreak usually wants to be certain that any person included in a study really had the disease. That investigator will prefer a specific or "strict" case definition. For instance, in an outbreak of *Salmonella Agona* infection, the investigators would be more likely to identify the source of the infection if they included only people who were confirmed to have been infected with that organism, rather than including anyone with acute diarrhea, because some people may have had diarrhea from a different cause. In this setting, the only disadvantages of a strict case definition is the equipment used to test everyone with symptoms and the underestimation of the total number of cases if some people with Salmonellosis are not tested.

Using Counts and Rates

As noted, one of the basic tasks in public health is identifying and counting cases. These counts, usually derived from case reports submitted by healthcare workers and laboratories to the health department, allow public health officials to determine the extent and patterns of disease occurrence by the time, place, and person. They may also indicate clusters or outbreaks of disease in the community.

Counts are also valuable for health planning. For example, a health official might use counts (i.e., numbers) to plan how many infection control isolation units or doses of vaccine may be needed.

However, simple counts do not provide all the information a health department needs. For some purposes, the counts must be put into context, based on the population in which they arise. Rates are measures that relate the number of cases during a certain period of time (usually per year) to the size of the population in which they occurred. For example, 42,745 new cases of AIDS were reported in the United States in 2001 (CDC, 2001b). This number, divided by the estimated 2001 population, results in a rate of 15.3 cases per 100,000 population. Rates are particularly useful for comparing the frequency of disease in different locations whose populations differ in size. For example, in 2003, Pennsylvania had nearly ten times the as many births (140,660) as its neighboring state, Delaware (11,264). However, Pennsylvania has nearly ten times the population of Delaware. So a more fair way to compare is to calculate rates. In fact, the birth rate was greater in Delaware (13.8 per 1,000 women aged 15 to 44 years) than in Pennsylvania (11.4 per 1,000 women aged 15 to 44 years) (Arias, 2003).

Important Point: Rate = Number of cases ÷ Size of the population/unit of time.

Rates are also useful for comparing disease occurrence during different periods of time. For example, 19.5 cases of chickenpox per 100,000 were reported in 2001 compared with 135.8 cases per 100,000 in 1991. In addition, rates of disease among different subgroups can be compared to identify those at increased risk of disease. These so-called high risk groups can be further assessed and targeted for special intervention. High risk groups can also be studied to identify risk factors that cause them to have increased risk of disease. While some risk factors such as age and family history of breast cancer may not be modifiable, others, such as smoking and unsafe sexual practices, are. Individuals can use knowledge of the modifiable risk factors to guide decisions about behaviors that influence their health.

Maquiladora: An Environmental Health Concern

(From NIH 2011. Tox Town—*Maquiladora.* Accessed 02/26/12 at http://toxtown.nim.nih.gov/ test_version/locations.php?id=35.) Maquiladoras are foreign-owned factories located in Mexico that are typically found along the U.S.-Mexico border. Maquiladoras produce a variety of products including electronic components, chemicals, clothes, machinery, and auto parts.

The maquiladora program began in 1965 as a part of the Mexican government's Border Industrialization Program. It was developed in response to the demise of the U.S. government's Bracero Program, which allowed Mexican farmworkers to legally perform seasonal work within the United States. The end of the Bracero Program caused an unemployment crisis in the border region. The Mexican government responded to this crisis by creating the maquiladora program, which provides an incentive to foreign manufacturers to move production to Mexico. This incentive was created by allowing duty-free import of raw materials and other supplies into the country with the stipulation that the manufactured goods and the resulting wastes would eventually be exported to another country. The passing of the North American Free Trade Agreement (NAFTA) in 1993 led to an increased number of maquiladoras in the border region. In 2003, there were 2,893 maquiladoras employing 1,063,123 people.

The high concentration of maquiladoras combined with less rigorous environmental regulations limited capacity to enforce environmental laws, and the expense of exporting hazardous waste has created an incentive for illegal dumping and has polluted the surrounding land, water, and air. Inside the maquiladoras, occupational hazards relating to toxic chemical exposure and workplace safety also affect human health. Occupational hazards are of particular concern in Mexico since first-time violators are rarely punished and since penalties are typically incurred only for imminent dangers and failures to address previously highlighted violations.

Descriptive Epidemiology

(Information in this section is adapted from NIH, 2011. Tox Town—*Maquiladora.* Accessed 02/26/12 at http://toxtown.nim.nih.gov/ test_version/locations.php?id=35.) As noted earlier, every novice newspaper reporter is taught that a story is incomplete if it does not describe the what, who, where, when, and why/how of a situation (see figure 4.3), whether it be a space shuttle launch or a house fire.

The 5 W's of descriptive epidemiology

- What = health issue of concern
- Who = person
- Where = place
- When = time
- Why/how = causes, risk factors, modes of transmission

FIGURE 4.3
The 5 Ws of descriptive epidemiology.

Epidemiologists strive for similar comprehensiveness in characterizing an epidemiologic event, whether it be a pandemic influenza or a local increase in all-terrain vehicle crashes. However, epidemiologists tend to use synonyms for the five Ws listed above: case definition, person, place, time, and causes/risk factors/modes of transmission. Descriptive epidemiology covers *time*, *place*, and *person*. Compiling and analyzing data by time, place, and person is desirable for several reasons:

- First, by looking at the data carefully, the epidemiologist becomes very familiar with the data. He or she can see what the data can or cannot reveal based on the variables available, its limitations (for example, the number of records with missing information of each important variable), and its eccentricities (for example, all cases ranging in age from two months to six years, plus one seventeen-year-old).
- Second, the epidemiologist learns the extent and pattern of the public health problem being investigated—which months, which neighborhoods, and which groups of people have the most and least cases.
- Third, the epidemiologist creates a detailed description of the health of a population that can be easily communicated with tables, graphs, and maps.
- Fourth, the epidemiologists can identify areas or groups within the population that have high rates of disease. This information in turn provides important clues to the causes of the disease, and these clues can be turned into testable hypotheses.

Time

The occurrence of disease changes over time. Some of these changes occur regularly, while others are unpredictable. Two diseases that occur during the same season each year include influenza (winter) and West Nile Virus infection (August–September). In contrast, diseases such as hepatitis B and Salmonellosis can occur at any time. For diseases that occur seasonally, health officials can anticipate their occurrence and implement control and prevention measures, such as an influenza vaccination campaign or mosquito spraying. For diseases that occur sporadically, investigators can conduct studies to indentify the causes and modes of spread, and then develop appropriately targeted actions to control or prevent further occurrence of the disease.

In either situation, displaying the patterns of disease occurrence by time is critical for monitoring disease occurrence in the community and for assessing whether the public health interventions made a difference.

Time data are usually displayed with a two-dimensional graph. The vertical or y-axis usually shows the number or rate of cases; the horizontal or x-axis shows the time periods, such as years, months, or days. The number or rate of cases is plotted over time. Graphs of disease occurrence over time are usually plotted as line graphs (see figure 4.4) or histograms (see figure 4.5)

Sometimes a graph shows the timing of events that are related to disease trends being displayed. For example, the graph may indicate the period of exposure or the date when control measures were implemented. Studying a graph that notes the period of exposure may lead to insights into what may have caused illness. Showing a graph that notes the timing of control measure shows what impact if any, the measures may have had on disease occurrence.

As noted above, time is plotted along the x-axis. Depending on the disease, the time scale may be as broad as years or decades, or as brief as days or even hours of the day. For some conditions—many chronic diseases, for example—epidemiologists tend to be interested in long-term trends or patterns in the number of cases or the rate. For other conditions, such as foodborne

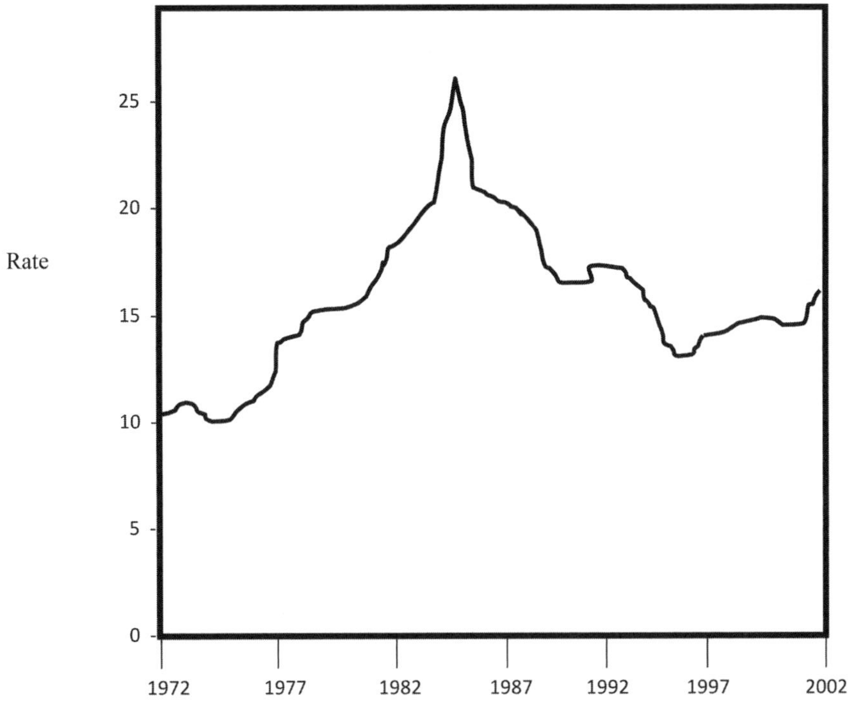

FIGURE 4.4
Reported cases of Salmonellosis per 100,000 population, by year—United States, 1972–2002. From the CDC, 2002, Summary of notifiable diseases—United States, 2002. Published April 30, 2004, for *MMWR* 2002; 51(53): 59.

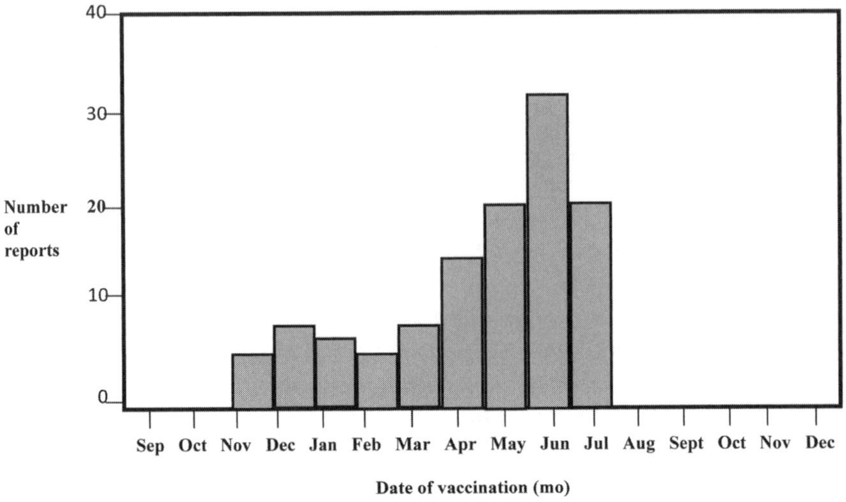

FIGURE 4.5
Number of intussusception reports after the rhesus rotavirus vaccine-tetravalent (RRV-TV), by vaccination date—United States, September 1998 to December 1999. From Zhou, W., V. Pool, K.K. Iskander, R. English-Bullard, R. Ball, and R. P. Wise et al., 2003. "Surveillance Summaries," January 24. *MMWR* 52(55-1): 1–26.

outbreaks, the relevant time scale is likely to be days or hours. Some of the common types of time-related graphs are further described below.

Secular (long-term) trends. Graphing the annual cases or rate of a disease over a period of years shows long-term or secular trends in the occurrence of the disease (see figure 4.3). Health officials use these graphs to assess the prevailing direction of disease occurrence (increasing, decreasing, or eventually flat), help them evaluate programs or make policy decisions, infer what caused an increase or decrease in the occurrence of a disease (particularly if the graph indicates when related events took place), and use past trends as a predictor of future incidence of disease.

Seasonality. Disease occurrence can be graphed by week or month over the course of a year or more to show its seasonal pattern, if any. Some diseases such as influenza and West Nile infection are known to have characteristic seasonal distributions. Seasonal patterns may suggest hypotheses about how the infection is transmitted, what behavioral factors increase risk, and other possible contributors to the disease or condition.

Day of week and time of day. For some conditions, displaying data by day of the week or time of day may be informative. Analysis at these shorter time periods is particularly appropriate for conditions related to occupational or environmental exposures that tend to occur at regularly scheduled intervals. For example, in a study by Goodman et al., (1985), they point out that farm tractor fatalities on Sundays were about half the number on the other days. The pattern of farm tractor injuries, by hour, peaked at 11:00 a.m., dipped at noon, and peaked again at 4:00 p.m. These patterns may suggest hypotheses and possible explanations that could be evaluated with further study.

Epidemic period. To show the time curve of a disease outbreak or epidemic, epidemiologists use a graph called an epidemic curve. As with the other graphs presented so far, an epidemic curve's y-axis shows the number of cases, while the x-axis shows time as either date of symptom onset or date of diagnosis. Depending on the incubation period (the length of time between exposure and onset of symptoms) and routes of transmission, the scale on the x-axis can be as broad as weeks (for a very prolonged epidemic) or as narrow as minutes (e.g., for food poisoning by chemicals that cause symptoms within minutes). Conventionally, the data are displayed as a histogram (which is similar to a bar chart but has no gaps between adjacent columns). Sometimes each case is displayed as a square. The shape and other features of an epidemic curve can suggest hypotheses about the time and source of exposure, the mode of transmission, and the causative agent.

Place

Describing the occurrence of disease by place provides insight into the geographic extent of the problem and its geographic variation. Characterization by place refers not only to place of residence but to any geographic location relevant to disease occurrence. Such locations include place of diagnosis or report, birthplace, site of employment, school district, hospital unit, or recent travel destinations. The unit may be as large as a continent or country or as small as a street address, hospital wing, or operating room. Sometimes place refers not to a specific location at all but to a place category such as urban or rural, domestic or foreign, and institutional or noninstitutional.

Analyzing data by place can identify communities at increased risk of disease. Even if the data cannot reveal why these people have an increased risk, it can help generate hypotheses to test with additional studies. For example, is a community at increased risk because of characteristics of the people in the community such as genetic susceptibility, lack of immunity, risky behaviors, or exposure to local toxins or contaminated food? Can the increased risk, particularly of a communicable disease, be attributed to characteristics of the causative agent such as a particularly virulent strain, hospitable breeding sites, or availability of the vector that transmits the organism to humans? Or can the increased risk be attributed to the environment that brings the agent and the host together, such as crowding in urban areas that increases the risk of disease transmission from person to person, or more homes being built in wooded areas close to deer that carry ticks infected with the organism that causes Lyme disease?

Person

Because personal characteristics may affect illness, organization and analysis of data by "person" may use inherent characteristics of people (for example, age, sex, race), biological characteristics (immune status), acquired characteristics (marital status), activities (occupation, leisure activities, use of medications/tobacco/drugs), or the conditions under which they live (socioeconomic status, access to medical care). Age and sex are included in almost all data sets and are the two most commonly analyzed "person" characteristics. However, depending on the disease and the data available, analyses of other person variables are usually necessary. Usually epidemiologists begin the analysis of person data by looking at each variable separately. Sometimes, two variables such as age and sex can be examined simultaneously. Person data are usually displayed in tables or graphs. It's important to note that "person" attributes include age, sex, ethnicity/race, and socioeconomic status.

Age. Age is probably the single most important "person" attribute, because almost every heath-related event varies with age. A number of factors that also vary with age include susceptibility, opportunity for exposure, latency or incubation period of the disease, and physiologic response (which affects, among other things, disease development). When analyzing by age, epidemiologists try to use age groups that are narrow enough to detect any age-related patterns that may be present in the data. For some diseases, particularly chronic diseases, ten-year age groups may be adequate. For other diseases, ten-year and even five-year age groups conceal important variations in disease occurrence by age.

Sex. Males have higher rates of illness and death than do females for many diseases. For some diseases, this sex-related difference is because of genetic, hormonal, anatomic, or other inherent differences between the sexes. These inherent differences affect susceptibility or physiologic response. For example, premenopausal women have a lower risk of heart disease than men of the same age. This difference has been attributed to higher estrogen levels in women.

Ethnic and racial groups. Sometimes epidemiologists are interested in analyzing person data by biological, cultural, or social groupings such as race, nationality, religion, or social groups such as tribes and other geographically or socially isolated groups. Differences in racial, ethnic, or other group variables my reflect differences in susceptibility or exposure, or differences in other factors that influence the risk of disease, such as socioeconomic status and access to health care.

Socioeconomic status. Socioeconomic status is difficult to quantify. It is made up of many variables such as occupation, family income, educational achievement or census track, living conditions, and social standing. The variables that are easiest to measure may not accurately reflect the overall concept. Nevertheless, epidemiologists commonly use occupation, family income, and educational achievement, while recognizing that these variables do not measure socioeconomic status precisely.

The frequency of many adverse health conditions increase with decreasing socioeconomic status. For example, tuberculosis is more common among people in lower socioeconomic strata. Infant mortality and time lost from work due to disability are both associated with lower income. These patterns may reflect more harmful exposures, lower resistance, and less access to health care. Or they may in part reflect an interdependent relationship that is impossible to untangle: Does low socioeconomic status contribute to disability, or does disability contribute to lower socioeconomic status, or both? What accounts for the disproportionate prevalence of diabetes and asthma in lower socioeconomic areas (Liao, et al., 2001; CDC, 1997b)?

A few adverse health conditions occur more frequently among people of higher socioeconomic status. Gout was known as the "disease of kings" because of its association with consumption of rich foods. Other conditions associated with higher socioeconomic status include breast cancer, Kawasaki syndrome, chronic fatigue syndrome, and tennis elbow. Differences in exposure account for at least some if not most of the difference in the frequency of these conditions.

Fish Farms: An Environmental Health Concern

(From NIH TOXTOWN. *Fish Farm.* Accessed 02/27/12 at http://toxtown.nlm.nih.gov/test_version/locations.php?id=25.) A fish farm is an aquaculture operation where fish are grown in a controlled water environment for commercial or recreational use. Aquaculture, a major global industry, is the growing of aquatic plants and animals for harvest. Fish aquaculture may take place in pens in rivers, lakes, streams, and coastal areas or on land using man-made tanks or ponds. Fish commonly produced in fish farms are catfish, salmon, trout, tilapia, shrimp, oysters, clams, and crawfish.

Fish farms have several benefits, such as maintaining a steady supply of fish and reducing the need for imported seafood. But they can also have harmful environmental effects. Fish farms that use natural waterways can introduce non-native species to the surrounding environments. Coastal areas may be damaged when turned into new production sites. Diseases can spread from fish in the growing pens to wild fish. The waste from all the fish concentrated in one area can pollute the water and settle in sediments. Fish may be fed antibiotics to keep them healthy, but these drugs may leak into the marine environment causing toxic conditions for other species. Fish farms that use man-made structures may be cleaned with chlorine.

Both farmed and wild-caught fish, especially salmon, may contain traces of mercury or PCBs. Recent studies, however, conclude that the health benefits of eating fish far outweigh any potential health risks from contaminants.

Analytic Epidemiology

As noted earlier, descriptive epidemiology can identify patterns among cases and in populations by time, place, and person. From these observations, epidemiologists develop hypotheses about the causes of these patterns and about the factors that increase risk of disease. In other words, epidemiologists can use descriptive epidemiology to generate hypotheses, but only rarely to test those hypotheses. For that, epidemiologists must turn to analytic epidemiology.

The key feature of analytic epidemiology is a comparison group. Consider a large outbreak of hepatitis A that occurred in Pennsylvania in 2003 (CDC, 2003c). Investigators found almost all of the case-patients had eaten at a particular restaurant during the two to six weeks before onset of illness (i.e., the typical incubation period for hepatitis A). While the investigators were able to narrow down their hypotheses to the restaurant and were able to exclude the food preparers and servers as the source, they did not know which particular food may have been contaminated. The investigators asked the case-patients which restaurant foods they had eaten, but that only indicated which foods were popular. The investigators, therefore, also enrolled and interviewed a comparison or control group—a group of persons who had eaten at the restaurant during the same period but who did not get sick. Of 133 items on the restaurant's menu, the most striking difference between the case and control groups was in the proportion that ate salsa (94 percent of case-patients ate, compared with 39 percent of controls). Further investigation of the ingredients in the salsa implicated green onions as the source of infection. Shortly thereafter, the Food and Drug Administration issued an advisory to the public about green onions and the risk of hepatitis A. This action was in direct response to the convincing results of the analytic epidemiology, which compared the exposure history of case-patients with that of an appropriate comparison group.

Important Point: Key feature of analytic epidemiology = Comparison group.

When investigators find that people with a particular characteristic are more likely than those without the characteristic to contract a disease, the characteristic is said to be associated with the disease. The characteristic may be a:

- Demographic factor such as age, race, or sex
- Constitutional factor such as blood group or immune status
- Behavior or act such as smoking or having eaten salsa
- Circumstance such as living near a toxic waste site

Identifying factors associated with disease helps health officials appropriately target public health prevention and control activities. It also guides additional research into the causes of disease.

Thus, analytic epidemiology is concerned with the search for causes and effects, or the why and the how. Epidemiologists use analytic epidemiology to quantify the association between exposures and outcomes and to test hypotheses about causal relationships. It has been said that epidemiology by itself can never prove that a particular exposure caused a particular outcome. Often, however, epidemiology provides sufficient evidence to take appropriate control and prevention measures.

Epidemiologic studies fall into two categories: *experimental* and *observational.*

Experimental Studies

In an experimental study, the investigator determines through a controlled process the exposure for each individual (clinical trial) or community (community trial), and then tracks the individuals or communities over time to detect the effects of the exposure. For example, in a clinical trial of a new vaccine, the investigator may randomly assign some of the participants to receive the new vaccine, while others receive a placebo shot. The investigator then tracks all participants, observes who gets the disease that the new vaccine is intended to prevent, and compares the two groups (new vaccine versus placebo) to see whether the vaccine group has a lower rate of disease. Similarly, in a trial to prevent onset of diabetes among high-risk individuals, investigators randomly assigned enrollees to one of three groups—placebo, an anti-diabetes drug, or lifestyle intervention. At the end of the follow-up period, investigators found the lowest incidence of diabetes in the lifestyle intervention group, the next lowest in the anti-diabetic drug group, and the highest in the placebo group (Knowler, 2002).

An example of an experimental study is where subjects were children enrolled in a health maintenance organization. At two months, each child was randomly given one of two types of a new vaccine against rotavirus infection. Parents were called by a nurse two weeks later and asked whether the children had experienced any of a list of side effects.

Observational Studies

In an observational study, the epidemiologist simply observes the exposure and disease status of each study participant. John Snow's studies of cholera in London were observational studies. The two most common types of observational studies are cohort studies and case-control studies; a third type is cross-sectional studies.

Cohort Study. A cohort study is similar in concept to the experimental study. In a cohort study the epidemiologist records whether each study participant is exposed or not, and then tracks the participants to see if they develop the disease. Note that this differs from an experimental study because, in a cohort study, the investigator observes rather than determines the participants' exposure status. After a period of time, the investigator compares the disease rate in the exposed group with the disease rate in the unexposed group. The unexposed group serves as the comparison group, providing an estimate of the baseline or expected amount of disease occurrence in the community. If the disease rate is substantively different in the exposed group compared to the unexposed group, the exposure is said to be associated with illness.

The length of follow-up varies considerably. In an attempt to respond quickly to a public health concern such as an outbreak, public health departments tend to conduct relatively brief studies. On the other hand, research and academic organizations are more likely to conduct studies of cancer, cardiovascular disease, and other chronic diseases that may last for years and even decades. The Framingham study is a well-known cohort study that has followed 5,000 residents of Framingham, Massachusetts, since the early 1950s to establish the rates and risk factors for heart disease (Kannel, 2000). The Nurses Health Study and the Nurses Health Study II are cohort studies established in 1976 and 1989, respectively, that have followed over 100,000 nurses each and have provided useful information on oral contraceptives, diet, and lifestyle risk factors (Colditz, 1997). These studies are sometimes called *follow-up* or *prospective* cohort studies, because participants are enrolled as the study begins and are then followed prospectively over time to identify occurrence of the outcomes of interest.

An alternative type of cohort study is a *retrospective* cohort study. In this type of study, both the exposure and the outcomes have already occurred. Just as in a prospective cohort study, the investigator calculates and compares rates of disease in the exposed and unexposed groups. Retrospective cohort studies are commonly used in investigations of disease in groups of easily identified people such as workers at a particular factory or attendees at a wedding. For example, a retrospective cohort study was used to determine the source of infection of cyclosporiasis, a parasitic disease that caused an outbreak among members of a residential facility in Pennsylvania in 2004 (CDC, 2004c). The investigation indicated that consumption of snow peas was implicated as the vehicle of the cyclosporiasis outbreak.

An example of an observational cohort study is where an occurrence of cancer was identified between April 1991 and July 2002 for 50,000 troops who served in the first Gulf War (ended April 1991) and 50,000 troops who served elsewhere during the same period.

Case-control study. In a case-control study, investigators start by enrolling a group of people with disease (at the CDC such persons are called case-patients rather than cases, because "case" refers to occurrence of disease, not a person). As a comparison group, the investigator then enrolls a group of people without disease (controls). Investigators then compare previous exposures between the two groups. The control group provides an estimate of the baseline or expected amount of exposure in that population. If the amount of exposure among the case group is substantially higher than the amount you would expect based on the control group, then illness is said to be associated with that exposure. The study of hepatitis A traced to green onions, described above, is an example of a case-control study. The key in a case-control study is to identify an appropriate control group, comparable to the case group in most respects, in order to provide a reasonable estimate of the baseline or expected exposure.

An example of an observational case-control study is where people diagnosed with new-onset Lyme disease were asked how often they walk through the woods, use insect repellant, wear short sleeves and pants, etc. Twice as many patients without Lyme disease from the same physician's practice were asked the same questions, and the responses in the two groups were compared.

Cross-sectional study. In this third type of observational study, a sample of persons from a population is enrolled and their exposures and health outcomes are measured simultaneously. The cross-sectional study tends to assess the presence (prevalence) of the health outcome at that point of time without regard to duration. For example, in a cross-sectional study of diabetes, some of the enrollees with diabetes may have lived with their diabetes for many years, while others may have been recently diagnosed.

From an analytic viewpoint the cross-sectional study is weaker than either a cohort or a case-control study because a cross-sectional study usually cannot disentangle risk factors for occurrence of disease (incidence) from risk factors for survival with the disease. On the other hand, a cross-sectional study is a perfectly fine tool for descriptive epidemiology purposes. Cross-sectional studies are used routinely to document the prevalence in a community of health behaviors (prevalence of smoking), health states (prevalence of vaccination against measles), and health outcomes, particularly chronic conditions (hypertension, diabetes).

An example of an observational cross-sectional study is a representative sample of residents being telephoned and asked how much they exercise each week and whether they currently have (ever been diagnosed with) heart disease.

The bottom line on analytic study in epidemiology is to identify and quantify the relationship between an exposure and a health outcome. The hallmark of such a study is the presence of at least two groups, one of which serves as a comparison group. In an experimental study, the investigator determines the exposure for the study subjects; in an observational study, the subjects are exposed under more natural conditions. In an observational cohort study, subjects are enrolled or grouped on the basis of their exposure, then are followed to document occurrence of disease. Difference in disease rates between the exposed and unexposed groups lead investigators to conclude that exposure is associated with disease. In an observational case-control study, subjects are enrolled according to whether they have the disease or not, then are questioned or tested to determine their prior exposure. Differences in exposure prevalence between the case and control groups allow investigators to conclude that the exposure is associated with the disease. Cross-sectional studies measure exposure and disease status at the same time, and are better suited to descriptive epidemiology than causation.

Tree Farm and Logging—An Environmental Health Concern

(From CDC 2011. TOXTOWN *Tree Farm and Logging*. Accessed 02/28/12 at http://toxtown.nlm.nih.gov/text_Version/locations.php?id=59.) A tree farm is a privately owned forest that produces renewable crops of trees and forest products. A tree farm may also provide wildlife habitat or recreational space. Tree farms may be a cause for concern because vehicles, machinery, and equipment used on the tree farm may run on diesel, gasoline, or propane for fuel. This equipment may emit carbon monoxide, particulate matter, and other air pollutants.

If pesticides and herbicides are used on the tree crops, they may end up in runoff that causes water quality problems in streams, rivers, and other bodies of water. They may also be emitted into the air.

Logging may cause soil erosion, air pollution, and destruction of wildlife and plant habitat. Soil erosion and sediment production can pollute streams and rivers and threaten aquatic life. Beneficial nutrients found in living trees, which help the soil and other plants, are removed from a forest when trees are cut down.

Tree farm workers may be exposed to pesticides, air pollution, and pests such as rodents and ticks. Safety concerns on a tree farm include vehicles, machinery, logging equipment, and tools.

Concepts of Disease Occurrence

A critical premise of epidemiology is that disease and other health events do not occur randomly in a population, but are more likely to occur in some members of the population than others because of risk factors that may not be distributed randomly in the population. As noted earlier, one important use of epidemiology is to identify the factors that place some members at greater risk than others.

Causation

A number of models of disease causation have been proposed. Among the simplest of these is the epidemiologic triad or triangle, the traditional model for infectious disease. The triad consists of an external *agent*, a susceptible *host*, and an *environment* that brings the host and agent together. In this model, disease results from the interaction between the agent and the susceptible host in an environment that supports transmission of the agent from a source to that host. Two ways of depicting this model are shown in Figure 4.6.

Agent, host, and environmental factors interrelate in a variety of complex ways to produce disease. Different diseases require different balances and interactions of these three components. Development of appropriate, practical, and effective public health measures to control or prevent disease usually requires assessment of all three components and their interactions.

Agent originally referred to an infectious microorganism or pathogen: a virus, bacterium, parasite, or other microbe. Generally, the agent must be present for disease to occur; however, presence of that agent alone is not

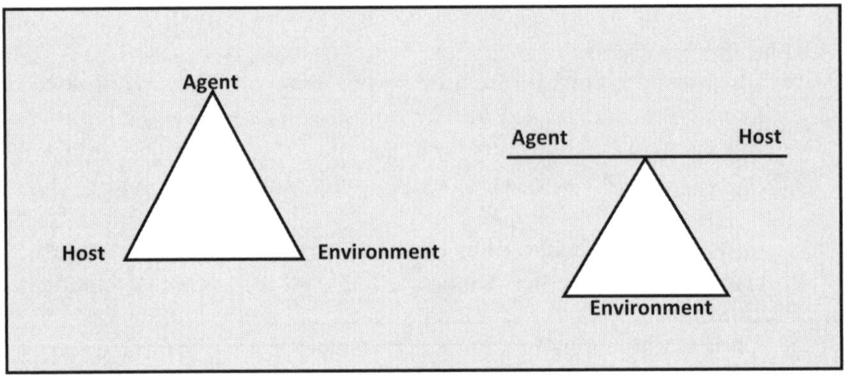

FIGURE 4.6
Epidemiologic triad. From the CDC, 2009. "Reproductive Health: Glossary 2009." Atlanta, GA: Centers for Disease Control and Prevention.

always sufficient to cause disease. A variety of factors influence whether exposure to an organism will result in disease, including the organism's pathogenicity (ability to cause disease) and dose.

Over time, the concept of agent has been broadened to include chemical and physical causes of disease or injury. These include chemical contaminants (such as the L-tryptophan contaminant responsible for eosinophilia-myalgia syndrome), as well as physical forces (such as repetitive mechanical forces associated with carpal tunnel syndrome). While the epidemiological triad serves as a useful model for many diseases, it has proven inadequate for cardiovascular disease, cancer, and other diseases that appear to have multiple contributing causes without a single necessary one.

Host refers to the human who can get the disease. A variety of factors intrinsic to the host, sometimes called risk factors, can influence an individual's exposure, susceptibility, or response to a causative agent. Opportunities for exposure are often influenced by behaviors such as sexual practices, hygiene, and other personal choices, as well as by age and sex. Susceptibility and response to an agent are influenced by factors such as genetic composition, nutritional immunologic status, anatomic structure, presence of disease or medications, and psychological makeup.

Environment refers to extrinsic factors that affect the agent and the opportunity for exposure. Environmental factors include physical factors such as geology and climate, biologic factors such as insects that transmit the agent, and socioeconomic factors such as crowding, sanitation, and the availability of health services.

Although causation is an essential concept in epidemiology, there is no single definition for the discipline. Based on experience and from a systematic review of the literature, we have surmised that whichever definition epidemiologists choose to use in defining causal factors related to their epidemiological research and/or practice, they likely relate to or use the terms and associated definitions described below.

Definitions of Causation

The following definitions are from Parascandola and Weed, 2001.

Production—causes are conditions that play essential parts in producing the occurrence of disease (MacMahon and Pugh, 1970; Susser, 1991).

Necessary causes—a condition without which the effect cannot occur. For example, HIV infection is a necessary cause of AIDS (Gordis, 2000; Charlton, 1996).

Sufficient-component causes—a sufficient cause guarantees that its effect will occur; when the cause is present, the effect *must* occur. A sufficient-component cause is made up of a number of components, no one of which is sufficient on its own, but which taken together make up a sufficient cause (Rothman, 1976; Rothman and Greenland, 1998).

Probabilistic cause—increases the probability of its effect occurring (Olsen, 1993; Karhausen, 1996; Elwood, 1994; Kleinbaum et al., 1982; Olsen, 1991; IARC, 1990). Such a cause would not be either necessary or sufficient.

Counterfactual causes—makes a difference in the outcome (or the probability of the outcome) when it is present, compared with when it is absent, while all else is held constant (Rubin, 1974; Holland, 1986; Maldonado and Greenland, 1998). The counterfactual approach also does not specifically require that causes must be necessary or sufficient for their effects.

Vineis and Kriebel (2006) point out that ". . . causation involves the relationship between at least two entities, an agent and a disease. Historically, at least two distinct eras of medical causality can be distinguished. The first era corresponds to the 'microbiological' revolution [the Pasteur/Koch era]. After the work of Pasteur and Koch, the agent of a disease came to be conceived of as a single necessary cause (e.g., *Mycobacterium* for tuberculosis). The concept of necessary cause means that the disease does not develop in the absence of exposure to the agent (p. 2). It is important to note that "such a view implies a) that the cause is, at least potentially, definable unequivocally and is easily identifiable unequivocally and is easily identifiable, and b) that the disease can be also defined unequivocally (i.e., it is not a complex and variable constellation of symptoms)" (Vineis and Kriebel, 2006). Obviously there are conditions in which the relationship between a necessary cause and the corresponding disease is indeed evident: for example, smallpox is easy to define and diagnose because it is a clear-cut entity; that is, no smallpox develops in the absence of the specific virus (due to a single necessary virus) and the causal link is clear and proven. This is clear and proven due to the disappearance of smallpox after large-scale vaccination (Vineis and Kriebel, 2006).

The ongoing study of chronic diseases like cancer and cardiovascular disease in medicine have ushered in the second era in the history of disease causation. In these diseases, the concept of a "necessary" condition is rarely, if ever, valid. Simply, we do not know what we do not know about any "necessary" cause of cancer (with the possible exception of cervical cancer). When we think about, discuss, or perform any research project to identify "the" absolute cause of cancer, we face a dilemma best characterized by uncertainty. We continue to collect data—we list known carcinogens; we track the prevalence of exposure to these agents; and we assemble and collate information on the magnitude of risk based on exposure response curves—and continue to conduct research. But, again, our current research efforts directed at the cause(s) of cancer simply lead to us to recognitions of additional uncertainties.

These additional uncertainties lead us to the idea of multiple causation, a metaphor for "causal web," or web of causation (Krieger, 1994) (in our usage in this text, the spider is metaphor for the disease, of course). The causal web reflects the fact that a concurrence of different "exposures," or conditions, is required to induce disease, none of which is in itself necessary. For example, lung cancer can be induced by a causal web, including tobacco smoking and individual predisposition from CYP1A1 (one of the main cytochrome P450 enzymes) and other high-risk genotypes (Vineis, 2003). Another causal web may be represented by asbestos exposure and low consumption of raw fruits and vegetables in the occurrence of mesothelioma. The idea of the web implies that while the disease is usually well-defined from a clinical point of view (e.g., lung cancer or mesothelioma), the etiologic perspective is more complex: not all lung cancer cases can be linked to the same exposures, but may instead share partially overlapping constellations of causes—an extensive web, but without the spider.

Did You Know?

The CYP1A1 gene encodes a member of the cytochrome P450 superfamily of enzymes. The cytochrome P450 proteins are monooxygenases that catalyze many reactions involved in drug metabolism and synthesis of cholesterol, steroids, and other lipids. This protein localizes to the endoplasmic reticulum and its expression is induced by some polycyclic aromatic hydrocarbons (PAHs), some of which are found in cigarette smoke. The enzyme's endogenous substrate is unknown; however, it is able to metabolize some PAHs to carcinogenic intermediates. The gene has been associated with lung cancer risk. A related family member, CYP1A2, is located approximately 25 kb away from CYP1A1 on chromosome 15 (NIH, 2012).

Guided by a multicausal view, the main causal model used by epidemiologists today is Rothman's "pies" (Rothman and Greenland, 2005). The idea is that a sufficient causal complex (a pie) is represented by the combination of several component causes (slices of the pie). A set of component causes occurring together may complete the "pie," creating a sufficient cause and thus initiating the disease process. Rothman's model has been useful on several accounts. For example, suppose three factors (A, B, and C) make a sufficient cause of disease X. Then, one can see that A will appear to be a stronger or weaker cause depending on how common the other "slices" B and C are. A will have a large impact on disease occurrence in a population in which B and C are common, but no effect at all (though being a sufficient cause) where B or C is absent. If it were true that the sufficient cause A+B+C were the *only* pathway to disease X, then it would follow that blocking or eliminating any of these three factors would prevent the disease. Thus, A and B and C would be necessary component causes. But if A, for example, also contributed to a sufficient cause with factors D, E, and F, then blocking B would not prevent disease X. This more complex view (many pies to which factors contribute) is supported by the epidemiologic evidence for most chronic diseases. There are only few examples of necessary component causes for cancer or heart disease.

Note that the above considerations concern our understanding of disease causality at the *individual level*. The model looks different if we shift from the individual to the *population*. Here, the idea of single "necessary" components makes sense. If we consider the current epidemic of lung cancer, for example, there is no doubt that it is attributable to the diffusion of the habit of tobacco smoking. For, although we cannot attribute any single case of lung cancer to the individual's smoking habits, there is no doubt that, on a population level, the epidemic would not have occurred without cigarette smoking. Notice that this assertion is not contradicted by the fact that lung cancer also occurs among non-smokers. Indeed, the evidence for cigarette smoking as a (population level) cause of lung cancer is quite strong: the risk of cancer in those who stop smoking decreases considerably, in comparison with continuing smokers, and, after a few years, approaches the risk of non-smokers (Vineis, 2004). It should be clear, then, that we have to apply different criteria of causation when considering the causes of disease at the individual or population level. We can say that for chronic diseases, the model of causal complexes in which there are necessary components is valid at the population level.

Epidemiology Case Example—Risk Factors and Health Outcomes

The example shown in Table 4.1 illustrates how classification of risk factors and health outcomes can be identified as necessary causes, sufficient causes, or component causes.

Epidemiology Case Example—Anthrax: Agent, Host, and Environment

Read the information in this section, below; then read how its causation is described in terms of agent, host, and environment (CDC, 2006).

What Is Anthrax?

Anthrax is an acute infectious disease that usually occurs in animals such as livestock, but can also affect humans. Human anthrax and intestinal anthrax symptoms usually occur within seven days after exposure.

Cutaneous. Most (about 95 percent) anthrax infections occur when the bacterium enters a cut or abrasion on the skin after handling infected livestock or contaminated animal products. Skin infection begins as a raised itchy bump that resembles an insect bite but within one to two days develops into a vesicle and then a painless ulcer, usually one to three centimeters in diameter, with a characteristic black necrotic (dying) area in the center. Lymph glands in the adjacent area may swell. About 20 percent of untreated cases of cutaneous anthrax will result in death. Deaths are rare with appropriate antimicrobial therapy.

Inhalation. Initial symptoms are like cold or flu symptoms and can include a sore throat, mild fever, and muscle aches. After several days, the symptoms may progress to cough, chest discomfort, severe breathing problems, and shock. Inhalation anthrax is often fatal. Eleven of the infamous mail-related cases in 2001 were inhalation; five (45 percent) of eleven patients died.

TABLE 4.1

Risk Factor	*Health Outcome*	*Type of Cause*
Hypertension	Stroke	Component
Treponema pallidum	Syphilis	Necessary
Type A personality	Heart disease	Component
Skin contact with a strong acid	Burn	Sufficient

Intestinal. Initial signs of nausea, loss of appetite, vomiting, and fever are followed by abdominal pain, vomiting of blood, and severe diarrhea. Intestinal anthrax results in death in 25 percent to 60 percent of cases.

While most human cases of anthrax result from contact with infected animals or contaminated animal products, anthrax also can be use as a biologic weapon. In 1979, dozens of residents of Sverdiovsk in the former Soviet Union are thought to have died of inhalation anthrax after an unintentional release of an aerosol from a biologic weapons facility. In 2001, twenty-two cases of anthrax occurred in the United States from letters containing anthrax spores that were mailed to members of Congress, television networks, and newspaper companies.

What Causes Anthrax?

Anthrax is caused by the bacterium *Bacillus anthracis*. The anthrax bacterium forms a protective shell called a spore. *Bacillus anthracis* spores are found naturally in soil, and can survive for many years.

How Is Anthrax Diagnosed?

Anthrax is diagnosed by isolating *Bacillus anthracis* from the blood, skin lesions, or respiratory secretions or by measuring specific antibodies in the blood of persons with suspected cases.

Is There a Treatment for Anthrax?

Antibiotics are used to treat all three types of anthrax. Treatment should be initiated early because the disease is more likely to be fatal if treatment is delayed or not given at all.

How Common Is Anthrax and Where Is It Found?

Anthrax is most common in agricultural regions of South and Central America, Southern and Eastern Europe, Asia, Africa, the Caribbean, and the Middle East, where it occurs in animals. When anthrax affects humans, it is usually the result of an occupational exposure to infected animals or their products. Naturally occurring anthrax is rare in the United States (twenty-eight reported cases between 1971 and 2000), but twenty-two mail-related cases were identified in 2001.

Infections occur most commonly in wild and domestic lower vertebrates (cattle, sheep, goats, camels, antelopes, and other herbivores), but it can also occur in humans when they are exposed to infected animals or tissue from infected animals.

How Is Anthrax Transmitted?

Anthrax can infect a person in three ways: by anthrax spores entering through a break in the skin, by inhaling anthrax spores, or by eating contaminated, undercooked meat. Anthrax is not spread from person to person. The skin ("cutaneous") form of anthrax is usually the result of contact with infected livestock, wild animals, or contaminated animal products such as carcasses, hides, hair, wool, meat, or bone meal. The inhalation form is contracted by breathing in spores from the same sources. Anthrax can also be spread as a bioterrorist agent.

Who Has an Increased Risk of Being Exposed to Anthrax?

Susceptibility to anthrax is universal. Most naturally occurring anthrax affects people whose work brings them into contact with livestock or products from livestock. Such occupations include veterinarians, animal handlers, abattoir workers, and laboratorians. Inhalation anthrax was once called Woolsorter's Disease because workers who inhaled spores from contaminated wool before it was cleaned developed the disease. Soldiers and other potential targets of bioterrorist anthrax attacks might also be considered at increased risk.

Is There a Way to Prevent Infection?

In countries where anthrax is common and vaccination levels of animal herds are low, humans should avoid contact with livestock and animal products and avoid eating meat that has not been properly slaughtered and cooked. Also, an anthrax vaccine has been licensed for use in humans. It is reported to be 93 percent effective in protecting against anthrax. It is used by veterinarians, laboratorians, soldiers, and others who may be at increased risk of exposure, but it is not available to the general public at this time.

For a person who has been exposed to anthrax but is not yet sick, antibiotics combined with the anthrax vaccine are used to prevent illness.

From the above information, we are able to describe its causation in terms of the following:

Agent. *Bacillus anthracis*, a bacterium that can survive for years in spore form, is a necessary cause.

Host. People are generally susceptible to anthrax. However, infection can be prevented by vaccination. Cuts or abrasions of the skin may permit entry of the bacteria.

Environment. Persons at risk for naturally acquired infection are those who are likely to be exposed to infected animals or contaminated animal products, such as veterinarians, animal handlers, abattoir workers, and laboratorians. Persons who are potential targets of bioterrorism are also at increased risk.

Colonia—An Environmental Health Concern

(From CDC 2011. *Tox Town—Colonia.* Accessed 03/03/12 at http://toxtown.nlh.nih.gov/text_version/locations.php?id=13.) *Colonia* is a Spanish word for neighborhood or community. Along the U.S.-Mexico border, the term *colonia* refers to low-income rural housing developments built on floodplains and other land with no agricultural value in the United States. *Colonias* are primarily located in the border region of Texas, but they can also be found near the borders of New Mexico, Arizona, and California. The border region contains over 1,400 colonias that house over 400,000 people. The majority of colonia residents are U.S. citizens, and the population is predominately Hispanic. Colonias share many of the environmental health concerns of the entire border region. However, since colonias often lack running water, sewage disposal, paved roads, and waste disposal, environmental health is a primary concern.

Natural History and Spectrum of Disease

(From CDC, 2006.) Natural history of disease refers to the progression of a disease process in an individual over time (see Figure 4.7), in the absence of treatment. For example, untreated infection with HIV causes a spectrum of clinical problems beginning at the time of seroconversion (i.e., development of antibodies in blood serum as a result of infection or immunization; primary HIV) and terminating with AIDS and usually death. It is now recognized that it may take ten years or more for AIDS to develop after seroconversions (Mindel and Tenant-Flowers, 2001). Many, if not most, diseases have a characteristic natural history, although the time frame and specific manifestations of disease may vary from individual to individual and are influenced by preventive and therapeutic measures.

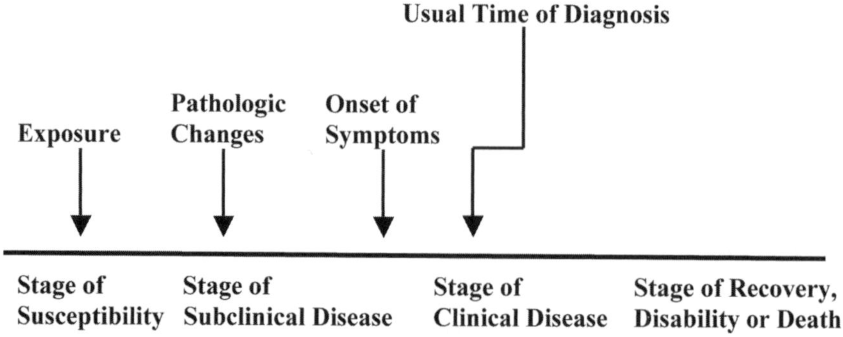

FIGURE 4.7
Natural history of disease timeline. From the CDC, 1992. *Principles of Epidemiology*, 2nd ed. Atlanta: U.S. Department of Health and Human Services.

The process beings with the appropriate exposure to or accumulation of factors sufficient for the disease process to begin in a susceptible host. For an infectious disease, the exposure is a microorganism. For cancer, the exposure may be a factor that initiates the process, such as asbestos fibers or components in tobacco smoke (for lung cancer), or one that promotes the process, such as estrogen (endometrial cancer).

After the disease process has been triggered, pathological changes then occur without the individual being aware of them. This stage of subclinical disease extends from the time of exposure to onset of disease symptoms (see figure 4.6) and is usually called the *incubation period* for infectious diseases and the *latency period* for chronic diseases. During this stage, disease is said to be asymptomatic (no symptoms) or inapparent. This period may be as brief as seconds for hypersensitivity and toxic reactions to as long as decades for certain chronic diseases. Even for a single disease, the characteristic incubation period has a range. For example, the typical incubation period for hepatitis A is as long as seven weeks. The latency period for leukemia to become evident among survivors of the atomic bomb blast in Hiroshima ranged from two to twelve years, peaking at six to seven years (Cobb, Miller, and Wald, 1959). Incubation periods of selected exposures and diseases varying from minutes to decades are displayed in Table 4.2.

Although disease is not apparent during the incubation period, some pathologic changes may be detectable with laboratory, radiographic, or other screening methods. Most screening programs attempt to identify the disease process during this phase of its natural history, since intervention at this early state is likely to be more effective than treatment given after the disease has progressed and become symptomatic.

TABLE 4.2
Incubation Periods of Selected Exposures and Diseases (CDC, 2006)

Exposure	*Clinical Effect*	*Incubation/Latency Period*
Saxitoxin and similar toxins from shellfish	Paralytic shellfish poisoning (tingling, numbness around lips and fingertips, giddiness, incoherent speech, respiratory paralysis, sometimes death)	few minutes to 30 minutes
Organophosphorus ingestion	Nausea, vomiting, cramps, headache, nervousness, blurred vision, chest pain, confusion, twitching, convulsions	few minutes to a few hours
Salmonella	Diarrhea, often with fever and cramps	usually 6–48 hours
SARS-associated	Severe Acute Respiratory Syndrome (SARS)	5–10, usually 4–6 days
Varicella-zoster virus	Chickenpox	10–21, usually 14–16 days
Treponema pallidum	Syphilis	10–90 days, usually 3 weeks
Hepatitis A virus	Hepatitis	14–50 days, average 4 weeks
Hepatitis B virus	Hepatitis	50–180 days, usually 2–3 months
Human immunodeficiency virus	AIDS	<1 to 15+ years
Atomic bomb radiation (Japan)	Leukemia	2–12 years
Radiation (Japan, Chernobyl)	Thyroid Cancer	3–20+ years
Radium (watch dial)	Bone cancer	8–40 years

The onset of symptoms marks the transition from subclinical to clinical disease. Most diagnoses are made during the stage of clinical disease. In some people, however, the disease process may never progress to clinically apparent illness. In others, the disease process may result in illness that ranges from mild to severe or fatal. This range is called the spectrum of disease. Ultimately, the disease process ends either in recovery, disability, or death.

For an infectious agent, *infectivity* refers to the proportion of exposed persons who become infected. *Pathogenicity* refers to the proportion of infected individuals who develop clinically apparent disease. *Virulence* refers to the proportion of clinically apparent cases that are severe or fatal.

Because the spectrum of disease can include asymptomatic and mild cases, the cases of illness diagnosed by clinicians in the community represent only the tip of the iceberg. Many additional cases may be too early to diagnose or may never progress to the clinical stage. Unfortunately, persons with inapparent to undiagnosed infections may nonetheless be able to transmit infection to others. Such persons who are infectious but have subclinical disease are called carriers. Frequently, carriers are persons with incubating disease or inapparent infection. Persons with measles, hepatitis A, and several other diseases become infectious a few days before the onset of symptoms. However, carriers may also be people who appear to have recovered from their clinical illness but remain infectious, such as chronic carriers of hepatitis B virus, or people who never exhibited symptoms. The challenge to public health workers is that these carriers, unaware that they are infected and infectious to others, are sometimes more likely to unwittingly spread infection than are people with obvious illness.

Chain of Infection

As described earlier, the traditional epidemiologic triad model holds that infectious diseases result from the interaction of agent, host, and environment. More specifically, transmission occurs when the agent leaves its *reservoir*, or host, through a *portal of exit*, is conveyed by some *mode of transmission*, and enters through an appropriate *portal of entry* to infect a *susceptible host*. This sequence is sometimes called the chain of infection (see figure 4.8).

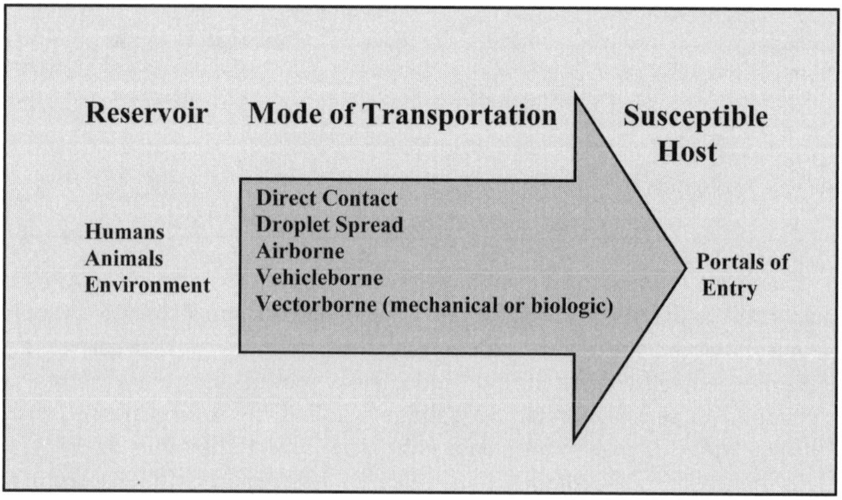

FIGURE 4.8
Chain of infection. From the CDC, 1992. *Principles of Epidemiology*, 2nd ed. Atlanta, GA: Centers of Disease Control and Prevention.

Reservoir

The reservoir of an infectious agent is the habitat in which the agent normally lives, grows, and multiplies. Reservoirs include humans, animals, and the environment. The reservoir may or may not be the source from which an agent is transferred to a host. For example, the reservoir of *Clostridium botulinum* is soil, but the source of most botulism infections is improperly canned food that contains *Clostridium botulinum* spores.

Human reservoirs. Many common infectious diseases have human reservoirs. Diseases that are transmitted from person to person without intermediaries include the sexually transmitted diseases, measles, mumps, strepotoccal infection, and many respiratory pathogens. Because humans were the only reservoir for the smallpox virus, naturally occurring smallpox was eradicated after the last human case was identified and isolated (Fenner et al., 1998).

Human reservoirs may or may not show the effects of illness. As noted earlier, a carrier is a person with inapparent infection who is capable of transmitting the pathogen to others. Asymptomatic or passive or healthy carriers are those who never experience symptoms despite being infected. Incubatory carriers are those who can transmit the agent during the incubation period before clinical illness begins. Convalescent carriers are those who have recovered from their illness but remain capable of transmitting to others. Chronic carriers are those who continue to harbor a pathogen such as hepatitis B virus or *Salmonella Typhi*, the causative agent of typhoid fever, for months or even years after their initial infection.

Did You Know?

One notorious carrier is Mary Mallon, or Typhoid Mary, who was an asymptomatic chronic carrier of *Salmonella* Typhi. As a cook in New York City and New Jersey in the early 1900s, she unintentionally infected dozens of people until she was placed in isolation on an island in the East River, where she died twenty-three years later (Leavitt, 1996).

Carriers commonly transmit disease because they do not realize they are infected, and consequently take no special precautions to prevent transmission. Symptomatic persons who are aware of their illness, on the other hand, may be less likely to transmit infection because they are either too sick to be out and about, take precautions to reduce transmission, or receive treatment that limits the disease.

Animal reservoirs. Humans are also subject to diseases that have animal reservoirs. Many of these diseases are transmitted from animal to animal, with humans as incidental hosts. The term **zoonosis** refers to an infectious disease that is transmissible under natural conditions from vertebrate animals to humans. Long-recognized zoonotic diseases include brucellosis (cows and pigs), anthrax (sheep), plague (rodents), trichinelloisis/trichinosis (swine), tularemia (rabbits), and rabies (bats, raccoons, dogs, and other mammals). Zoonoses newly emergent in North America include West Nile encephalitis (birds), and monkeypox (prairie dogs). Many newly recognized infectious diseases in humans, including HIV/AIDS, Ebola infection, and SARS, are thought to have emerged from animal hosts, although those hosts have not yet been identified.

Environmental reservoirs. Plants, soil, and water in the environment are also reservoirs for some infectious agents. Many fungal agents, such as those that cause histoplasmosis, live and multiply in the soil. Outbreaks of Legionnaires' disease are often traced to water supplies in cooling towers and evaporative condensers, reservoirs for the causative organism *Legionella pneumophila.*

Portal of Exit

Portal of exit is the path by which a pathogen leaves its host. The portal of exit usually corresponds to the site where the pathogen is localized. For example, influenza viruses and *Mycobacterium tuberculosis* exit the respiratory tract, schistosomes though urine, choler vibrios in feces, *Sarcoptes scabiei* in scabies skin lesions, and Enterovirus 70, a cause of hemorrhagic conjunctivitis, in conjunctival secretions. Some bloodborne agents can exit by crossing the placenta from mother to fetus (rubella, syphilis, toxoplasmosis), while others exit through cuts or needles in the skin (hepatitis B) or blood-sucking arthropods (malaria).

Modes of Transmission

An infectious agent may be transmitted from its natural reservoir to a susceptible host in different ways. There are different classifications for modes of transmission. Here is one classification:

- Direct contact: droplet spread
- Indirect contact: airborne transmission (vehicleborne or vectorborne: mechanical or biological)

In *direct transmission*, an infectious agent is transferred from a reservoir to a susceptible host by direct contact or droplet spread.

Direct contact occurs through skin-to-skin contact, kissing, and sexual intercourse. Direct contact also refers to contact with soil or vegetation harboring infectious organisms. Thus, infectious mononucleosis ("kissing disease") and gonorrhea are spread from person to person by direct contact. Hookworm is spread by direct contact with contaminated soil.

Droplet spread refers to spray with relatively large, short-range aerosols produced by sneezing, coughing, or even talking. Droplet spread is classified as direct because transmission is by direct spray over a few feet, before the droplets fall to the ground. Pertussis and meningococcal infection are examples of diseases transmitted from an infectious patient to a susceptible host by droplet spread.

Indirect transmission refers to the transfer of an infectious agent from a reservoir to a host by suspended air particles, inanimate objects (vehicles), or animate intermediaries (vectors).

Airborne transmission occurs when infectious agents are carried by dust or droplet nuclei suspended in air. Airborne dust includes materials that have settled on surfaces and become resuspended by air currents as well as infectious particles blown form the soil by the wind. Droplet nuclei are dried residue of less than five microns in size. In contrast to droplets that fall to the ground within a few feet, droplet nuclei may remain suspended in the air for long periods of time and may be flown over great distances. Measles, for example, has occurred in children who came into a physician's office after a child with measles had left, because the measles virus remained suspended in the air (Remington, 1985).

Vehicles that may indirectly transmit an infectious agent include food, water, biologic products (blood), and fomites (inanimate objects such as handkerchiefs, bedding, or surgical scalpels). A vehicle may passively carry a pathogen—as food or water may carry the hepatitis A virus. Alternatively, the vehicle may provide an environment in which the agent grows, multiplies, or produces toxin—as improperly canned foods provide an environment that supports production of botulinum toxin by *Clostridium botulinum.*

Vectors, such as mosquitoes, fleas, and ticks, may carry an infectious agent through purely mechanical means or may support growth or changes in the agent. Examples of mechanical transmission are flies carrying *Shigella* on their appendages and fleas carrying Yersinia *pestis*, the causative agent of plague, in their gut. In contrast, in biologic transmission, the causative agent of malaria or guinea worm disease undergoes maturation in an intermediate host before it can be transmitted to humans.

Portal of Entry

The portal of entry refers to the manner in which a pathogen enters a susceptible host. The portal of entry must provide access to tissues in which the pathogen can multiply or a toxin can act. Often, infectious agents use the same portal to enter a new host that they used to exit the source host. For example, influenza virus exits the respiratory tract of the source host and enters the respiratory tract of the new host. In contrast, many pathogens that cause gastroenteritis follow a so-called "fecal-oral" route because they exit the source host in feces, are carried on inadequately washed hands to a vehicle such as food, water, or a utensil, and enter a new host through the mouth. Other portals of entry include the skin (hookworm), mucous membranes (syphilis), and blood (hepatitis B, human immunodeficiency virus).

Host

The final link in the chain of infection is a susceptible host. Susceptibility of a host depends on genetic or constitutional factors, specific immunity, and nonspecific factors that affect an individual's ability to resist infection or to limit pathogenicity. An individual's genetic makeup may either increase or decrease susceptibility. For example, people with sickle cell trait seem to be at least partially protected from a particular type of malaria. Specific immunity refers to protective antibodies that are directed against a specific agent. Such antibodies may develop in response to infection, vaccine, or toxoid (toxin that has been deactivated but retains its capacity to stimulate production of toxin antibodies) or may be acquired by transplacental transfer from mother to fetus or by injection of antitoxin or immune globulin. Nonspecific factors that defend against infection include the skin, mucous membranes, gastric acidity, cilia in the respiratory tract, the cough reflex, and nonspecific immune response. Factors that may increase susceptibility to infection by disrupting host defense include malnutrition, alcoholism, and disease or therapy that impairs the nonspecific immune response.

Implications for Public Health

Knowledge of the portals of exit and entry and modes of transmission provides a basis for determining appropriate control measures. In general, control measures are usually directed against the segment in the infection chain that is most susceptible to intervention, unless practical issues dictate otherwise.

For some diseases, the most appropriate intervention may be directed at controlling or eliminating the agent at its source. A patient sick with a communicable disease may be treated with antibiotics to eliminate the infection. An asymptomatic but infected person may be treated both to clear the infection and to reduce the risk of transmission to others. In the community, soil may be decontaminated or covered to prevent escape of the agent.

Some interventions are directed at the mode of transmission. Interruption of direct transmission may be accomplished by isolation of someone with infection, or counseling persons to avoid the specific type of contact associated with transmission. Vehicle-borne transmission may be interrupted by elimination or decontamination of the vehicle. To prevent fecal-oral transmission, efforts often focus on rearranging the environment to reduce the risk of contamination in the future and on changing behaviors, such as promoting hand washing. For airborne diseases, strategies may be directed at modifying ventilation or air pressure, and filtering or treating the air. To interrupt vector-borne transmission, measures may be directed toward controlling the vector population, such as spraying to reduce the mosquito population.

Some strategies that protect portals of entry are simple and effective. For example, bed nets are used to protect sleeping persons from being bitten by mosquitoes that may transmit malaria. A dentist's mask and gloves are intended to protect the dentist from a patient's blood, secretions, and droplets, as well to protect the patient from the dentist. Wearing of long pants and sleeves and use of insect repellent are recommended to reduce the risk of Lyme disease and West Nile Virus infection, which are transmitted by the bites of ticks and mosquitoes, respectively.

Some interventions aim to increase a host's defenses. Vaccinations promote development of specific antibodies that protect against infection. On the other hand, prophylactic use of antimalarial drugs, recommended for visitors to malaria-endemic areas, does not prevent exposure through mosquito bites, but does prevent infection from taking root.

Finally, some interventions attempt to prevent a pathogen from encountering a susceptible host. The concept of *herd immunity* (*community immunity*; see figure 4.9) suggests that if a high enough proportion of individuals in a population are resistant to an agent, then those few who are susceptible will be protected by the resistant majority, since the pathogen will be unlikely to "find" those few susceptible individuals. The degree of herd immunity necessary to prevent or interrupt an outbreak varies by disease. In theory, herd immunity means that not everyone in a community needs to be resistant (immune) to prevent disease spread and occurrence of an outbreak. In practice, herd immunity has not prevented outbreaks of measles and rubella in populations with immunization levels as high as 85 percent to 90 percent. One problem is that, in highly immunized populations, the relatively few susceptible people are often clustered in subgroups defined by socioeconomic or cultural factors. If the pathogen is introduced into one of these subgroups, an outbreak may occur.

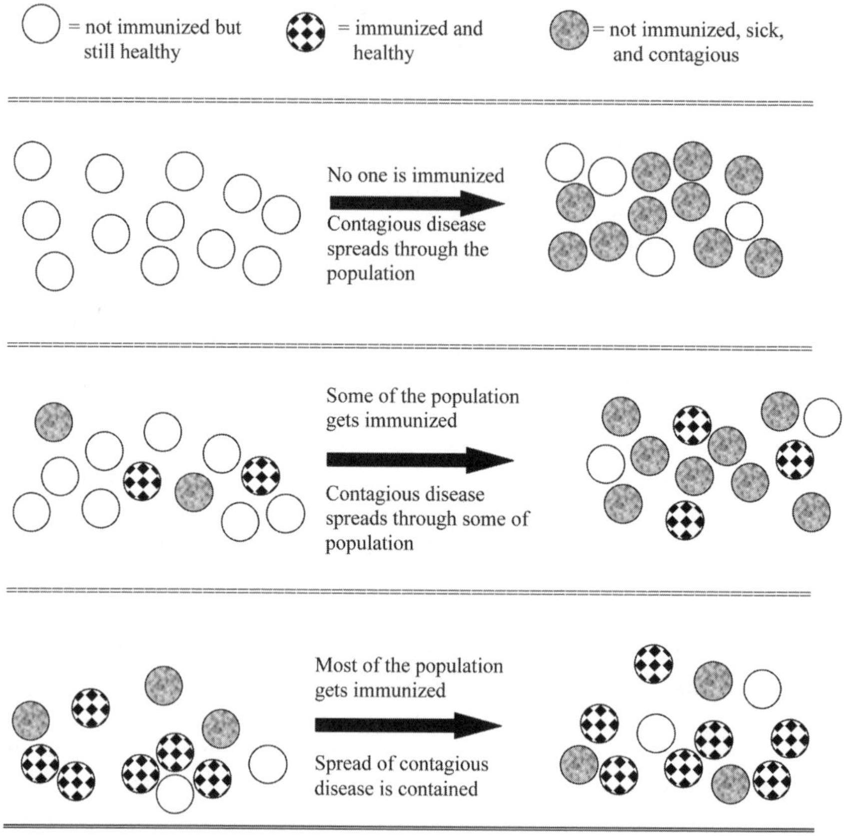

FIGURE 4.9
Herd community. From NIAID, 2010, accessed 03/05/12 at http://www.niaid.nih.govtopics/Pages/communityImmunity.aspx.

Epidemiology Case Example—Dengue Chain of Infection

(CDC, 2005. *Dengue Fever*. Accessed 03/05/12 at http://www.cod.gov/dvbid/dengue/index.htm.) Basic information, reservoirs, portal of exit, modes of transmission, portal of entry, and factors in host susceptibility are discussed.

What Is Dengue?

Dengue (pronounced *den-gee*) is an acute infectious disease that comes in two forms: dengue and dengue hemorrhagic fever. The principal symptoms of dengue are high fever, severe headache, backache, joint pains, nausea and vomiting, eye pain, and rash. Generally, younger children have a milder illness than older children and adults.

Dengue hemorrhagic fever is a more severe form of dengue. It is characterized by a fever that lasts for two to seven days, with general signs and symptoms that could occur with many other illnesses (e.g., nausea, vomiting, abdominal pain, and headache). This stage is followed by hemorrhagic manifestations, tendency to bruise easily or other types of skin hemorrhages, bleeding nose or gums, and possibly internal bleeding. The smallest blood vessels (capillaries) become excessively permeable ("leaky"), allowing the fluid component to escape from the blood vessels. This may lead to failure of the circulatory system and shock, followed by death, if circulatory failure is not corrected. Although the average case-fatality rate is about 5 percent, with good medical management, mortality can be less than 1 percent.

What Causes Dengue?

Dengue and dengue hemorrhagic fever are caused by any one of four closely related flaviviruses, designated DEN-1, DEN-2, DEN-3, or DEN-4.

How Is Dengue Diagnosed?

Diagnosis of dengue infection requires laboratory confirmation, either by isolating the virus from serum within five days after onset of symptoms, or by detecting convalescent-phase specific antibodies obtained at least six days after onset of symptoms.

What Is the Treatment for Dengue or Dengue Hemorrhagic Fever?

There is no specific medication for treatment of a dengue infection. People who think they have dengue should use analgesics (pain relievers) with acetaminophen and avoid those containing aspirin. They should also rest, drink plenty of fluids, and consult a physician. Persons with dengue hemorrhagic fever can be effectively treated by fluid replacement therapy if an early clinical diagnosis is made, but hospitalization is often required.

How Common Is Dengue and Where Is It Found?

Dengue is endemic in many tropical countries in Asia and Latin America, most countries in Africa, and much of the Caribbean, including Puerto Rico. Cases have occurred sporadically in Texas. Epidemics occur periodically. Globally, an estimated 50 to 100 million cases of dengue and several hundred thousand cases of dengue hemorrhagic fever occur each year, depending on epidemic activity. Between 100 and 200 suspected cases are introduced into the United States each year by travelers.

How Is Dengue Transmitted?

Dengue is transmitted to people by the bite of an Aedes mosquito that is infected with a dengue virus. The mosquito becomes infected with dengue virus when it bites a person who has dengue or DHF and after about a week can transmit the virus while biting a healthy person. Monkeys may serve as a reservoir in some parts of Asia and Africa. Dengue cannot be spread directly from person to person.

Who Has an Increased Risk of Being Exposed to Dengue?

Susceptibility to dengue is universal. Residents of or visitors to tropical urban areas and other areas where dengue is endemic are at highest risk of becoming infected. While a person who survives a bout of dengue caused by one serotype develops lifelong immunity to that serotype, there is no cross-protection against the three other serotypes.

What Can Be Done to Reduce the Risk of Acquiring Dengue?

There is no vaccine for prevention dengue. The best preventive measure for residents living in areas infested with Aedes aegypti is to eliminate the places where the mosquito lays her eggs, primarily artificial containers that hold water.

Items that collect rainwater or are used to store water (for example, plastic containers, fifty-five-gallon drums, buckets, or used automobile tires) should be covered or properly discarded. Pet and animal watering containers and vases with fresh flowers should be emptied and scoured at least once a week. This will eliminate the mosquito eggs and larvae and reduce the number of mosquitoes present in these areas. For travelers to areas with dengue, as well as people living in areas with dengue, the risk of being bitten by mosquitoes indoors is reduced by utilization of air-conditioning or windows and doors that are screened. Proper application of mosquito repellents containing 20 percent to 30 percent DEET as the active ingredient on exposed skin and clothing decreases the risk of being bitten by mosquitoes. The risk of dengue infection for international travelers appears to be small, unless an epidemic is in progress.

Can Epidemics of Dengue Hemorrhagic Fever Be Prevented?

The emphasis for dengue prevention is on sustainable, community-based, integrated mosquito control, with limited reliance on insecticides (chemical larvicides and adulticides). Preventing epidemic disease requires a coordinated community effort to increase awareness about dengue/DHF, how to recognize it, and how to control the mosquito that transmits it. Residents are responsible for keeping their yards and patios free of sites where mosquitoes can be produced.

After studying the above information, we can outline the chain of infection by identifying the following:

Reservoirs: Humans and possibly monkeys

Portals of exit: Skin (via mosquito bite)

Modes of transmission: Indirect transmission to humans by mosquito vector

Portals of entry: Skin to blood (via mosquito bite)

Host susceptibility: Universal, except for survivors of dengue infection who are immune to subsequent infection from the same serotype

Epidemic Disease Occurrence

Level of Disease

The amount of a particular disease that is usually present in a community is referred to as the baseline level of the disease. This level is not necessarily the desired level, which may, in fact, be zero, but rather is the observed level. In the absence of intervention and assuming that the level is not high enough to deplete the pool of susceptible persons, the disease may continue to occur at this level indefinitely. Thus, the baseline level is often regarded as the expected level of the disease.

While some diseases are so rare in a given population that a single case warrants an epidemiologic investigation (e.g., rabies, plague, polio), other diseases occur more commonly so that only deviations from the norm warrant investigation. *Sporadic* refers to a disease that occurs infrequently and irregularly. *Endemic* refers to the constant presence and/or usual prevalence of a disease or infectious agent in a population within a geographic area. *Hyperendemic* refers to persistent, high levels of disease occurrence.

Occasionally, the amount of disease in a community rises above the expected level. *Epidemic* refers to an increase, often sudden, in the number of cases of a disease above what is normally expected in that population in that area. *Outbreak* carries the same definition of epidemic, but is often used for a more limited geographic area. *Cluster* refers to an aggregation of cases grouped in place and time that are suspected to be greater than the number expected, even though the expected number may not be known. *Pandemic* refers to an epidemic that has spread over several countries or continents, usually affecting a large number of people.

Epidemics occur when an agent and susceptible hosts are present in adequate numbers, and the agent can be effectively conveyed from a source to the susceptible hosts. More specifically, an epidemic may result from:

- A recent increase in amount or virulence of the agent.
- The recent introduction of the agent into a setting where it has not been before.
- An enhanced mode of transmission so that more susceptible persons are exposed.
- A change in the susceptibility of the host response to the agent.
- Factors that increase host exposure or involve introduction through new portals of entry (Kelsey et al., 1986).

Note that the previous description of epidemics presumes only infectious agents, but non-infectious diseases such as diabetes and obesity exist in epidemic proportions in the United States (CDC, 2004d; NCHS, 2005).

Example Situations

Sporadic disease: Single case of histoplasmosis was diagnosed in a community.

Endemic disease: About sixty cases of gonorrhea are usually reported in this region per week, slightly less than the national average.

Hyperendemic disease: Average annual incidence was 364 cases of pulmonary tuberculosis per 100,000 population in one area, compared with the national average of 134 cases per 100,000 population.

Pandemic disease: Over 20 million people worldwide died from influenza in 1918–1919.

Epidemic disease: Twenty-two cases of legionellosis occurred within three weeks among residents of a particular neighborhood (usually they would have zero to one per year).

Discussion Questions

1. Describe the difference between distribution and determinants. What does each cover?
2. In epidemiology, public health action is not included. Why?
3. Why is John Snow's investigation of cholera considered a model for epidemiologic field investigations?
4. Why is the use of an appropriate comparison group the hallmark feature of an analytic epidemiologic study?
5. What is the key difference between a cohort and a case-control study?

References and Recommended Reading

Arias, E., R.N. Anderson, K. Hsiang-Ching, S.L. Murphy, and D.D. Kovhanek, 2003. Deaths: final data for 2001. *National vital statistics reports*; vol. 52, no. 3. Hyattsville, MD: National Center for Health Statistics.

Beaglehole, R., R. Bonita, and T. Kjellstrom, 1993. *Basic epidemiology*. Geneva: World Health Organization.

Cates, W., 1982. Epidemiology: Applying principles to clinical practice. *Contemp ObGyn* 20:147–161.

CDC, 1997a. Case definition for infectious conditions under public health surveillance. *MMWR Recomm Rep* 46(RR-10):1–55.

CDC, 1997b. Asthma mortality—Illinois, 1979–1994, *MMWR* 46(MM37):877–80.

CDC, 2001a. Updated guidelines for evaluating public health surveillance systems: recommendations from the guidelines Working Group. *MMWR Recommendations and Reports* 50(RR13).

CDC, 2001b. Summary of notifiable disease—United States, *MMWR 2001* 50(53).

CDC, 2003a. Outbreak of severe acute respiratory syndrome—worldwide. *MMWR 2003* 52226–52228.

CDC, 2003b. Revised U.S. surveillance case definition for severe acute respiratory syndrome (SARS) and update on SARS cases—United States and Worldwide. *MMWR 2003* 52:1202–1206.

CDC, 2003c. Hepatitis A outbreak associated with green onions at a restaurant—Monaca, Pennsylvania, 2003. *MMWR 2003* 52(47):1155–1157.

CDC, 2004a. Framework for revaluating public health surveillance systems for early detection of outbreaks: recommendations for the CDC Working Group. *MMWR* 53(RR05):1–11.

CDC, 2004b. Indicators for chronic disease surveillance. *MMWR Recomm Rep 2004* 53(RR-11):1–6.

CDC, 2004c. Outbreak of Cyclosporiasis associated with snow peas—Pennsylvania. *MMWR 2004* 53:576–578.

CDC, 2004d. Prevalence of overweight and obesity among adults with diagnosed diabetes—United States, 1988–1994 and 1999–2002. *MMWR* 53(45):1066–1068.

CDC, 2006. *Principles of Epidemiology in Public Health Practice*, 3rd edition. Atlanta, GA: Centers for Disease Control and Prevention.

CDC, 2012. Interim guidance on infection control precautions for patients with suspected severe acute respiratory syndrome (SARS) and close contacts in households. Available from: http:llwww.cdc.gov/ncidod/sars/ic-closecontacts.htm.

Charlton, B.G., 1996. Attribution of causation in epidemiology: chain or mosaic *J Clin Epidemiol* 49:105–107.

Cobb, S., M. Miller, and N. Walk, 1959. On the estimation of the incubation period in malignant disease. *J Chron Dis* 9:385–393.

Colditz, G.A., J.E. Manson, and S.E. Hankinson, 1997. The Nurses' Health Study: 20-year contribution to the understanding of health among women. *Women's Heath* 49–62.

Derelanko, M.J., 2012. Pollutant Source Pathways. Accessed 03/07/12 at http://www.jhsph.edu/courses/Environmentalhealth/lecturenotes.cfm.

Eidson, M., R.M. Philen, C.M. Sewell, R. Voorhees, and E.M. Kilborne, L-tryptophan and eosinophilia-myalgia syndrome in New Mexico. *Lancet* 335:645–648.
Elwood, M., 1998. *Casual relationships in medicine: a practical system for critical appraisal*. Oxford: Oxford University Press.
Fenner, R., D.A. Henderson, I. Arita, Z. Jezek, and I.D. Landnyi, 1988. *Smallpox and its eradication*. Geneva: World Health Organization.
Fraser, D.W., T.R. Tsai, W. Orenstein, W.E. Parkin, H.J. Beecham, R.G. Sharrar, et al., 1977.
Legionnaires' disease: description of an epidemic of pneumonia. *New Engl J Med* 297:1189–1197.
Goldsmith, J.R., 1986. *Environmental Epidemiology: Epidemiological Investigation of Community Environmental Health Problems*. Boca Raton, FL: CRC Press.
Goodman, R.A., J.D. Smith, R.K. Sikes, D.L. Rogers, and J.L. Mickey, 1985. Fatalities associated with farm tractor injuries: an epidemiologic study. *Public Health Rep* 100:29–33.
Gordis, L., 2000. *Epidemiology*, 2nd ed. Philadelphia: W.G. Saunders.
Greenwood, M., 1935. *Epidemics and crowd-diseases: an introduction to the study of epidemiology*. New York: Oxford University Press.
Holland, P.W., 1986. Statistics and causal inference. *J Am Stat Assoc* 81:945–960.
IARC, 1990. *Scientific Publications no. 100*. Lyon, France: International Agency for Research on Cancer.
Kamps, B.S., and C. Hoffmann, eds., 2003. *SARS Reference*, 3rd ed. Flying Publisher, accessed 03/06/12 at http://www.sarsreference.comindex.htm.
Kannel, W.B., 2000. The Framingham study: its 50-year legacy and future promise. *J Atheroscler Thromb* 6:60–66.
Karhausen, L.R., 1996. The logic of causation in epidemiology. *Scand J Soc Med* 24:8–13.
Kelsey, J.L., W.D. Thompson, and A.S. Evans, 1986. *Methods in observational epidemiology*. New York: Oxford University Press.
Kleinbaum, D.G., L. Kupper, and H.I. Morgenstern, 1983. *Epidemiologic research: principles and quantitative methods*. Belmont, CA: Lifetime Learning Publishing.
Knowler, W.C., E. Barrett-Connor, S.E. Fowler, R.F. Hamman, J.M. Lachin, E.A. Walker, et al., 2002. Diabetes Prevention Program Research Group. Reduction in the incidence of type 2 diabetes with lifestyle intervention or metformin. *N Engl J Med* 346:393–403.
Krieger, N. 1994. Epidemiology and the web of causation: Has anyone seen the spider? *Soc. Sci. Med* 39:887–903.
Last, J.M., 2001. *Dictionary of Epidemiology*, 4th ed. New York: Oxford University Press.
Leavitt, J.W., 1996. *Typhoid Mary: captive to the public's health*. Boston: Beacon Press.
Liao, Y., P. Tucker, C.A. Okoro, W.H. Giles, A.H. Mokdad, V.B. Harris, et al., REACH 2010 surveillance for health status in minority communities—United States, 2001–2001. *MMWR* 53:1–3.
MacDonald, P., J. Boggs, R. Whitwam, M. Beatty, S. Hunter, N. MacCormack, et al., 2000. Listeria-associated birth complications linked with homemade Mexican-style

cheese. North Carolina, October 200 [abstract]. 50th Annual Epidemic Intelligence Service Conference; Atlanta, GA.
MacMahon, B., and T. Pugh, 1970. *Epidemiology: Principles and Methods.* Boston: Little, Brown.
Maldonado, G., and S. Greenland, 1998. Estimating causal effects. Presented at the Society for Epidemiological Research 31st Annual Meeting, Chicago, Illinois. *Am J Epidemiolog* 147:S80.
Mindel, A., and M. Tenant-Flowers, 2001. Natural history and management of early HIV infection. *BMJ 2001* 332:1290–1293.
Moeller, D.W., 2011. *Environmental Health*, 4th ed. Cambridge, MA: Harvard University Press.
Monson, R., 1990. *Occupational Epidemiology*, 2nd ed. Boca Raton, FL: CRC Press.
Morris, J.N., 1957. *Uses of epidemiology*. Edinburg: Livingstone.
Murphy, T.V., P.M. Gargiullo, M.S. Massoudi, et al., 2001. Intussusception among infants given an oral rotavirus vaccine. *N Eng J Med* 344:564–572.
NCHS, 2005. National Center for Health Statistics. Accessed 03/07/12 at http://www.cdc.gov/nchs/productspubs/pubd/hestatsoverw*ght99.htm.*
NIH, 2012. *CYP1A! Cytochrome P450, family 1, subfamily A, polypeptide 1.* Accessed 03/01/12 at http://www.ncbi.nlm.nih.gove/gene/1543.
Olsen, J., 1991. Causes and prevention. *Scand J Soc Med* 19:1–6.
Olsen, J., 1993. Some consequences of adopting a conditional deterministic causal model in epidemiology. *Eur J Public Health* 3:204–209.
Orenstein, W.A., and R.H. Bernier, 1990. Surveillance: Information for action. *Pediatr Clin North Am* 37709–377034.
Parascandola, M., and D.L. Weed, 2001, Causation in epidemiology. *J Epidemiol Community Health* 55:905–912.
Remington, P.L., W.N. Hall, I.H. Davis, A. Herald, and R.A. Gunn, 1985. Airborne transmission of measles in a physician's office. *JAMA* 253:1575–1577.
Rothman, K.J., 1976. Causes. *Am J Epidemiolo* 104:587–592.
Rothman, K.J., 1993. Policy recommendations in epidemiology research papers. *Epidemiol* 4:94–99.
Rothman, K.H., and S. Greenland, 2005. Causation and casual inference in epidemiology. *Am J Public Health* 98:S144–150.
Rothman, K.H., and S. Greenland, 1998. *Modern epidemiology*, 2nd ed. Philadelphia: Lippincott, Williams and Wilkins.
Rubin, D.B., 1974. Estimating cause effects of treatments in randomized and nonrandomized studies. *Journal of Educational Psychology* 66:688–701.
Snow, J., 1936. *Snow on Cholera*. London: Humphrey Milford, Oxford University Press.
Susser, M., 1991. What is a cause and how do we know one? *Am J Epidemiolo* 133:636–648.
Tyler, C.W., and J.M. Last, 1992. Epidemiology. In: Last, J.M., R.B. Wallace, eds. Maxcy-Rosenau-Last Public health and preventive medicine, 14th ed. Norwalk, CT: Appleton and Lange.
U.S. Department of Health and Human Services (HHS), 1991. *Healthy people 2000; national health promotion and disease prevention objectives.* Washington, D.C.: HHS.

U.S. Department of Health and Human Services (HHS), 2000a. *Healthy people 2010*, 2nd ed. Washington, D.C.: U.S. Government Printing Office.

U.S. Department of Health and Human Services (HHS), 2000b. *Tracking healthy people 2010*. Washington, D.C.: GPO.

U.S. Department of Health and Human Services, 2004. Public Health Service, National Toxicology Program. *Current directions and evolving strategies good science for good decisions*. Accessed 03/08/2011 at http://ntp.niehs.nih.gov/ntp/Main_Pages/PUBS/2004CurrentDirections_Press.pdf.

Vineis, P., and D. Kriebel, 2006. Causal Models in Epidemiology: Past Inheritance and Genetic Future. Accessed 02/29/12 at http://www.ncbi.nlm.nih.gov/pmc/articles/PMC1557493/.

Vineis, P., F. Vegilia, S. Benhamou, D. Bulkiewicz, I. Cascorbi, M. Clapper, et al., 2003. CYP1A1 T3801 C polymorphism and lung cancer: a pooled analysis of 2451 cases and 3358 controls. *Int J Cancer* 104:650–657.

Vineis, P., M. Alavanja, P. Buffler, E. Fontham, S. Franceshi, Y.T. Gao, et al., 2004. Tobacco and cancer: recent epidemiological evidence. *J Natl Cancer Inst* 96:99-106.

Wagner, M.M., F.C. Tsui, J.U. Espino, V.M. Dato, D.F. Sittig, F.A. Caruana, et al., 2001. The emerging science of very early detection of disease outbreaks. *J. Pub Health Mgmt Pract* 6:51–59.

5

Foodborne Disease

Food is our common ground, a universal experience.

—James Beard

USDA FSIS Reports

(Food and Safety Bulletin, 02/07/2012, Possible Chicken Salad Products Listeria Contamination; Food and Safety Bulletin, 03/02/2012, Ready-to-Eat Pizza Stick Products Recalled; and Food and Safety Bulletin, 03/06/2012, Kansas Firm Recalls Pizza Topping Products Nationwide.)

North Carolina Firm Recalls Chicken Products Due to Possible Listeria Contamination. Bost Distributing Company, doing business as Harold Food Company, a Bear Creek, North Carolina, establishment, is recalling approximately 1,200 pounds of chicken salad products. The products contain eggs that are the subject of a Food and Drug Administration (FDA) recall due to concerns about contamination with *Listeria monocytogenes.*

Indiana Firm Recalls Ready-to-Eat Pizza Stick Products Produced with Benefit of Inspection. Pasou Foods, a Syracuse, Indiana, establishment, is recalling approximately 147 pounds of frozen, fully cooked, ready-to-eat pizza stick products because a meat ingredient used in the product may have been produced without the benefit of federal inspection.

Kansas Firm Recalls Pizza Topping Products Nationwide Due to Misbranded and Undeclared Allergen. Tyson Prepared Foods, Inc., a South Hutchinson, Kansas, establishment, is recalling approximately 12,060 pounds of pizza topping products because of misbranding and an undeclared allergen. The packaging identifies the products as beef, but they actually contain pork. Additionally, the pork products contain soy, a known potential allergen, which is not declared on the label.

Poisoned Picnic

(From F. R. Spellman and Joni Price-Bayer, 2011. *In Defense of Science*. Latham, MD: Government Institutes; based on CDC, 2009. *Poisoned Picnic*. Accessed 03/07/2012 at www.bam.gov/teachers/activities/epi_6_picnic.pdf.) The picnic started at 6:00 p.m. Tuesday evening at Grand City Park. The park is located by the Grand River and contains several gazebos and picnic areas. The administration and faculty of Grand City Middle School organized the picnic as a relaxing event to be held before the faculty meeting. Many faculty and staff brought members of their family.

Mrs. Smith and Ms. Johnston arrived at 5:30 to set up. Mr. Albert arrived next to set up the grill. He brought his grill from home and had to take a few minutes to clean it off because it had not been used since last summer. Mr. Drake arrived next, after having bought the hamburgers at the supermarket. After the charcoal was lit and aluminum foil was placed over the grills, Mr. Albert began to cook.

At 5:55, Mrs. Smith realized that there was only one serving spoon. At that point, she left to get some more spoons. The other teachers waited for a while, but finally decided to start eating at about 6:20.

When all of the food arrived there was a full menu that included baked beans, chicken, ham, green bean casserole, tuna casserole, cherry pie, pudding, potato salad, macaroni salad, corn, and hamburgers. Drinks included soda, water, coffee, and tea.

Mr. Drake was first through the line. He tried green bean casserole, ham, and a hamburger.

Ms. Cummings was next. She ate potato salad, ham, and a hamburger.

The third person through the line was Mr. Carlson. He ate green bean casserole, potato salad, and a hamburger.

Mrs. Albert was next in line. She sampled potato salad, a hamburger, and cherry pie.

At this point, Mrs. Smith returned with more serving spoons. Mrs. Bell came at the same time. She was a little late because she had to be sure that her chicken was done.

Mrs. Wolfe went through the line next. She ate green bean casserole, chicken, a hamburger, and pudding.

Next was Mr. Lewis, who ate baked beans, green bean casserole, macaroni salad, and corn.

The line became a little unorganized at this point and it is not clear who went through next. Mrs. Smith and Ms. Johnston were two of the last people through since they helped to serve.

Mrs. Smith ate green bean casserole, potato salad, a hamburger, and pudding.

Others in attendance included Mr. Harvey, Ms. Jackson, Mr. Dooley, Mrs. Jones, and Mrs. Darwin. A lot of the guests said they could not remember exactly what they ate, but Mr. Harvey, Mr. Dooley, Mrs. Jones, and Mrs. Bell all had hamburgers, baked beans, and macaroni salad.

Ms. Jackson and Mrs. Darwin had ham, baked beans, corn, and some pudding for dessert.

Ms. Cain, Mrs. Williams, Dr. Oakton, Mrs. Corning, and Mrs. Reid have not yet been interviewed. Some other staff members arrived just in time for the faculty presentations, which started at 7:45. These included Mrs. Robinson, Mrs. Brown, and Mrs. Wright.

Some of the faculty and staff walked around while they ate but most sat in one of the gazebos. The presentations were held in the main gazebo, which was a relief for some of the faculty because it seemed to be one of the few places free of duck droppings.

Even during the meeting, some of the kids chased ducks with their water guns. These kids never seemed to run out of water because the guns held almost a gallon each, but even if they did run out, they quickly refilled them from the river. Just about everyone at the picnic, except for those who came only for the meeting, were soaked. Since it was a hot day, the only time anyone seemed to mind the soaking was when one of the kids missed their intended target and almost put out the grill. After this incident, which happened about 6:10, the kids stayed away from the main gazebo where the food was located, and turned their attention to the ducks and teachers walking around.

Grand City officials were alarmed by the illnesses and deaths that seemed to be associated with the event. They have promised a full investigation. Even the wastewater treatment plant just a few hundred yards up the river will have to submit a report on their procedures for water treatment. This is the first time anything like this has happened at the park and officials want to be sure that it does not happen again.

Park managers said that most of the symptoms such as dehydration, stomach cramps, nausea, and vomiting indicate some type of food poisoning. However, at this point they cannot be certain.

You are now part of a team of epidemiologists that has been called in to get to the bottom of this mystery. You will need to identify the cause of the disease and prevent any further outbreaks. Time is of the essence. The first thing you will want to do is meet with your team members and outline the information you have been given and then decide what additional information you need. Grand City authorities have promised complete cooperation in this matter. Good luck!

The first order of business for the team of epidemiologists is to determine the goal of their investigation and to compile data related to the picnic. The goal is to determine the cause of the mysterious disease and how to prevent future outbreaks. Item #1, the picnic menu, and other collected items are shown below.

Item #1: Grand City Middle School Faculty Picnic Menu

Baked beans: Simply purchased two large cans of baked beans and heated on stovetop to boiling.

Pudding: Mixed four packets of chocolate pudding with four cups of milk. Heated and then refrigerated.

Chicken: Baked chicken legs for 1 hour.

Ham: Baked ham for 2 hours 30 minutes until thermometer read 150 degrees for 20 minutes.

Green Bean Casserole: Cracker crust covered with two cans of cream of mushroom soup and two jars of green beans. Topped with 2 cans of small onions. Baked for 20 to 25 minutes to warm.

Potato salad: One jar of salad dressing (mayonnaise), assorted diced vegetables, 2 tablespoons sugar, ½ cup mustard, 6 cups diced and cooked potatoes.

Macaroni salad: One box elbow macaroni, 3 tablespoons mustard, one jar salad dressing (mayonnaise), various diced vegetables.

Tuna casserole: Cracker crust, 3 cans tuna, one can cream of mushroom soup, one can cream of chicken soup. Mixed and topped with parmesan cheese topping.

Hamburgers: Purchased at the supermarket just before the picnic (receipt showed time was 12:25).

Corn: 2 large cans of corn heated to simmering.

Cherry pie: Mountaintop cherry pie, baked 40 minutes, premade.

Pause #1: It's Got to Be the Potato Salad or Macaroni Salad!

We feel it is important to pause for a moment. We want to point out to readers that we have presented this particular case to undergrad environmental health students in a few of our 300 level college courses for several years. We know this is the right place to pause in our presentation because right after we introduce the above picnic menu to our students, many of their faces light up and beam that look of knowing the answers. We are used to this occurrence and always cease our presentation to allow the students the chance to comment on what they think. Invariably their statements seem to relate the following: "Ah, they all got sick because of the potato and/or macaroni salad . . . it was probably the mayonnaise that poisoned 'em." We allow the students to banter round and round with their statements without our interjection. But as with all bubbles that eventually burst into nothingness, the time soon arrives to burst a few bubbles and put the students back on the correct scientific approach to finding the picnic poison.

We explain that commercially prepared mayonnaise gets a bad rap when it comes to making picnickers sick from eating potato salad and/or macaroni salad. A commercially made jar of mayonnaise is set with a pH level point that bacteria can't survive. The pH level makes mayonnaise safe to eat even if not refrigerated.

When people become sick from eating these salads, the causal factors are related to how long the salads have sat in the sun, along with the potatoes and onions. Bacteria have never found a freshly cut potato or onion that they did not immediately attack. Bacteria can't resist either one and it is these potatoes and onions that are the culprits—not mayonnaise. However, in this case the potato and macaroni salads are not the culprits.

Item #2: Poisoned Picnic Faculty Information Cards

Create cards with the following information on them:

Mrs. Cain: Brought plates and cups to the picnic. Had chicken, potato salad, pudding, green bean casserole. Became sick Tuesday evening. Symptoms include nausea, vomiting, and dizziness.

Mr. Lewis: Organized a game of volleyball set up by the gazebo. The players were a favorite target for the water guns! The only foul was when Mrs. Cain stepped on a duck going after the ball. Mr. Lewis became ill Tuesday evening. He was treated and released from the hospital Wednesday morning.

Mrs. Williams: Recovering. Became ill Tuesday night and was rushed to the hospital by her husband. Her son enjoyed his water gun, dousing teachers with river water. She loved the burgers made by Mr. Albert. She also tried some green bean casserole, chicken, and pudding. Her son did not become ill.

Mrs. Reid: Sampled a little bit of everything. She became ill Tuesday night and finally went to the hospital Wednesday morning. She complained of stomach cramps and nausea. Doctors quickly began an IV to help replenish lost fluids. She briefly went into a coma then slowly recovered.

Dr. Oakton: Recovering. Had a great time except for when she stepped in duck droppings, which seemed to be everywhere. She didn't even mind being soaked. She tried a little bit of everything to eat.

Mr. Albert: Mr. Albert took control of the grill. Mr. Drake soon showed up with the hamburger meat and started making the burgers. Mr. Albert had some potato salad, green bean casserole, a hamburger, and pudding for desert. Mr. Albert became ill, suffering from numbness, disorientation, nausea, and vomiting. He was treated and released after several days in the hospital.

Mrs. Corning: Arrived late, just in time to grab a burger and some green bean casserole. Most of the utensils and food were already put away. She became ill Wednesday morning and had to leave work around 8:30. She suffered from nausea, dizziness, and was so disoriented that she could not drive home.

Mrs. Smith: She arrived early with her son and helped to set up for the picnic. After many of the staff arrived, she realized that there was only one serving spoon so she went home to get some more. She returned about thirty minutes later with spoons (after several faculty had gone through the line) to find her son chasing ducks with the water guns. Both Mrs. Smith and her son became ill.

Mrs. Johnston: Helped to set up for the picnic. She had a hamburger, baked beans, pudding, and corn. She and several other teachers spent their time sitting in one of the gazebos talking and watching the children dash about after the ducks. Ms. Johnston is lactose-intolerant. She became ill just a couple of hours after the picnic suffering from severe stomach pains. She went to bed and recovered overnight.

Mrs. Albert: Complained of stomach cramps early Tuesday night. Her condition continued to worsen until she finally had to be taken to the hospital. She was given massive doses of antibiotics. Her condition became worse as symptoms began to include vomiting and disorientation. She soon found that she could not remember much about the picnic. After some time her condition improved.

Item #3 consists of specific, additional information about the poisoned picnic event.

Item #3: Picnic Information Card

It was determined that there was only one burger on the grill when it was soaked. Mr. Albert decided to throw it away because he had to lift up the grill and add more charcoal. Many times he would walk away from the grill to talk to someone and return to some very well-done burgers. No one seemed to mind; that's the way they wanted them.

It was also learned that the wastewater treatment plant performed several tests on the water coming from the plant (effluent). The effluent was virtually void of any bacteria. The plant was doing a good job. They also did tests on the water around the park and found no notable bacterial contamination.

The epidemiological team obtained item 4, the pathology reports on victims.

Item #4: Pathology Report

Victim: Mrs. Wolfe. Admitted to hospital suffering from abdominal pain and vomiting. Began diagnostic tests but patient's condition deteriorated. Death due to respiratory and heart failure. Time of death: 3:30 a.m., 9/21/05.

Victim: Mr. Carlson. Paramedic response to home. Pronounced dead on arrival. Attempts to revive failed. Time of death: 11:30 p.m., 9/20/05.

Victim: Mr. Drake. Admitted to hospital suffering from abdominal pain, headache, and paralysis of extremities. Lapsed into shock. Pulmonary failure followed. Time of death: 2:30 a.m., 9/21/05.

Victim: Mrs. Cummings. Admitted to hospital suffering paralysis. Unable to communicate to hospital staff. Died of heart and respiratory failure. Time of death: 1:20 a.m., 9/21/05.

Item #5: Round Up the Usual Suspects

Because the epidemiological investigative team determined the culprit in the poisoned picnic episode was likely food poisoning, they listed specific disease-causing microbes that could be responsible. The suspect list is provided in table 5.1.

From the toxin suspect list shown above and the information and clues gathered from other sources, the team was able to select the correct food item and toxin responsible for the incident. The results of their findings are detailed in the incident explanation below.

Pause #2: The Culprit Identified?

Before presenting the poisoned picnic team's findings, we thought we would ask: Have you identified the food and the toxin? Let's see if your decision matches the team's findings.

The Explanation

In determining the food item and correct toxin that caused the poisoned picnic event, the investigation team pieced together the following clues from a variety of sources:

- The opening information states that there was only one serving spoon when the teachers began going through the line. Moreover, the first person

TABLE 5.1
Poisoned Picnic: List of Suspect Disease-Causing Microbes

Microbe Name	*Description*
Staphylococcus aureus	*Staph* bacteria are very common on the skin of most animals including humans. The bacteria can, in rare instances, be very dangerous when ingested. Food poisoning can result if the bacteria are allowed to multiply and produce toxins on handled food before eating. Foods commonly contaminated in the United States include turkey, ham, processed meats, chicken salad, pastries, and ice cream in which staph has grown. Infection is characterized by sudden nausea, vomiting, diarrhea, and often shock within a few hours of eating contaminated food. Usually other bacteria in the body help to keep staph at bay, but if something happens to upset this balance, such as the introduction of massive doses of antibiotics, infection can occur with fatal results.
Bacillus cereus	*Bacillus cereus* is one of the many types of bacteria that cause food poisoning. General symptoms of stomach cramps, dizziness, and vomiting can be attributed to bacteria but the infection is rarely fatal. The illnesses caused by this unsightly intestinal intruder are actually caused by toxins it produces.
Escherichia coli	*E. coli* are found in the intestines of healthy animals, including humans. There are many different strains or types of this bacterium and most are harmless to humans. However, some strains have been known to cause E. coli food poisoning. People can become infected by eating undercooked meat products (especially ground meat) that have been contaminated with animal feces. E. coli also can be contracted by ingesting other types of food, water, or anything else that has been contaminated with human or animal waste. Upon infection, the bacteria multiply and produce toxins. Victims usually suffer severe bloody diarrhea and stomach cramps. Some suffer vomiting. Infection can be fatal, especially for the elderly and the young since their immune systems are not as strong as those in other age groups. Death can result from dehydration or damage to the red blood cells and kidneys.
Salmonella typhimurium	This bacterium causes salmonella food poisoning. It is usually caused by eating undercooked fowl (such as chicken or turkey) but can also be contracted from a contaminated water supply. Symptoms include headache, chills, and stomach pain and are usually followed by nausea, vomiting, diarrhea, and fever. Symptoms usually last 3 to 4 days. After this time, an individual may become a carrier of the disease. Rarely fatal if treated quickly.
Clostridium botulinum	This bacterium causes botulism. Botulism is actually caused by the toxins produced by the bacteria and not the presence of the bacteria itself. This organism reproduces by forming spores, which can be found in soil and therefore on vegetation. The bacteria are often found in meats and improperly canned foods. Infected individuals initially suffer from nausea, vomiting, and diarrhea. If enough toxin is ingested or produced, death can result from a general breakdown of the nervous system. Individuals may become paralyzed and death usually occurs as result of respiratory paralysis.

Microbe Name	*Description*
Clostridium perfringens	This bacterium causes gas gangrene. The condition can lead to a buildup of gases in the muscles of the body rendering them useless. Infection also causes stomach pain and cramping (gastroenteritis). Disease is usually caused by eating meat that has been stewed or boiled and then set aside before reheating and serving.
Streptococcus pyogenes	This bacterium is widely distributed among humans. Infected individuals may spread this pathogen through respiratory droplets. This organism can cause sharp outbreaks of sore throats and scarlet fever. High fever and skin lesions characterize infection by this pathogen. Individuals with known infection should be isolated.

through the line had green bean casserole. That person was one of the four fatalities. The next person through the line was also a fatality. This person did not eat green bean casserole. But the toxin was on the serving spoon after being used for the casserole and then for the potato salad. The second person through the line got it from the spoon! After this, the other serving spoons arrived and the cross-contamination soon ended.

- Team members realized that everyone who had the green bean casserole contracted the illness. Two people became ill, one of whom died, who did not have the green bean casserole. The second person through the line, as stated above, contracted the disease from the serving spoon. The other person was lactose-intolerant and became ill because of the pudding, which was made with milk.
- From the list of usual suspects, team members were able to link the symptoms of the disease to botulism. From this and other research, team members knew that botulism can come from improperly canned vegetables. The menu states that the green bean casserole was made using canned green beans. Botulism is easily destroyed by high heat. However, the menu also states that the casserole was only heated to warm. This would not have provided enough heat to destroy the toxin.
- In addition to identifying the cause of the disease, team members were also to outline a strategy to prevent future outbreaks of the disease. Because this outbreak was due to food poisoning, team members realized that they needed to educate the teachers, students, and parents about proper food handling techniques. For this specific contaminant, information should include inspection of canned foods for bulges (a result of gas buildup from growth of the bacteria), and also stress the importance of sufficient heat to destroy the toxin. Other general information should include cleanliness in food preparation areas, the washing of hands, and

the use of clean utensils in food preparation. This information could be presented in the form of a brochure or a school auditorium skit that would be broadcast on CCTV, local television, or radio stations.

Final Pause

So, did you determine that the green bean casserole was the bad food, that the spoon was the source of cross-contamination, and that the deadly toxin was *Clostridium botulinum*? If so, great! If not, no problem. The point is, the poisoned picnic event demonstrates the need for a step-by-step approach based on actual facts in determining causal factors related to an event such as this one.

For those of you who immediately jumped on the potato salad or macaroni salad (laced with copious amounts of mayonnaise, of course) as being the bad food, as mentioned, you are not alone in your assumptions. And this is the point about the need for science and the scientists' toolbox. Investigative opinions should be based on facts and good science—peer-reviewed and verified facts and the practice of good science only, please!

Setting the Stage

(From CDC, 2012. Questions and Answers about Foodborne Illness [sometimes called "Food Poisoning"]. Accessed 03/08/2012 at http://www.cdc.gov/foodsafety/facts.html.) The focus of this chapter is on foodborne disease, including foodborne infections, food safety, and foodborne outbreaks. Foodborne disease (sometimes called "foodborne illness," "foodborne infection," or "food poisoning") is a common, costly—yet preventable—public health problem. CDC (2012) estimates that each year roughly one in six Americans (or 48 million people) gets sick, 128,000 are hospitalized, and 3,000 die of foodborne diseases. The 2011 estimates provide the most accurate picture yet of which foodborne bacteria, viruses, and microbes ("pathogens") are causing the most illnesses in the United States. According to the 2011 estimates, the most common foodborne illnesses are caused by norovirus and by the bacteria *Salmonella*, *Clostridium perfringens*, and *Campylobacter*. In addition to the many different foodborne illnesses caused by bacteria, viruses, and microbes, poisonous chemicals or other harmful substances can also cause foodborne diseases if they are present in food.

- More than 250 different foodborne disease have been described. Most of these diseases are infections, caused by a variety of bacteria, viruses, and parasites that can be foodborne.

- Other diseases are poisonings, caused by harmful toxins or chemicals that have contaminated the food, for example, poisonous mushrooms.
- These different diseases have many different symptoms, so there is no one "syndrome" that encompasses foodborne illness. However, the microbe or toxin enters the body through the gastrointestinal tract and causes the first symptoms there, so nausea, vomiting, abdominal cramps, and diarrhea are common symptoms in many foodborne diseases.

Many microbes can spread in more than one way, so we cannot always know that a disease is foodborne. The distinction matters, because public health authorities need to know how a particular disease is spreading to take the appropriate steps to stop it.

For example, *Escherichia coli* 0157:H7 infections can spread through contaminated food, contaminated drinking water, contaminated swimming water, and from toddler to toddler at a day-care center. Depending on which means of spread caused a case, the measures to stop other cases from occurring could range from removing contaminated food from stores, chlorinating a swimming pool, or closing a child day-care center.

Food Services—An Environmental Health Concern

(Information in this section adapted from NIH, 2012. *Food Services: Why are food services a concern?* Accessed 03/08/12 at http://toxtown.nlm.hih.gov/text_version/locations.php?id=26.)

Why Are Food Services a Concern?

Food services include restaurants, delis, and companies that prepare food for others. Food service workers and the people they serve may be exposed to chemicals used in food processing, cooking, or storage, or used in operating a facility. Basic food safety issues are also very important for public health.

Chemicals that may be used in maintaining a restaurant include bleach, household chemicals, and pesticides. Food service operations may use natural gas or propane as cooking fuels. Food processors may use sulfur dioxide as a food preservative for fruits and vegetables or use food packaging that contains phthalates (esters used as plasticizers to increase flexibility).

Observing basic food safety and meeting health department standards are very important in preventing the spread of infectious diseases. Workers need to follow good personal hygiene such as handwashing and not working when sick. Proper handling of food is also critical. Food handling safety issues include exposure to or eating of raw or undercooked foods, including raw shellfish that may be contaminated. Prepared food that has not been refrigerated for more than two hours may pose health risks. All food needs to be maintained at the proper cold or hot temperature and kitchen surfaces kept clean.

If smoking is allowed in a restaurant, exposure to secondhand smoke is a health issue for workers and customers.

Foodborne Disease Outbreaks

An outbreak of foodborne illness occurs when a group of people consume the same contaminated food and two or more of them come down with the same illness.

- It may be a group that ate a meal together somewhere, or it may be a group of people who do not know each other at all, but who all happened to buy and eat the same contaminated item from a grocery store or restaurant. For an outbreak to occur, something must have happened to contaminate a batch of food that was eaten by a group of people.
- Often, a combination of events contributes to the outbreak. A contaminated food may be left out at room temperature for many hours, allowing the bacteria to multiply to high numbers, and then be insufficiently cooked to kill the bacteria.

Many outbreaks are local in nature. They are recognized when a group of people realize that they all became ill after a common meal, and someone calls the local health department. This classical local outbreak might follow a catered meal at a reception, a potluck supper, or a meal at an understaffed restaurant on a particularly busy day. However, outbreaks are increasingly being recognized that are more widespread, that affect persons in many different places, and that are spread out over several weeks.

- For example, a recent outbreak of *Salmonellosis* was traced to persons eating a breakfast cereal produced at a factory in Minnesota, and marketed under several different brand names in many different states. No one county or state had very many cases and the cases did not know each other. The outbreak was recognized because it was caused by an unusual strain of salmonella, and because state public health laboratories type salmonella strain and they noticed a sudden increase in this one rare strain.
- In another recent outbreak, a particular peanut snack food caused the same illness in Israel, Europe, and North America. Again, this was recognized as an increase in infections caused by a rare strain of salmonella.

Did You Know?

The vast majority of reported cases of foodborne illness are not part of recognized outbreaks, but occur as individual or "sporadic" cases. It may be that many of these cases are actually part of unrecognized widespread or diffused outbreaks. Detecting and investigating such widespread outbreaks is a major challenge to our public health system. This is the reason that new and more sophisticated laboratory methods are being used at the CDC and in state public health department laboratories.

Investigating Foodborne Disease Outbreaks

A foodborne outbreak is an indication that something needs to be improved in our food safety system. Public health scientists investigate outbreaks to control them, and also to learn how similar outbreaks can be prevented in the future. Just as when a fire breaks out in a large building or when an airliner crashes, two activities are critical when an outbreak occurs.

- First, emergency action is needed to keep the immediate danger from spreading, and second, a detailed objective scientific investigation is needed to learn what went wrong, so that future similar events can be prevented.
- Much of what we know about foodborne disease and its prevention comes from detailed investigation of outbreaks. This is often how a new pathogen is identified, and this is how the critical information linking a pathogen to a specific food and animal reservoir is first gathered. The full investigation can require a team with multiple talents, including the epidemiologist, microbiologist, food sanitarian, food scientist, veterinarian, and factory process engineer.

Detecting Outbreaks of Foodborne Disease

The initial clue that an outbreak is occurring can come in various ways.

- It may be when a person realizes that several other people who were all together at an event have become ill and he or she calls the local health department.
- It may be when a physician realizes she has seen more than the usual number of patients with the same illness.
- It may be when a county health department gets an unusually large number of reports of illness.

- The hardest outbreaks to detect are those that are spread over a large geographic area, with only a few cases in each state. These outbreaks can be detected by combining surveillance reports at the regional or national level and looking for increase in infections of a specific type. This is why state public health laboratories determine the serotype of salmonella bacteria isolated from people.
- New "DNA fingerprinting" technologies can make detecting outbreaks easier, too. For example, the new molecular subtyping network, PulseNet, allows state laboratories and the CDC to compare strains of *E. coli* 0157:H7 and an increasing number of other pathogens from all across the United States to detect widespread outbreaks.

After an apparent cluster of cases is detected, it is important to determine whether these cases represent a real increase above the expected number of cases and whether they really might be related. Sometimes a cluster of reported cases is caused by something other than an actual outbreak of illness. For example, if the person responsible for reporting has just returned from a vacation and is clearing up a backlog of cases by reporting them all at once, the sudden surge of reports is just a false cluster.

Did You Know?

After *Salmonella, Staphylococcus aureus* is the second most frequent cause of foodborne disease. *Staphylococci* are responsible for approximately 25 percent of all cases of food poisoning. *Staphylococcus* (staph) infection can manifest itself in a variety of ways, from food poisoning to skin infections to septicemia (blood infection). In severe cases, it can be life-threatening (Balch, 2010).

Foodborne Disease Outbreak Investigation Procedure

- Once an outbreak is strongly suspected, an investigation begins.
- A search is made for more cases among persons who may have been exposed. The symptoms, time of onset, and location of possible cases is determined, and a "case definition" is developed that describes these typical cases.
- The outbreak is systematically described by time, place, and person. A graph is drawn of the number of people who fell ill on each successive day to show pictorially when it occurred. A map of where the ill people live, work, or eat may be helpful to show when it occurred. Calculating the distribution of causes by age and sex shows who is affected. If the causative microbe is not known, samples of stool or blood are collected

from ill people and sent to the public health laboratory to make the diagnosis.

- To identify the food or other source of the outbreak, the investigators first interview a few persons with the most typical cases about exposures they may have had in the few days before they got sick. In this way, certain potential exposures may be excluded while others that are mentioned repeatedly emerge as possibilities.
- Combined with other information, such as the likely sources for the specific microbe involved, these hypotheses are then tested in a formal epidemiologic investigation.
- The investigators conduct systematic interviews about a list of possible exposures with the ill people, and with a comparable group people who are not ill. By comparing how often an exposure is reported by ill people and by well people, investigators can measure the association of the exposure with illness. Using probability statistics, similar to those used to describe coin flips, the probability of no association is directly calculated.

For example, imagine that an outbreak has occurred after a catered event. Initial investigation suggested that Hollandaise sauce was eaten by a least some of the attendees, so it is on the list of possible hypotheses.

- We interview twenty persons who attended the affair, ten of whom became ill and ten who remained well. Each ill or well person is interviewed about whether or not they ate the Hollandaise sauce, as well as various other food items.
- If half the people ate the sauce, but the sauce was not associated with the illness, then we would expect each person to have a 50/50 chance of reporting that they ate it, regardless of whether they were ill or not.
- Suppose, however, that we find that ten ill people but none of the well people reported eating Hollandaise sauce at the event? This would be very unlikely to occur by chance alone if eating the Hollandaise sauce were not somehow related to the risk of illness. In fact, it would be about as unlikely as getting heads ten times in a row by flipping a coin (that is 50 percent multiplied by itself ten times over, or a chance of just under 1 in 1,000).
- The epidemiologist concluded that eating the Hollandaise sauce was very likely to be associated with the risk of illness. Note that the investigator can draw this conclusion even though there is no Hollandaise sauce left to test in a laboratory.
- The association is even stronger if she can show that those who ate second helpings of Hollandaise were even more likely to become ill, or that

people who ate leftover Hollandaise sauce that went home in doggie bags also became ill.

Once a food item is statistically implicated in this manner, further investigation into its ingredients and preparation, and microbiologic culture of leftover ingredients or the food itself (if available), may provide additional information about the nature of contamination.

- Perhaps the Hollandaise sauce was made using raw eggs. The source of the raw eggs can be determined, and it may even be possible to trace them back to the farm and show that chickens on the farm are carrying the same strain of *Salmonella* in their ovaries. If so, the eggs from that farm can be pasteurized to prevent them from causing other outbreaks.
- Some might think that the best investigation method would just be to culture all the leftover foods in the kitchen, and conclude that the one that is positive is the one that caused the outbreak. The trouble is that this can be misleading, because it happens after the fact. What if the Hollandaise sauce is all gone, but the spoon that was in the sauce got placed in potato salad that was not served at the function? Now, cultures of the potato salad yield a pathogen, and the unwary tester might call that the source of the outbreak, even though the potato salad had nothing to do with it. This means that laboratory testing without epidemiologic investigation can lead to the wrong conclusion.

Even without isolating microbes from food, a well-conducted epidemiologic investigation can guide immediate efforts to control the outbreak. A strong and consistent statistical association between illness and a particular food item that explains the distribution of the outbreak in time, place, and person should be acted upon immediately to stop further illness from occurring. An outbreak ends when the critical exposure stops.

- This may happen because all the contaminated food is eaten or recalled, because a restaurant is closed or a food processor shuts down or changes its procedures, or because an infected food handler is no longer infectious or is no longer working with food.
- An investigation that clarifies the nature and mechanism of contamination can provide critical information even if the outbreak is over.
- Understanding the contamination event well enough to prevent it can guide the decision to resume usual operations, and lead to more general prevention measures that reduce the risks of similar outbreaks happening elsewhere.

Did You Know?

The U.S. Department of Agriculture told *The Daily* online newspaper that it is buying 7 million pounds of pink slime for school lunch programs across the country. Pink slime consists of beef by-products, cow intestines, connective tissue, and other parts that are not used in traditional beef cuts. Those parts are susceptible to *E. coli* and salmonella contamination, so the last ingredient to the "pink slime" is ammonium hydroxide, which kills the bacteria (USAToday.com, 2012).

Foods Most Associated with Foodborne Illness

- Raw foods of animal origin are the most likely to be contaminated, that is, raw meat and poultry, raw eggs, unpasteurized milk, and raw shellfish.
- Because filter-feeding shellfish strain microbes from the sea over many months, they are particularly likely to be contaminated if there are any pathogens in the seawater.
- Foods that mingle the products of many individual animals, such as bulk raw milk, pooled raw eggs, or ground beef, are particularly hazardous because a pathogen present in any one of the animals may contaminate the whole batch. A single hamburger may contain meat from hundreds of animals; a single restaurant omelet may contain eggs from hundreds of chickens; a glass of raw milk may contain milk from hundreds of cows; and a broiler chicken carcass can be exposed to the drippings and juices of many thousands of other birds that went through the same cold water tank after slaughter.
- Fruits and vegetables consumed raw are a particular concern. Washing can decrease but not eliminate contamination, so the consumers can do little to protect themselves. Recently, a number of outbreaks have been traced to fresh fruits and vegetables that were processed under less than sanitary conditions. These outbreaks show that the quality of the water used for washing and chilling the produce after it is harvested is critical. Using water that is not clean can contaminate many boxes of produce. Fresh manure used to fertilize vegetables can also contaminate them. Alfalfa sprouts and other raw sprouts pose a particular challenge, as the conditions under which they are sprouted are ideal for growing microbes as well as the sprouts, and because they are eaten without further cooking. That means that a few bacteria present on the seeds can grow to high numbers of pathogens on the sprouts. Unpasteurized fruit juice can also be contaminated if there are pathogens in or on the fruit that is used to make it.

Food Contamination

We live in a microbial world, and there are many opportunities for food to become contaminated as it is produced and prepared.

- Many foodborne microbes are present in healthy animals (usually in their intestines) raised for food. Meat and poultry carcasses can become contaminated during slaughter by contact with small amounts of intestinal contents.
- Similarly, fresh fruits and vegetables can be contaminated if they are washed or irrigated with water that is contaminated with animal manure or human sewage.
- Some types of salmonella can infect a hen's ovary so that the internal contents of a normal-looking egg can be contaminated with salmonella even before the shell is formed.
- Oysters and other filter-feeding shellfish can concentrate *Vibrio* bacteria that are naturally present in seawater, or other microbes such as norovirus that are present in human sewage dumped into the sea.

Later in food processing, other foodborne microbes can be introduced from infected humans who handle the food, or by cross-contamination from some other raw agricultural product.

- For example, *Shigella* bacteria, hepatitis A virus, and norovirus can be introduced by the unwashed hands of food handlers who are themselves infected.
- In the kitchen, microbes can be transferred from one food to another food by using the same knife, cutting board, or other utensil to prepare both, without washing the surface or utensil in between.
- A food that is fully cooked can become recontaminated if it touches other raw foods or drippings from raw foods that contain pathogens.

The way that food is handled after it is contaminated can also make a difference in whether or not an outbreak occurs.

- Many bacterial microbes need to multiply to a larger number before enough are present in food to cause disease. Given warm, moist conditions and ample supply of nutrients, on bacterium that reproduced by dividing itself every half hour, can produce 17 million progeny in twelve hours.
- As a result, lightly contaminated food left out overnight can be highly infectious by the next day. If the food were refrigerated promptly, the bacteria would not multiply at all.

- In general, refrigeration or freezing prevents virtually all bacteria from growing but generally preserves them in a state of suspended animation. This general rule has a few surprising exceptions. Two foodborne bacteria, *Listeria monocytogenes* and *Yesinia enterocolitica*, can actually grow at refrigerator temperatures.
- High salt, high sugar, or high acid levels keep bacteria from growing, which is why salted meats, jam, and pickled vegetables are traditional preserved foods.

Microbes are killed by heat:

- If food is heated to an internal temperature above 1600F, or 780C, for even a few seconds, this is sufficient to kill parasites, viruses, or bacteria, except for the *Clostridium* bacteria, which produce a heat-resistant form called a spore. *Clostridium* spores are killed only at temperatures above boiling. This is why canned foods must be cooked to a high temperature under pressure as part of the canning process.
- The toxins produced by bacteria vary in their sensitivity to heat. The *staphylococcal* toxin that causes vomiting is not inactivated even if it is boiled. Fortunately, the protein toxin that causes botulism is completely inactivated by boiling.

Shellfishing—An Environmental Health Concern

Shellfish are marine animals that have a shell. The Food and Drug Administration defines shellfish as clams, mussels, oysters, and scallops. Shellfish provide food for humans and many other species. Some are found close to shore in tidal flats where people may dig or gather them for personal consumption. Offshore shellfish beds or reefs are more likely to be harvested by commercial fishermen using boats equipped with dredges or tongs.

Shellfish must grow and live in clean water to be safely eaten. Shellfish are "filter feeders," which means they pump and filter large amounts of water through their bodies every day as they eat. As the water filters through, the shellfish strain out particles for their food. These particles can include harmful chemicals, waste, bacteria, viruses, and marine toxins, which can contaminate the shellfish.

Shellfish can become contaminated because of poorly treated sewage from wastewater treatment plants, cesspools, and septic systems. Contamination may also come from polluted water runoff from marinas, farms, and wildlife waste. They can be contaminated by marine toxins that are produced by algae blooms. Some marine toxins are dangerous and can cause extremely serious shellfish poisoning and even death in humans who eat contaminated shellfish.

Local and state health departments monitor shellfish for contamination and will ban harvesting in contaminated shellfish beds and coastal areas. Health departments also warn of the increased health risks from eating raw shellfish and may advise cooking shellfish to kill bacteria.

Food Workers' Safety and Health

(Based on information from F. R. Spellman, 2008. *Occupational Safety and Health Simplified for the Food Manufacturing Industry*. Lanham, MD: Government Institutes Press.)

OSHA Regional News Release: Region 6, December 15, 2003

OSHA Cites Austin Food Manufacturing Company for Worker Death: Michael Angelo's Gourmet Foods, Inc., Will Pay $140,220 Fine

Austin, Texas—Michael Angelo's Gourmet Foods, Inc., in Austin, Texas, has agreed to pay $140,220 in penalties for citations issued by the U.S. Department of Labor's Occupational Safety and Health Administration (OSHA) for failure to provide employees with adequate protection and training to prevent machines from starting up during cleaning operations.

A worker for the frozen food manufacturer was killed in June when he was pulled into a meat mixer.

OSHA began its investigation on June 13 after the employee was found dead by a coworker at the company's headquarters. Michael Angelo Gourmet Foods employs about 440 workers at that location.

The company was cited for ten safety and health violations for exposing employees to electrical hazards such as defective electrical cords, failing to properly guard machinery that could cause amputations and other injuries, failing to ensure locks were provided on machinery requiring maintenance, and failing to train its employees to perform lockout procedures. Lockout/tagout involves shutting off and locking out the energy source to a machine that is undergoing maintenance or repair to prevent an accidental startup.

OSHA has inspected Michael Angelo twice since 2000, and the company has paid more than $63,000 in penalties for similar violations.

Michael Angelo's Gourmet Foods has agreed to fully comply with OSHA's standards by providing lockout/tagout training for its employees, managers, and supervisors. The company also has agreed to abate the violations and certify their abatement within ten days. In addition, the company will retain the services of an outside consultant to review its safety and health program.

Did You Know?

Food manufacturing has one of the highest incidences of injury and illness among all industries; animal slaughtering plants have the highest incidence among all food manufacturing industries.

BLS (2005) reports that workers in the food manufacturing industry link farmers and other agricultural products with consumers. They do this by processing raw fruits, vegetables, grains, meats, and dairy products into finished goods ready for the grocer or wholesaler to sell to households, restaurants, or institutional food services.

Food manufacturing workers perform tasks as varied as the many foods we eat. For example, they slaughter, dress, and cut meat or poultry; process milk, cheese, and other dairy products; can and preserve fruits, vegetables, and frozen specialties; manufacture flour, cereal, pet foods, and other grain mill products; make bread, cookies, cakes, and other bakery products; manufacture sugar and candy and other confectionery products; process shortening, margarine, and other fats and oils; and prepare packaged seafood, coffee, potato and corn chips, and peanut butter. Although this list is long, it is not exhaustive: food manufacturing workers also play a part in delivering numerous other food products to our tables.

Quality control and quality assurance (QC & QA) are vital to this industry. The U.S. Department of Agriculture (USDA) oversees all aspects of food manufacturing. In addition, other food safety programs have been adopted recently as issues of chemical contamination, and the growing number of new foodborne pathogens remains a public health concern. For example, by applying science-based controls from raw materials to finished products, a program called Hazard Analysis and Critical Control Point (HACCP; see figure 5.1) is an evaluation system to identify, to monitor, and to control contamination risks in food-service establishments. Simply, it focuses on identifying critical hazards and preventing them from contaminating food. Note that General Mills, one of the leaders in HACCP implementation, defines a "critical hazard" as an imminent health hazard or a guest dissatisfaction. Since incorporating this system totally into its operations, General Mills has increased food quality substantially.

Thirty-four percent of all food manufacturing workers are employed in plants that slaughter and process animals, and another 19 percent work in establishments that make bakery goods (table 5.2). Seafood product preparation and packaging, the smallest sector of the food manufacturing industry, accounts for only 3 percent of all jobs.

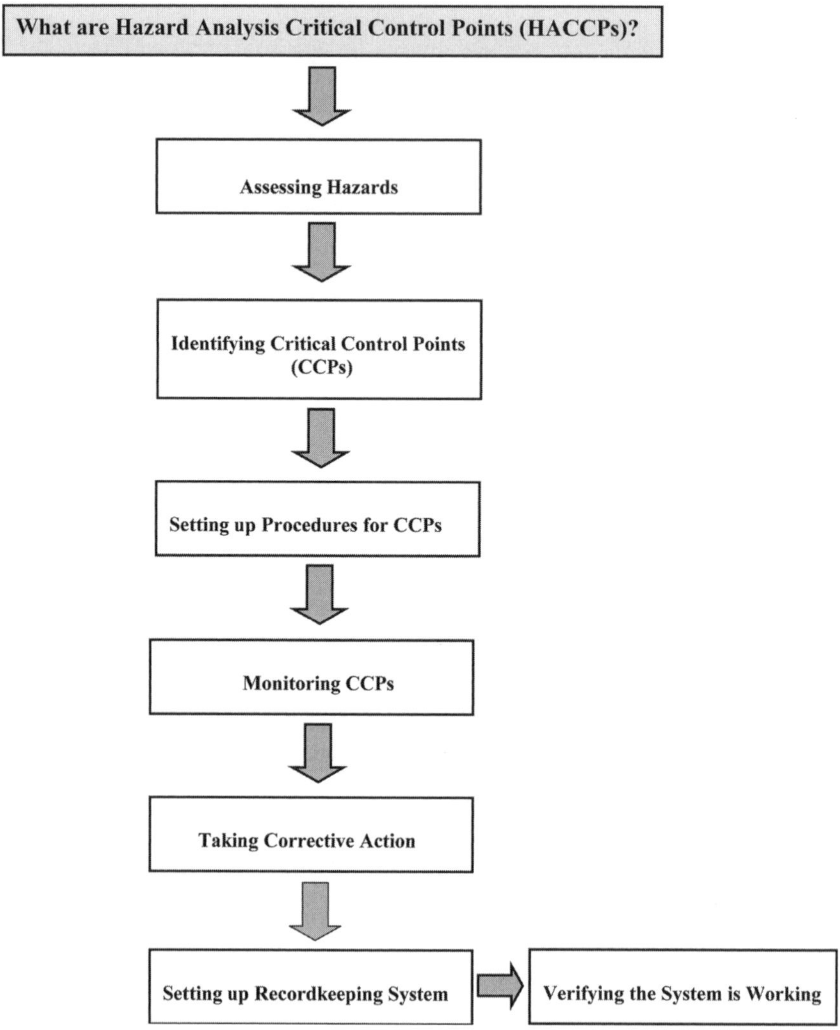

FIGURE 5.1.
HACCPs.

Farm-to-Fork Continuum: Agriculture

Although this book is not intended to be a detailed description and accounting of agricultural safety and safe work practices per se, the authors feel it is important to include a brief overview of farm or barnyard safety in this discussion.

Some of us might imagine living and working on a farm in the stereotypical mode portrayed by Grant Wood (1930) in his classic painting, *American*

TABLE 5.2
Distribution of Wage and Salary Employment in Food Manufacturing by Industry Segment, 2004, Employment in Thousands (From BLS, 2005)

Industry Segment	*Employment, 2004*	*2004–2014 % Change*
Total Employment	1497.5	3.81
Animal slaughtering and processing	505.3	12.82
Bakeries and tortilla manufacturing	287.8	3.79
Fruit and vegetable preserving and specialty food manufacturing	181.7	–1.49
Other food manufacturing	154.1	6.42
Dairy product manufacturing	132.0	–0.61
Sugar and confectionary product manufacturing	83.7	–4.42
Grain and oilseed milling	60.6	–6.60
Animal food manufacturing	50.7	–4.93
Seafood product preparation and packaging	41.6	–3.85

Source: BLS 2005.

Gothic, where a Midwest pitchfork-holding farmer stands next to his daughter before a large Gothic window. Most of us, however, probably imagine that living and working on a farm out in the country as a safe, healthy, quiet, peaceful, outdoorsy, get-back-to-nature activity; for the most part, this is the case.

Most people in risky occupations, including farming, live to a ripe old age. However, like other risky occupations, the farm environment has many hazards, the full extent of which the average person has no knowledge or no appreciation. This lack of knowledge or appreciation of potential safety problems can be extremely hazardous, costing fingers, hands, arms, and/or lives. Another unavoidable potentially dangerous circumstance in farming has to do with the occupation's need for farmers to perform many solitary tasks way out in the back forty, so to speak. Working alone, far from other people who could help in an emergency, adds to the severity of many accidents.

Did You Know?

In any given year, the number of farm fatalities is usually exceeded only by those related to mining.

OSHA fully understands that, contrary to the popular image of fresh air and peaceful surroundings, a farm is not a hazard-free work setting. As mentioned, every year thousands of farm workers are injured and hundreds more die in farming accidents. According to the National Safety Council (2007), agriculture is the most hazardous industry in the nation.

Farm workers—including farm families and migrant workers—are exposed to hazards such as the following:

- Chemicals/pesticides
- Cold
- Dust
- Electricity
- Grain bins
- Hand tools
- Highway traffic
- Lifting
- Livestock handling
- Machinery/equipment
- Manure pits
- Mud
- Noise
- Ponds
- Silos
- Slips/trips/falls
- Sun/heat
- Toxic gases
- Tractors
- Wells

What Are the High Risk Factors on Farms?

The following factors may increase risk of injury or illness for farm workers:

Age—Injury rates are highest among children age fifteen and under and adults over sixty-five.

Equipment and Machinery—Most farm accidents and fatalities involve machinery. Proper machine guarding and performing equipment maintenance according to manufacturers' recommendations can help prevent accidents.

Protective Equipment—Using protective equipment (such as seat belts on tractors) and personal protective equipment (such as safety gloves, coveralls, boots, hats, aprons, goggles, and face shields) could significantly reduce farming injuries.

Medical Care—Hospitals and emergency medical care are typically not readily accessible in rural areas near farms.

How Can We Improve Farm Safety?

OSHA (2007) points out that farmers can start by increasing their awareness of farming hazards and making a conscious effort to prepare for emergency situations including fires, vehicle accidents, electrical shocks from equipment and wires, and chemical exposures. They should be especially alert to hazards that might affect children and the elderly. They should minimize hazards by carefully selecting the products they buy to ensure that they provide good tools and equipment. They should always use seat belts when operating tractors, and establish and maintain good housekeeping practices. Here are some other steps farmers should take to reduce illnesses and injuries on the farm:

- Read and follow instructions in equipment operator's manuals and on product labels.
- Inspect equipment routinely for problems that might cause accidents.
- Discuss safety hazards and emergency procedures with their workers.
- Install approved rollover protective structures, protective enclosures, or protective frames on tractors.
- Make sure that guards on farm equipment are replaced after maintenance.
- Review and follow instructions in material safety data sheets (MSDSs) and on labels that come with chemical produces, and communicate information on these hazards to your workers.
- Take precautions to prevent entrapment and suffocation caused by unstable surfaces of grain storage bins, silos, or hoppers. Never "walk the grain."
- Be aware that methane gas, carbon dioxide, ammonia, and hydrogen sulfide can form in unventilated grain silos and manure pits and can suffocate or poison workers or explode.
- Take advantage of safety equipment, such as bypass starter covers, power take-off master shields, and slow-moving vehicle emblems.

Did You Know?

"Walking down grain" or similar practices, where employees walk on grain to get grain to flow out of a grain storage structure or where employees are on top of moving grain, are not permitted.

The 29 CFR 1928 Agriculture Standard

If federal OSHA administers your state and your employee workers are outside your immediate family, you must post a copy of the appropriate volume or volumes of the Code of Federal Regulations, 29 CFR 1910 (General Industry Standards; see table 5.3) and 29 CFR 1928 (Agriculture Standards; see table 5.4). There must be a copy physically present at locations where people-for-hire work.

Did You Know?

Most food production jobs require little formal education or training; many can be learned in a few days.

Food Manufacturing Working Conditions

Many production jobs in food manufacturing involve repetitive, physically demanding work. Food manufacturing workers are highly susceptible to repetitive-strain injuries to their hands, wrists, and elbows. This type of injury is especially common in meat-processing and poultry-processing plants. Production workers often stand for long periods and may be required to lift heavy objects or use cutting, slicing, grinding, and other dangerous tools and machines. To deal with difficult working conditions, ergonomic programs have been introduced to cut down on work-related accidents and injuries.

In 2003, there were 8.6 cases of work-related injury or illness per 100 full-time food manufacturing workers, *much higher than the rate of 5.0 cases for the private sector as a whole.* Injury rates vary significantly among specific food manufacturing industries, ranging from a low of 1.8 per 100 workers in retail bakeries to 12.9 per 100 in animal slaughtering plants, the highest rate in food manufacturing.

Eighty-nine percent of the establishments in food manufacturing employ fewer than 100 workers.

In an effort to reduce occupational hazards, many plants have redesigned equipment, increased the use of job rotation, allowed longer or more frequent breaks, and developed training programs in sage work practices. Furthermore, meat and poultry plants must comply with a wide array of OSHA regulations ensuring a safe work environment. Although injury rates remain high, training and other changes have reduced those rates. Some workers wear protective hats, gloves, aprons, and shoes. In many industries, uniforms and protective clothing are changed daily for reasons of sanitation.

Because of the considerable mechanization in the industry, most food manufacturing plants are noisy, with limited opportunities for interaction among workers. In some highly automated plants, "hands-on" manual work has been replaced by computers and factory automation, resulting in less waste and higher productivity. While much of the basic production—such

TABLE 5.3
29 CFR 1928—Agriculture Standards

Subpart C—Employee Operating Instruction	
Section	*Designation*
1928.51	Roll-over protective structures (ROPs) for tractors used in agricultural operations
1928.52	Protective frames for wheel-type agricultural tractors—test procedures and performance requirements
1928.53	Protective enclosures for wheel-type agricultural tractors—test procedures and performance requirements
Subpart D—Safety for Agricultural Equipment	
Section	*Designation*
1928.57	Guarding of farm field equipment, farmstead equipment, and cotton gins
Subpart I—General Environment Controls	
Section	*Designation*
1928.110	Field Sanitation
Subpart M—Occupational Health	
Section	*Designation*
1928.1027	Cadmium

TABLE 5.4
29 CFR 1910—General Industry Standards

Subpart J—General Environmental Controls	
Section	*Designation*
1910.142	Temporary Labor Camps
1910.145	Specifications for accident prevention signs and tags
Subpart H—Hazardous Materials	
Section	*Designation*
1910.111	Storage and handling of anhydrous ammonia
Subpart R—Special Industries	
Section	*Designation*
1910.266	Logging Operations
Subpart Z—Toxic and Hazardous Substances	
Section	*Designation*
1910.1200	Hazard Communication
1910.1201	Retention of DOT markings, placards, and labels
1910.1027	Cadmium

as trimming, chopping, and sorting—will remain labor-intensive for many years to come, automation is increasingly being applied to various functions, including inventory management, product movement, and quality control issues such as packing and inspection.

Did You Know?

Automation and increasing productivity will limit employment growth, but unlike many other industries, food manufacturing is not highly sensitive to economic conditions.

Working conditions also depend on the type of food being processed. For example, some bakery employees work at night or on weekends and spend much of their shifts near ovens that can be uncomfortably hot. In contrast, workers in dairies and meat-processing plants typically work daylight hours and may experience cold and damp conditions. Some plants, such as those producing processed fruits and vegetables, operate on a seasonal basis, so workers are not guaranteed steady, year-round employment and occasionally travel from region to region seeking work. These plants are increasingly rare, however, as the industry continues to diversify and manufacturing plants are producing alternative foods during otherwise inactive periods.

Food Manufacturing Employment

In 2004, the food manufacturing industry provided 1.5 million jobs. Almost all employees were wage- and salaried-workers, but a few food manufacturing workers were self-employed and unpaid family workers. In 2004, about 29,000 establishments manufactured food, with 8 percent employing fewer than 100 workers. Nevertheless, establishments employing 500 or more workers accounted for 36 percent of all jobs.

The employment distribution in this industry varies widely. Animal slaughtering and processing employs the largest proportion of workers. Economic changes in livestock farming and slaughtering plants have changed the industry. Increasingly, fewer, but larger, farms are producing the vast majority of livestock in the United States. Similarly, there are now fewer, but much larger, meat-processing plants, owned by fewer companies—a development that has tended to concentrate employment in a few locations.

Food manufacturing workers are found in all states, although some sectors of the industry are concentrated in certain parts of the country. For example, in 2004, California, Illinois, Iowa, Pennsylvania, and Texas employed 24 percent of all workers in animal slaughtering and processing. That same year,

Wisconsin employed 33 percent of all cheese manufacturing workers, and California accounted for 20 percent of fruit and vegetable preserving and specialty food manufacturing workers.

Occupations in the Food Manufacturing Industry

The food manufacturing industry employs many different types of workers. More than half are production workers, including skilled precision workers and less skilled machine operators and laborers. Production jobs require manual dexterity, good hand-eye coordination, and, in some sectors of the industry, strength.

Red Meat Production Workers—Red meat production is the most labor-intensive food-processing operation. Animals are not uniform in size, and slaughterers and meatpackers must slaughter, skin, eviscerate, and cut each carcass into large pieces. They usually do this work by hand, using large, suspended power saws. They also clean and salt hides and make sausage. Meat, poultry, and fish cutters and trimmers use hand tools to break down the large primary cuts into smaller sizes for shipment to wholesalers and retailers. These workers use knives and other hand tools to eviscerate, split, and bone chickens and turkeys.

Most butchers and meat, poultry, and fish cutters and trimmers frequently work in cold, damp rooms, which are refrigerated to prevent meat from spoiling. Workrooms are often damp because meat cutting generates large amounts of blood and condensation. These occupations require physical strength to lift and carry large cuts of meat and the ability to stand for long periods. Butchers and meat, poultry, and fish cutters and trimmers work in clean and sanitary conditions; however, their clothing is often soiled with animal blood and the air may smell unpleasant. They work with powerful cutting equipment and are susceptible to cuts on the fingers or hands. Risks are minimized with the proper use of equipment and hand and stomach guards. The repetitive nature of the work, such as cutting and slicing, may lead to wrist damage (carpal tunnel syndrome) (CALMIS, 2007).

Bakers—Bakers mix and bake ingredients according to recipes to produce breads, cakes, pastries, and other goods. Bakers produce goods in large quantities, using mixing machines, ovens, and other equipment.

Did You Know?

On January 19, 2007, a wide range of safety and health hazards at a Syracuse bakery has resulted in $120,600 in proposed fines from OSHA. Penny Curtiss Baking Co., Inc., which manufactures bread and other bakery products, was cited for a total of forty-two alleged serious safety and health hazards at its production plant following an OSHA inspection that began in July in response to an employee complaint.

Well-managed bakeries are generally kept spotlessly clean, and personal cleanliness is very important. However, work areas can be uncomfortably hot and noisy. Many employers who require uniforms furnish and launder employee uniforms. Oven mitts are also usually supplied to employees when necessary. Bakery production jobs are usually performed at a fast, steady pace while standing. Many plant jobs involve strenuous physical work, including heavy lifting, despite the use of machinery.

Hands-on Food Production Workers—Many food manufacturing workers use their hands or small hand tools to do their jobs. Cannery workers perform a variety of routine tasks—such as sorting, grading, washing, trimming, peeling, or slicing—in the canning, freezing, or packing of food products. Hand food decorators apply artistic touches to prepared foods. Candy molders and marzipan (confection) shapers form sweets into fancy shapes by hand.

Did You Know?

On March 11, 2004, because of a fatality at a Mississippi cannery, OSHA proposed citation penalties totaling $229,000.

Food Manufacturing Machine Operators—With increasing levels of automation in the food manufacturing industry, a growing number of workers are operating machines. For example, food batchmakers operate equipment that mixes, blends, or cooks ingredients used in manufacturing various foods, such as cheese, candy, honey, and tomato sauce. Dairy processing equipment operators process milk, cream, cheese, and other dairy products. Cutting and slicing machine operators slice bacon, bread, cheese, and other foods. Mixing and blending machine operators produce dough batter, fruit juices, or spices. Crushing and grinding machine operators turn raw grains into cereals, flour, and other milled-grain products, and they produce oils from nuts or seeds. Extruding and forming machine operators produce molded food and candy, and casing finishers and stuffers make sausage links and similar products. Bottle packers and bottle fillers operate machines that fill bottles and jars with preserves, pickles, and other foodstuffs.

Food Cooking Machine Operators—Food cooking machine operators and tenders steam, deep-fry, boil, or pressure-cook meats, grains, sugar, cheese, or vegetables. Food and tobacco roasting, baking, and drying machine operators and tenders operate equipment that roasts grains, nuts, or coffee beans and tend ovens, kilns, dryers, and other equipment that removes moisture from macaroni, coffee beans, cocoa, and grain. Baking equipment operators tend ovens that bake bread, pastries, and other products. Some foods—ice cream, frozen specialties, and meat, for example—are placed in freezers or refrigerators by cooling and freezing equipment operators. Other workers tend machines and equipment that clean and wash food or food processing equipment. Some machine operators also clean and maintain machines and perform duties such as checking the weight of foods.

Food Manufacturing Maintenance Workers—Many other workers are needed to keep food manufacturing plants and equipment in good working order. Industrial machinery mechanics repair and maintain production machines and equipment. Maintenance repairers perform routine maintenance on machinery, such as changing and lubricating parts. Specialized mechanics include heating, air-conditioning, and refrigeration mechanics, farm equipment mechanics, and diesel engine specialists.

QA/QC—Still other workers directly oversee the quality of the work and of final products. Supervisors direct the activities of production workers. Graders and sorters of agricultural products, production inspectors, and quality control technicians evaluate foodstuffs before, during, or after processing.

Packagers/Transporters—Food may spoil if not packaged properly or delivered promptly, so packaging and transportation employees play a vital role in the industry. Among those are freight, stock, and material movers, who manually move materials; hand packers and packagers, who pack bottles and other items as they come off the production line; and machine feeders and offbearers, who feed materials into machines and remove goods from the end of the production line. Industrial truck and tractor operators drive gasoline- or electric-powered vehicles equipped with forklifts, elevated platforms, or trailer hitches to move goods around a storage facility. *Truck drivers* transport and deliver livestock, materials, or merchandise and may load and unload trucks. Driver/sales workers drive company vehicles over established routes to deliver and sell goods, such as bakery items, beverages, and vending-machine products.

Managers—The food manufacturing industry also employs a variety of managerial and professional workers. Managers include top executives, who make policy decisions; industrial production managers, who organize, direct, and control the operation of the manufacturing plant; and advertising, marketing, promotions, public relations, and sales managers, who direct advertising, sales promotion, and community relations programs.

Professional Staff—Engineers, scientists, and technicians are becoming increasingly important as the food manufacturing industry implements new automation and food safety processes. These workers include industrial engineers, who plan equipment layout and workflow in manufacturing plants, emphasizing efficiency and safety. Also, mechanical engineers plan, design, and oversee the installation of tools, equipment, and machines. Chemists perform tests to develop new products and maintain the quality of existing products. Computer programmers and systems analysts develop computer systems and programs to support management and scientific research. Food scientists and technologists work in research laboratories or on production lines to develop new products, test current ones, and control food quality, including minimizing foodborne pathogens.

Sales/Marketing and Other Workers—Finally, many sales workers, including sales representatives, wholesale and manufacturing, are needed to sell the manufactured goods to wholesale and retail establishments. Bookkeeping, accounting, and auditing clerks and procurement clerks keep track of the food products going into and out of the plant. Janitors and cleaners keep buildings clean and orderly.

Did You Know?

The injury and illness rate for the food manufacturing industry is significantly higher than that for the manufacturing sector as a whole.

The Jungle Revisited I (2008)

The American author and socialist Upton Sinclair published his eye-opening, heart-throbbing, vomit-generating, horrific account of the meat-packing industry in his bestselling book *The Jungle*. Sinclair's account of workers falling into meat processing tanks and being ground, along with animal parts, into "Durham's Pure Leaf Lard," and the morbidity of the working conditions, gripped public attention.

Because of public outcry and new regulations, Sinclair's "Jungle" no longer exists to the extent he described it in 1906. However, consider today and those workers who labor in a world of long knives and huge saws, blood and bone, sweltering heat and arctic chill. It is no place for the squeamish; some workers can't stomach the gore—chopping up the meat and bones of hundreds of animals, day after day. Additionally, a rite of passage for many meatpackers is the slashes, burns, and scars received as part of their daily duties.

Meatpacking Industry—An Environmental Health Concern

According to the U.S. Department of Labor (1988), the meatpacking industry, which employs over 1,000,000 workers, is considered to be one of the most hazardous industries in the United States. According to the Bureau of Labor Statistics (BLS), this industry has had the highest injury rate of any industry in the country for five consecutive years (1980–1985) with a rate three times that of other manufacturing industries.

BLS studies have also shown that for 1985, 319 workers were injured during the first month of employment in the industry. Of those workers, 29 percent were cut by knives or machinery and 30 percent of all injuries occurred to workers twenty-five years of age or younger. Younger new workers are at the highest occupational risk and suffer a significant proportion of all injuries.

Workers can be seriously injured by moving animals prior to stunning, and by stun guns that may prematurely or inadvertently discharge while they try to still the animal. During the hoisting operation, it is possible for a 2,000-pound carcass to fall on workers and injure them if faulty chains break or slip off the carcass's hind leg. Workers can suffer from crippling arm, hand, or wrist injuries. For example, carpal tunnel syndrome, caused by repetitive motion, can literally wear out the nerves running through one section of the wrist. Workers can be cut by their own knives or by other workers' knives during the butchering process. Back injuries can result from overhead rails. Workers can be severely burned by cleaning solvents and burned by heat sealant machines when they wrap meat. It is not uncommon for workers to sever fingers or hands on machines that are improperly locked out or inadequately guarded. For example, in 1985, BLS studies also reported 1,748 cases of injuries to the fingers, including 76 amputations. Many workers can also injure themselves by falling on treacherously slippery floors and can be exposed to extremes of heat and cold.

Potential Hazards

Machinery such as head splitters, bone splitters, snout pullers, and jaw pullers, as well as band saws and cleavers, pose potential hazards to workers during the various stages of processing animal carcasses. A wide variety of other occupational safety and health hazards exists in the industry. These hazards are identified and discussed in the following paragraphs.

Knife Cuts—Knives are the major causes of cuts and abrasions to the hands and the torso. Although modern technology has eliminated a number of hand knife operations, the hand knife remains the most commonly used tool and causes the most frequent and severe accidents.

- A worker used a knife to pick up a ham prior to boning; the knife slipped out of the ham striking him in the eye and blinding him.
- Another worker was permanently disfigured when his knife slipped out of a piece of meat and struck his nose, upper lip, and chin.

- Workers have also been cut by other workers as they remove their knives from a slab of meat. These "neighbor cuts" are usually the direct result of overcrowded working conditions.

Falls—Falls also represent one of the greatest sources of serious injuries. Because of the nature of the work, floor surfaces throughout the plants tend to be wet and slippery. Animal fat, when allowed to accumulate on floors to dangerous levels, and blood, leaking pipes, and poor drainage are the major contributors to treacherously slippery floors.

Back Injuries—Back injuries tend to be more common among workers in the shipping department. These employees, called "luggers," are required to lug or carry on their shoulders carcasses (weighing up to 300 pounds) to trucks or railcars for shipment.

Toxic Substances—Workers are often exposed to ammonia. Ammonia is a gas with a characteristic pungent odor and is used as a refrigerant, and occasionally, as a cleaning compound. Leaks can occur in the refrigeration pipes carrying ammonia to coolers. Contact with anhydrous liquid ammonia or with aqueous solution is intensely irritating to the mucous membranes, eyes, and skin. There may be corrosive burns to the skin or blister formation. Ammonia gas is also irritating to the eyes and to moist skin. Mild to moderate exposure to the gas can produce headaches, salivation, burning of the throat, perspiration, nausea, and vomiting. Irritation from ammonia gas to the eyes and nose may be sufficiently intense to compel workers to leave the area. If escape is not possible, there may be severe irritation of the respiratory tract with the production of cough, pulmonary edema, or respiratory arrest. Bronchitis or pneumonia may follow a severe exposure.

On some occasions, employees have been exposed to unsafe levels of carbon dioxide from the dry ice used in the packaging process. When meat is ready to be frozen for packaging, it is put into vats where dry ice is stored. During this process, carbon dioxide gas may escape from these vats and spread throughout the room. Breathing high levels of this gas causes headaches, dizziness, nausea, vomiting, and even death. Workers are also exposed to carbon monoxide. Carbon monoxide is a colorless, odorless gas that is undetectable by the unaided senses and is often mixed with other gases. Workers are exposed to this gas when smokehouses are improperly ventilated. Overexposed workers may experience headaches, dizziness, drowsiness, nausea, vomiting, and death. Carbon monoxide also aggravates other conditions, particularly heart disease and respiratory problems.

Workers are also exposed to the thermal degradation products of polyvinyl chloride (PVC) food-wrap film. PVC film used for wrapping meat is cut on a hot wire, wrapped around the package of meat, and sealed by the use of a heated pad. When the PVC film is heated, thermal degradation products irritate workers' eyes, noses, and throats or cause more serious problems such as wheezing, chest pains, coughing, difficulty in breathing, nausea, muscle pains, chills, and fever.

Cumulative Trauma Disorders—Cumulative trauma disorders are widespread among workers in the meatpacking industry. Cumulative trauma disorders such as tendonitis (inflammation of a tendon sheath), and carpal tunnel syndrome are very serious diseases that often afflict workers whose jobs require repetitive hand movement and exertion.

Carpal tunnel syndrome is the disorder most commonly reported for this industry and is caused by repeated bending of the wrist combined with gripping, squeezing, and twisting motions. A swelling in the wrist joint causes pressure on a nerve in the wrist. Early symptoms of the disease are tingling sensations in the thumbs and in the index and middle fingers. Experience has shown that if workers ignore these symptoms, sometimes misdiagnosed as arthritis, they could experience permanent weakness and numbness in the hand coupled with severe pain in the hands, elbows, and shoulders.

Infectious Diseases—Workers are also susceptible to infectious disease such as brucellosis, erysipeloid, leptospirous, dermatophytoses, and warts. Brucellosis is caused by a bacterium and is transmitted by the handling of cattle or swine. Persons who suffer from this bacterium experience constant or recurring fever, headaches, weakness, joint pain, night sweats, and loss of appetite.

Erysipeloid and leptospirosis are also caused by bacteria. Erysipeloid is transmitted by infection of skin puncture wounds, scratches, and abrasions; it cause redness and irritation around the site of infection and can spread to the bloodstream and lymph nodes. Leptospirosis is transmitted through direct contact with infected animals or through water, moist soil, or vegetation contaminated by the urine of infected animals. Muscular aches, eye infections, fever, vomiting, chills, and headaches occur, and kidney and liver damage may develop.

Dermatophytosis, on the other hand, is a fungal disease and is transmitted by contact with the hair and skin of infected persons and animals. Dermatophytosis, also known as ringworm, causes the hair to fall out and small yellowish cuplike crusts to develop on the scalp.

Meatpacking Operational Hazards

Many of the operational hazards in the meatpacking industry are listed in table 5.5.

TABLE 5.5
Operational Hazards in the Meatpacking Industry (USDOL, 2007)

Operation Performed	*Equipment/Substances*	*Accidents/Injuries*
Stunning	Knocking gun	Severe shock, body punctures
Skinning/Removing front legs	Pincher device	Amputations, eye injuries, cuts, falls
Splitting animal	Splitter saws	Eye injury, carpal tunnel syndrome, amputations, cuts, falls
Removing brain	Head splitter	Cuts, amputations, eye injury, falls
Transporting products	Screw conveyors, screw conveyors	Fractures, cuts, amputations, falls
Cutting/trimming/boning	Hand knives, saws—circular saw, band saw	Cuts, eye injuries, carpal tunnel syndrome, falls
Removing jaw bone/snout	Jaw bone, snout puller	Amputations, falls
Preparing bacon for slicing	Bacon/belly press	Amputations, falls
Tenderizing	Electrical meat tenderizers	Severe shock, amputations, cuts, eye injuries
Cleaning equipment	Lock-out, tag-out	Amputations, cuts
Hoisting/shackling	Chain/dolly assembly	Falls, falling carcasses
Wrapping meat	Sealant machine/ polyvinyl chloride, meat	Exposure to toxic substances, severe burns to hands/arms, falls
Lugging meat	Carcasses	Severe back/shoulder injuries, falls
Refrigeration/curing cleaning and wrapping	Ammonia, carbon dioxide, carbon monoxide, polyvinyl chloride	Upper respiratory irritation and damage

Source: USDOL 2007.

The Jungle Revisited II

The American public paid little attention to the treatment of the central character of Sinclair's novel, a Lithuanian immigrant named Jurgis Rudkus who worked in the meatpacking plants. Sinclair reportedly lamented, "I aimed at the public's heart, and I hit it in the stomach" (Rasmussen, 2003). Inside the plants, workers faced constant wage cuts, production line speedups, injuries and disease, and instant dismissal and blacklisting if they protested conditions. Outside, they lived with no medical care, no education, and no decent housing. Rudkus and his coworkers were "aliens" both legally and culturally; not citizens, unable to speak good English, ignorant of their rights, and afraid to turn to governmental authorities for help. A century later, abusive working conditions and treatment still torment a mostly immigrant labor force in the American meatpacking industry.

Poultry Processing

Inferno in Hamlet, North Carolina

In September 1991, twenty-five people died and fifty-four were injured as a result of a fire in the Imperial Food Products, Inc., poultry plant in Hamlet, North Carolina. The cause of the fire was the ignition of hydraulic oil from a ruptured line only a few feet from a natural-gas-fired cooker. The cooker was used to cook chicken pieces for distribution to restaurants.

Many OSHA violations were uncovered after the fire. The basic OSHA exit and fire safety violations that contributed to the deaths and injuries were:

- Locked doors
- No marking of exits or non-exits
- Excessive travel distances to exits
- No fire alarms
- Obstructed doors
- No emergency action plan or fire prevention plan
- No automatic fire suppression plan

The tragic Hamlet fire received a lot of publicity. In spite of this publicity, blocked exits continue to be found in poultry processing facilities. OSHA cited a plant in Hudson, Missouri, for blocking fire and emergency exits in July 1997.

Best Food Nation (2007) points out that in America's poultry industry today, family farmers work with production and processing companies to provide consumers with tasty, nutritious, and economical food. Poultry is the number-one protein purchased by American consumers, at more than 100 pounds per year for every man, woman, and child.

Chicken is relatively lean and has fewer calories and is thus viewed as a healthy alternative to other food types. Moreover, American consumers have demanded the convenience of "fast foods," precut and packaged meats and boneless chicken pieces. The poultry industry (along with others) has had to institute changes to meet these public demands. Changes in the industry have heightened to the need for increased safety vigilance and implementation of safe work practices. For example, appropriate guards around the moving parts of machinery and the blades of saws and knives; adequate ventilation; the use of personal protective equipment; and good housekeeping practices are just a few of the safe practices required.

The poultry industry can include hatcheries and farms where chicks are grown; feed mills where grains are stored, selected, and mixed for hatcheries; and processing plants. All fowl (turkey, chicken, duck, capon, quail, etc.) that are processed and made available for consumption could be considered part of the poultry industry. This text focuses on the processing of chicken, but also applies to the processing of other poultry.

OSHA (2007) points out that the poultry industry can be divided into two stages, each with its own particular hazards. The first stage is the raising of live birds to the desired size, with delivery to the processing plant and preparing the live birds for slaughtering. The second stage is the slaughtering, processing, and packaging of the birds. Major hazards in the first stage are generally respiratory hazards resulting from exposure to organic dusts (litter, manure, dander) and ammonia. These are controlled using ventilation and personal protective equipment (PPE). Additional hazards include those associated with agricultural machinery, feed delivery systems, waste removal systems, and ergonomic hazards, especially as birds are prepared for slaughtering. A potentially significant hazard is the presence of microbiologicals and endotoxins in the organic dusts. In the second stage, common elements in an effective safety and health programs include control of ergonomic hazards (discussed in detail later) to prevent cumulative trauma disorders, machine guarding and PPE to prevent cuts, care of walking/working surfaces to reduce trips and falls, design and maintenance of electrical systems, and lockout/tagout procedures to prevent accidental startup of machinery.

An Environmental Health Hazard

As is clearly evident from figure 5.2, poultry processing is accomplished through a set of step-by-step stages, starting with and continuing with various ongoing sanitation tasks (accomplished within/around each procedure) through the receiving and killing, evisceration, cutting and deboning, packout, and warehousing of product.

Sanitation Worker—The job of the sanitation worker is one of the most hazardous jobs in the poultry processing industry. Sanitation workers may work a regular production shift, or they may be part of a special sanitation or cleaning crew.

The focus of sanitation workers who work a regular production shift is cleaning the machinery and floors. They move product to allow cleaning and

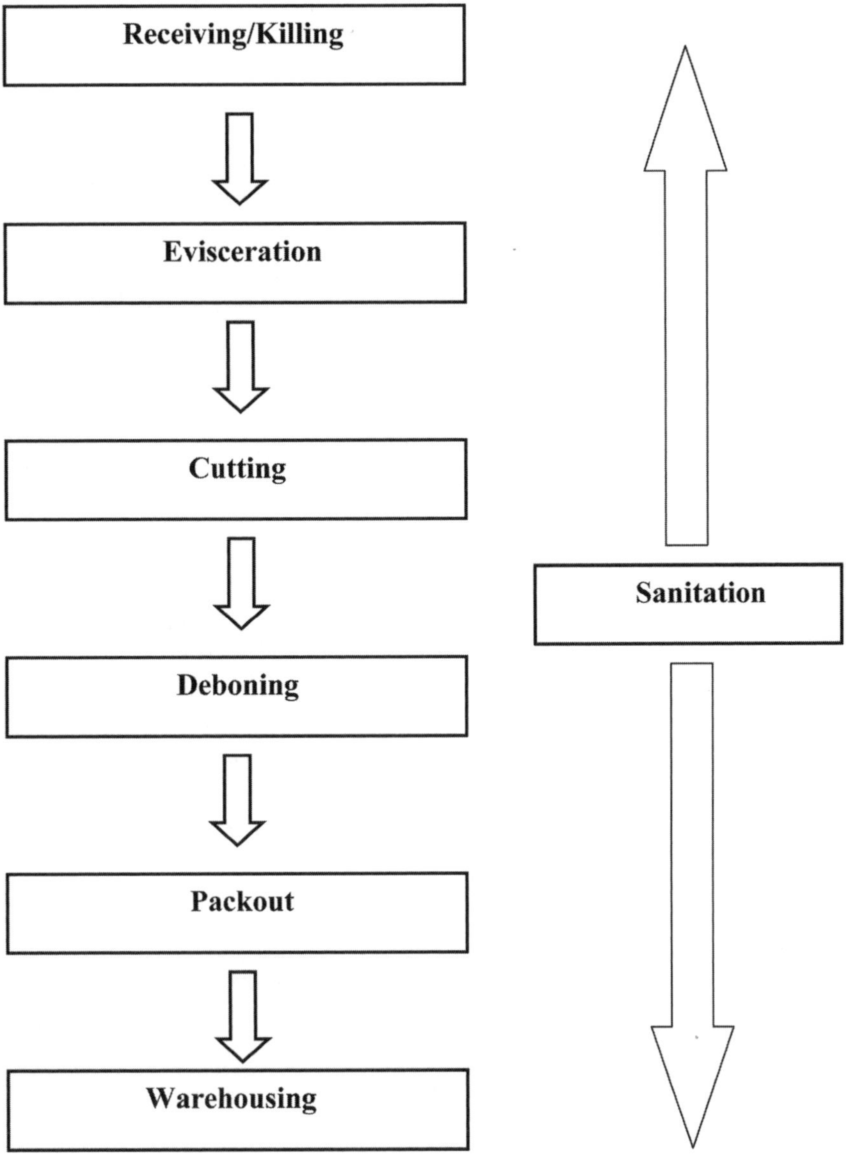

FIGURE 5.2
Poultry processing procedure.

use high-pressure water hoses and squeegees to clean the floors. This type of job is frequently an entry-level position. Workers who hold these jobs do not have the experience needed to be familiar with the many hazards of the equipment and the environment in which they work. They need extensive training.

Sanitation Worker Fatality Incident (#1)

A sanitation worker at a poultry plant was cleaning out a chiller. The motor that powers the paddles inside the chiller was accidentally turned on by a coworker who was cleaning a different chiller. Neither the main power control nor the chiller control box had been locked or tagged out in accordance with 29 CFR 1910.147, the control of hazardous energy (lockout/tagout). The employee was struck by and caught between the rotating paddle blades and the interior wall of the chiller. He died of severe chest injuries approximately twenty-eight hours after the accident.

The daily sanitation or cleanup crew has the responsibility of cleaning all product contact surfaces throughout the plant to comply with requirements of the Food Safety and Inspection Service (FSIS), USDA. If the cleanup crew has not done a satisfactory job, the FSIS inspector will not allow the plant to begin a production shift.

When the sanitation crew must remove guards or components to effectively clean processing equipment, and this action (or any other action) exposes crew members to hazardous energy, the equipment must be isolated from its energy source(s), and the energy isolation devices must be locked out or tagged out. In some situations, the equipment must be re-energized for a limited period of time for testing or repositioning purposes. During the testing or positioning period, a sequence of steps must be followed for the integrity of employee protection, and alternative protection, such as removing workers from the machine area, must be provided to prevent employee exposure to machine hazards. Once the testing/positioning activity is completed, the equipment again must be de-energized and locked or tagged out before undertaking further cleaning activities.

Depending on the part of the country in which the plant is located, the sanitation crew may be plant employees or they may work for a contractor. Again, they may be entry-level employees who need extra training to become familiar with the hazards of their work and ways to lessen the hazards.

An additional condition that may contribute to the hazardous nature of the work is that the crew may receive eight hours' pay regardless of how early they finish the job. This gives them an incentive to work as fast as possible and may even contribute to taking shortcuts, such as not locking or tagging out equipment.

The sanitation worker is exposed to most of the safety and health hazards throughout the plant, including:

- Cuts, lacerations, and amputations when removing blades from equipment.
- Being struck by, struck against, or caught in equipment, such as chiller paddles, when climbing over or under equipment.
- Slips, trips, and falls, whether from ladders or climbing on equipment, slipping on wet surfaces, or tripping over drain covers that have been removed and not replaced. Strains, sprains, and/or fractures may result.
- Electric shock, which is an increased risk in poultry processing plants because of the wet environment.
- Chemical hazards, such as cleaners, that can cause skin or eye irritation or burns.

Receiving and Killing—The receiving and killing operation is a largely automated process in most poultry plants and includes receiving live birds, killing, scalding, defeathering, and removing feet. This operation includes the following tasks:

Task 1: Forklift Operator—Moves the poultry cages from trucks to conveyor/dumping areas. This process moves from the outside to the inside of the dock area. Hazards of this task may include:

- **Collision**—forklifts are driven in heavily congested areas where many other operations are being performed. Visibility and sightlines can be limited, increasing the chance of collision between machinery and employees.
- **Defective forklift**—over a long period of time the solid rubber wheel on the front of the forklift may develop a flat spot, resulting in unstable loads and poor handling.
- **Slips and falls**—the employee is outside and is exposed to all weather conditions: rain, snow, ice, and heat, resulting in possible slips or falls from working/walking on slippery surfaces.
- **Strains, reaches, and cuts**—employees must release pull chains to remove chicken cages from trucks, resulting in strains from reaching and possible cuts.

Task 2a: Automatic Dumper Operator (Automatic Dumper System)—An employee operates the machinery that mechanically dumps live birds from catch-cages. After being removed from cages, the birds are moved by conveyor to the live hang area. Hazards of this task may include:

- **Awkward machine controls**—controls may be difficult to manipulate and/or improperly located. This may cause ergonomic stress on the hands, arms, shoulders, and upper back, and may result in injury.
- **Falls and catches by machinery**—frequent and rapid movement near heavy machinery increases the chance that employees may slip into the machine or be caught by or between moving parts. This is especially true if machinery is outside or exposed to wet or icy conditions.

Task 2b: Manual Back Dock Worker—Some facilities still use smaller catch-cages holding ten to twelve birds each that are normally emptied manually by the truck to the dock area by forklift. Workers then annually lift and remove cages from stacks and empty poultry out of cages. Birds are transported by conveyor from this station to the live hang area. Hazards of this task may include:

- **Lifting, bending, and reaching**—the worker must manually remove cages from stacks, and lift and tilt them to empty chickens from cages. The lifting, bending, and reaching causes stress to the back and shoulders.
- **Disease exposure**—as birds are dumped, feather danger and fecal debris may become airborne and inhaled by employees. Diseases associated with handling live chickens and contact with feces and dust include allergic alveolitis, cryptosporidiosis, histoplasmosis, hypersensitivity pneumonitis, psittacosis, and Newcastle disease.

Psittacosis

Psittacosis is caused by a bacterium, *Chlamydia psittaci*, which is transmitted to humans from birds. "Psittacine" birds, like parrots and parakeets, are classically responsible, although pigeons, chickens, and turkeys may carry the disease, as well. An infected bird may appear to have red, watery eyes; nasal discharge; diarrhea; and a poor appetite. After a bird recovers from infection, the bacteria may remain in the blood, feathers, and droppings for many weeks. Humans may acquire psittacosis by inhaling infected particles from bird droppings. Symptoms begin one to three weeks after exposure, and usually include headache, fever, and cough. A "flu-like" syndrome of nausea/vomiting, joint aches, and muscle aches is also common. Severe infection may develop into pneumonia that requires hospitalization. Psittacosis is treated with common antibiotics (doxycycline or erythromycin), though recovery may take several weeks. Sustained immunity to infection does not develop; some people have been reported to get the disease more than once. Less than 1 percent of all cases are fatal.

Task 3: Live Hang—Takes live birds from conveyors and hangs them in shackles. Hazards of this task may include:

- **Reaching to conveyor and shackles**—ergonomic hazards of reaching down to access birds on supply conveyor and reaching up to hang them on the shackle conveyor can lead to shoulder, back, and neck strain because of awkward postures and repetitive motion. The employees at the beginning of the line often work faster than those near the end of the line causing fatigue. Workers stand for long periods of time.
- **Disease exposure**—workers get covered with poultry mess and dust that can expose them to diseases associated with handling live chickens and contact with poultry feces and dust, such as allergic alveolitis, cryptosporidiosis, histoplasmosis, hypersensitivity pneumonitis, psittacosis, and Newcastle disease.
- **Poor lighting**—it is difficult for workers to see when lighting is reduced to calm the chickens. This lack of illumination contributes to slips, falls, and cuts, and makes inadequately guarded fans even more dangerous.
- **Standing for a long time**—standing for a long time can cause pain and strains in the legs and lower back. Common types of footwear worn in this area (e.g., rubber boots) do not provide much arch support.

Task 4: Kill Room Attendant (Backup Killer)—The kill room attendant monitors the automated poultry killing process and uses a knife to kill any birds missed by the machine. Hazards of this task may include:

- **Standing for a long time**—standing for a long time can cause pain and strains in the legs and lower back. Common types of footwear worn in this area (e.g., rubber boots) do not provide much arch support.
- **Falls, back injuries, and cuts**—employee stands in two to three inches of blood, which creates slippery floor conditions, sometimes resulting in a worker falling while holding a knife.
- **Blood on employee**—during the processing of birds in this area, blood may get in the worker's face and eyes, creating a hazard of infection and disease.
- **Restricted area**—employee is surrounded by equipment and product that may block sightlines where access from and to work areas is

restricted. Access points may not be obvious in cases of fire or other emergencies.
- **Ergonomic hazards from use of knives**—workers use a knife to perform this cutting task. Factors such as poorly fitting gloves, slick handles, inappropriately sized handles, or dull blades can increase the force that must be used. Minimize repetitive or prolonged exertion of finger force when performing cutting tasks, which can stress the tendons and tendon sheath of the hand.

Task 5: Picking Room Operator—A picking room operator ensures that equipment (e.g., stunners, scalders, picks, and conveyors) function properly. Most of the time on the job is spent walking around the equipment, performing a quality assessment of its operation. Hazards of this task may include:

- **Electrical shock**—workers in wet areas may contact electrical wires, causing electrical shock.
- **Strangulation and amputation**—employees may need to work around moving and unguarded equipment where an accident may result in possible amputations or strangulation.
- **Noise**—workers in this area are exposed to high noise levels from the surrounding machinery and processing equipment, which can result in hearing loss.

Task 6: Paw Room Grader—The paw room grader inspects and sorts product (feet) on conveyor. This is sometimes used as light duty position. Hazards of this task may include:

- **Standing for a long time**—standing for a long time reduces blood flow to the legs, forces isolated muscles to work for an extended period of time, and increases the risk of fatigue and varicose veins.
- **Prolonged bending of neck**—employees spend a long time looking down at a conveyor belt, which moves product past them. This can cause neck and shoulder pain and potentially carpal tunnel–like symptoms.

Evisceration—Evisceration processes remove the internal organs of the poultry. Hearts, livers, gizzards, and necks may also be cleaned and packaged in evisceration. This operation includes the following tasks:

Task 1: Rehang—After the carcass has been removed from the kill line by cutting off the feet, it is lifted from a conveyor shelf and rehung on shackles on the evisceration line for further processing. Hazards of this task may include:

- **Reaching to the conveyor and shackles**—employees bend and reach to lift chickens from supply conveyor and then reach out and away, sometimes above shoulder height, to place them on a shackle conveyor. Injuries to the shoulder, back, and neck are common due to awkward postures and high repetition. Employees at the beginning of the line often work faster than those near the end of the line because there is always a full supply of birds and all shackle positions are open.
- **Standing for a long time**—standing for a long time reduces blood flow to the legs, forces isolated muscles to work for an extended time, and increases the risk of fatigue and varicose veins.

Task 2: Opener (Vent Opener)—The opener uses scissors to manually cut open the bird. Most companies have eliminated this position by installing an automatic vent opener machine. Employees that serve as backup to the machine monitor the birds coming out of the machine and manually open any birds that may have been missed. Hazards of this task may include:

- **Ergonomic hazards from the use of scissors**—workers often use manual scissors that can cause ergonomic stress on the arms, hands, and fingers. Repeated opening of the jaws can irritate and inflame the tendons and sheaths of the hand. This is especially a problem if employees are positioned either too high or low in relation to the bird, such that the wrist is bent while finger force is exerted. The tendon and sheath can experience contact damage as they are pulled across the bones and ligaments of the wrists. Contact between the loop handles of the scissors and the sides of the fingers can damage nerves and blood vessels.
- **Reaching to the shackles**—workers are required to repeatedly reach to the shackles to access the bird so that various tasks can be performed. Reaching creates stress on the arms, shoulders, neck, and back because the weight of the arm and scissors must be supported.
- **Standing for a long time**—standing for a long time reduces blood flow to the legs, forces isolated muscles to work for an extended time, and increases risk of fatigue and varicose veins.

Task 3: Neck Breaker—The neck breaker uses a knife to cut the neck of the bird. Most companies have eliminated this position by installing an automatic neck breaking machine. Employees serve as backup to this machine. Hazards of this task may include:

- **Standing for a long time**—standing for a long time reduces blood flow to the legs, forces isolated muscles to work for an extended time, and increases risk of fatigue and varicose veins.

- **Reaching to the shackles**—neck breakers perform various tasks by reaching repeatedly to the shackles. Reaching creates stress on the arms, shoulders, neck, and back.
- **Ergonomic hazards from use of knives**—workers use a knife to cut the neck away from the body. The cutting motion may entail some bending of the wrist. Factors such as poorly fitting gloves, slick handles, inappropriately sized handles, or dull knives increase the force that must be used. Finger force and bending the wrist are recognized risk factors for developing many hand injuries; minimize these factors when performing cutting tasks.

Task 4: Oil Sack Cutter—The oil sack cutter cuts the oil sack from the birds. Most companies have eliminated this position by installing an automatic opening machine. Employees that serve as backup to the machine walk back and forth monitoring the procedure. Hazards of this task may include:

- **Standing for a long time**—standing for a long time reduces blood flow to the legs, forces isolated muscles to work for an extended time, and increases risk of fatigue and varicose veins.
- **Reaching to the shackles**—workers perform various oil sack cutter tasks by reaching repeatedly to the shackles. Reaching creates stress on the arms, shoulders, neck, and back.
- **Ergonomic hazards form use of knives**—workers use a knife to cut the oil sack. The cutting motion may entail bending the wrist. Factors such as poorly fitting gloves, slick handles, inappropriately sized handles, or dull knives increase the force that must be used. Finger force and bending of the wrist are recognized risk factors for developing many hand injuries; minimize these factors when performing cutting tasks.

Task 5: Arranger—The arranger, also called the presenter, removes the viscera from the body cavity and arranges them for USDA inspection. The initial removal is often accomplished by the automatic vent opening machine. Hazards of this task may include:

- **Pulling and turning viscera**—workers repeatedly pull the viscera from the body cavity with fingers and twist the forearm to present them for inspection. This process causes potential injury to both the wrist and elbow. The more the wrist is bent during this process, the greater the risk of injury.
- **Reaching to the shackles**—workers perform various arranger tasks reaching repeatedly to the shackles. Reaching creates stress on the arms, shoulders, neck, and back.

- **Standing for a long time**—standing for a long time reduces blood flow to the legs, forces isolated muscles to work for an extended time, and increases risk of fatigue and varicose veins.

Task 6: Giblet Harvester—The giblet harvester separates the heart, liver, and gizzard from the rest of the viscera and positions them to be cut by a saw. The heart, liver, and gizzard then fall to a wash table where an initial cleaning is performed and they are directed for further processing. Hazards of this task may include:

- **Reaching to the shackles**—workers are required to repeatedly reach to the shackles to access the bird so various tasks can be performed. Reaching creates stress on the arms, shoulders, neck, and back.
- **Standing for a long time**—standing for a long time reduces blood flow to the legs, forces isolated muscles to work for extended periods of time, and increases risk of fatigue and varicose veins.

Task 7: Gizzard Harvester—A gizzard harvester separates gizzards from other items in the viscera and directs the gizzard to the gizzard table. Hazards of this task may include:

- **Reaching across high and/or wide work surface**—employees repeatedly reach across a conveyor or worktable to obtain product for processing. Repetitive reaching stresses the shoulder and upper back and may require bending at the waist, which can stress the lower back.
- **Ergonomic hazards from use of knives**—workers use a knife to perform some trimming and cleaning functions. Most knives have a straight, in-line design. Using this type of knife on a horizontal cutting surface forces employees to bend their wrists to perform the cut. Bending the wrist while exerting finger force is stressful to the tendons and muscles of the hand and forearm. Factors such as poorly fitting gloves, slick handles, inappropriately sized handles, or dull knives increase the force that must be used. Finger force and bending the wrist should be minimized when performing cutting tasks.
- **Ergonomic hazards from use of scissors**—workers often use manual scissors that can cause ergonomic stress on the arms, hands, and fingers. Repeated opening of the jaws can irritate and inflame the tendons and sheaths of the hand. This is especially a problem if employees are positioned either too high or low in relation to the bird, such that the wrist is bent while finger force is exerted. The tendon and sheath can experience contact damage as they are pulled across the bones and ligaments of the

wrist. Contact between the loop handles of the scissors and the sides of the fingers can damage nerves and blood vessels.
- **Standing for a long time**—standing for a long time reduces blood flow to the legs, forces isolated muscles to work for an extended time, and increases risk of fatigue and varicose veins.

Task 8: Gizzard Table Operator—Gizzard table operators manually trim and clean gizzards. They then place gizzards in an automatic splitting machine so they are opened up and washed when they reach the gizzard peeler station. Hazards of this area may include:

- **Reaching across high and/or wide work surface**—employees repeatedly reach across a conveyor or worktable to obtain product for processing. Repetitive reaching stresses the shoulder and upper back and may require bending at the waist that can stress the lower back.
- **Ergonomic hazards from use of knives**—workers use a knife to perform some trimming functions. Most knives have a straight, in-line design. Using this type of knife on a horizontal cutting surface forces employees to bend their wrists to perform the cut. Bending the wrist while exerting finger force is stressful to the tendons and muscles of the hand and forearm. Factors such as poorly fitting gloves, slick handles, inappropriately sized handles, or dull knives increase the force that must be used. Finger force and bending of the wrist should be minimized when performing cutting tasks.
- **Ergonomic hazards from use of scissors**—workers often use manual scissors that can cause ergonomic stress on the arms, hands, and fingers. Repeated opening of the jaws can irritate and inflame the tendons and sheaths of the hand. This is especially a problem if employees are positioned either too high or low in relation to the bird, such that the wrist is bent while finger force is exerted. The tendon and sheath can experience contact damage as they are pulled across the bones and ligaments of the wrist. Contact between the loop handles of the scissors and the sides of the fingers can damage nerves and blood vessels.
- **Standing for a long time**—standing for a long time reduces blood now to the legs, forces isolated muscles to work for an extended time, and increases risk of fatigue and varicose veins.

Task 9: Gizzard Table-Peeler Operator—The employee presses the inside of gizzard against a rotating drum with a raspy surface that peels the inner lining from the gizzard. The employee feeds peeled gizzards to bagging area. Hazards of this task may include:

- **Hands/fingers getting caught by rollers**—as workers move gizzards over peeler, fingers may get caught in rollers.
- **Reaching across high and/or wide work surface**—employees repeatedly reach to pull product to the peeler. Repetitive reaching stresses the shoulder and upper back.
- **Standing for a long time**—standing for a long time reduces blood flow to the legs, forces isolated muscles to work for an extended time, and increases risk of fatigue and varicose veins.
- **Prolonged and forceful finger exertion**—employees use finger force to press gizzards against rotating drums for the majority of the task. Significant finger force must be exerted since the product is small and slippery. The wrist is usually bent during this process to place gizzards in the proper alignment. Exerting finger force of a prolonged time can stretch and fray tendons. Bending the wrist while exerting finger force can create further damage to tendons and their sheath, which can lead to injuries of the hand, wrist, and elbow.

Task 10: Heart and Liver Cutter/Inspector—An employee washes and visibly inspects hearts and livers before they are sent to the bagging station. Hazards of this area may include:

- **Reaching across high and/or wide work surface**—employees repeatedly reach across a conveyor or worktable to obtain product for processing. Repetitive reaching stresses the shoulder and upper back and may require bending at the waist, which can stress the lower back.
- **Ergonomic hazards from use of knives**—employees use in-line, straight knives to clean and trim. Using this type of knife on a horizontal cutting surface forces the employees to bend their wrists to perform the cut. Bending the wrist while exerting finger force is stressful to the tendons and muscles of the hand and forearm. Factors such as poorly fitting gloves, slick handles, inappropriately sized handles, or dull knives increase the force that must be used. Finger force and bending of the wrist should be minimized when performing cutting tasks.
- **Ergonomic hazards from use of scissors**—workers often use manual scissors that can cause ergonomic stress on the arms, hands, and fingers. Repeated opening of the jaws can irritate and inflame the tendons and sheaths of the hand. This is especially a problem if employees are positioned either too high or low in relation to the bird, such that the wrist is bent while finger force is exerted. The tendon and sheath can experience contact damage as they are pulled across the bones and ligaments of the wrist. Contact between the loop handles of the scissors and the sides of the fingers can damage nerves and blood vessels.

- **Standing for a long time**—standing for a long time reduces blood flow to the legs, forces isolated muscles to work for an extended time, and increases risk of fatigue and varicose veins.

Task 11: Bagger—A bagger bulk packs hearts, livers, or gizzards into various-sized bags before the bags are placed into boxes for shipment. Baggers may also repack giblets (heart, liver, gizzard, neck) into small paper bags for reinsertion into body cavity of whole birds. Some operations also require these employees to move the filled bag to a sealer where the bag is closed with a clip or heat seal. Hazards of this task may include:

- **Repetitive pinch grips**—employees secure and hold bags using a one- or two-finger pinch grip when removing it from the bagging fixture, transporting it to the bag sealer, and feeding it into the bag sealer. Using pinch grip places significant stress on the tendons of the fingers, which can lead to injuries of the hand, wrist, and forearm.
- **Reaching across or up to high work surfaces**—employees repeatedly reach to bins or across tabletops to obtain product for bagging and place product in bags. Repetitive reaching stresses the shoulder and upper back.
- **Standing for a long time**—standing for a long time reduces blood flow to the legs, forces isolated muscles to work for an extended time, and increases risk of fatigue and varicose veins.

Task 12: Lung Vacuumer—a lung vacuumer uses a small suction device to remove the lungs and the kidneys from the body cavity. Hazards of this task may include:

- **Awkward hand/arm postures**—worker must bend wrist and/or pull elbow away from the body to position vacuum into body cavity to remove the lungs and kidneys.
- **Reaching to the shackles**—workers are required to reach to the shackles to access the bird, thus causing ergonomic stress on the arms, shoulders, neck, and back.
- **Standing for a long time**—standing for a long time reduces blood flow to the legs, forces isolated muscles to work for an extended time, and increases risk of fatigue and varicose veins.

Task 13: Backup Eviscerator/Inspector—the backup eviscerator is a final product inspector who feels inside the carcass and looks for any remaining pieces of viscera before removing the carcass from the shackle. Hazards of this task may include:

- **Reaching to the shackles**—workers access birds by reaching to the shackles, thus causing ergonomic stress on the arms, shoulders, neck, and back.
- **Wrist deflection**—when employees are too low in relation to the bird, they must reach up to access the body cavity resulting in wrist bending. This can result in tendon and nerve damage, leading to pain and numbness in the hand, wrist, or elbow.
- **Standing for a long time**—standing for a long time reduces blood flow to the legs, forces isolated muscles to work for an extended time, and increases risk of fatigue and varicose veins.

Support Task: Rework Floor Person—A rework floor person manually reworks damaged or improperly processed items. This may include trimming, washing, and salvaging parts. Often, employees receive work from tubs and then replace them onto the shackle. Hazards of this task may include:

- **Bending at the waist to reach into tub**—repeatedly bending forward and reaching out away from the body stresses the back even if there is little being lifted because the upper body must be supported. When loads are being lifted, bending over at the waist increases the distance the load is held away from the body and increases the stress placed on the back.
- **Reaching to the shackles**—workers must reach to the shackles to place reworked produce for further processing. Reaching creates stress on the arms, shoulders, neck, and back.
- **Ergonomic hazards from use of knives**—workers use a knife to perform some trimming and cleaning functions. Most knives have a straight, in-line design. Using this type of knife on a horizontal cutting surface forces the employees to bend their wrists to perform the cut. Bending the wrist while exerting finger force is stressful to the tendons and muscles of the hand and forearm. Factors such as poorly fitting gloves, slick handles, inappropriately sized handles, or dull knives increase the force that must be used. Minimize finger force and bending of the wrist when performing cutting tasks.
- **Ergonomic hazards from use of scissors**—workers may use scissors to trim and clean product. Scissors can cause ergonomic stress to the hands and fingers, which results in nerve and tendon damage to the hand and forearm.
- **Reaching across high and/or wide work surface**—employees repeatedly reach across a conveyor or worktable to obtain product for processing. Repetitive reaching stresses the shoulder and upper back, and may require bending at the waist, which can stress the lower back.

Support Task: Ice Attendant—The ice attendant manually brings ice from the icehouse to the packing line, paw room, and other areas as needed. Usually the ice is transported in tubs. Hazards of this task may include:

- **Slips, trips, and falls**—workers are standing on wet floors that may have bird skin, bird parts, and ice on them, creating a slipping hazard. Metal drain covers on the floor are also very slippery and pose a hazard. A falling worker may contact dangerous equipment.
- **Moving heavy tubs of ice**—employees manually push tubs of ice. Pushing tubs, especially when on slick or icy floors, stresses the back, shoulders, ankles, and knees.
- **Shoveling loads of ice**—employees support a load that can easily weigh fifteen pounds from the end of a shovel handle. In a manner similar to that encountered on a child's teeter-totter, leverage can increase the effect of this load by two to four times, depending on the length of the shovel handle. Additionally employees may need to repeatedly bend at the waist to scoop from the bottom of the tubs and may need to lift ice above head height. The back and shoulders can be negatively affected by these motions.

Cutting and Deboning—After a chicken has been eviscerated and cleaned, it is prepared for packaging as a whole bird, or it may enter one of the two processes: 1) the cutting process for preparation of a bone-in product, or 2) the cutting and deboning process for preparation of bone-out products.

Process 1: Cutting—In the cutting process, the wings and legs/thighs are removed from the carcass and the back is cut away from the breast. Bones are not removed. At this point parts can be packaged as a consumer product, bulk-packed for delivery to other processors, or shipped to other parts of the plant for further processing.

Task 1: Line Loader—Birds are often transported from the evisceration line to the cone conveyor or line in a tub. Line loaders grasp two to three birds in each hand and lift them from the tub and place them on a conveyor or staging shelf, which is generally waist- to shoulder-height. Other personnel usually place the birds on the cone or shackle. Hazards of the task may include:

- **Bending at the waist to reach into tubs**—repeatedly bending forward and reaching out away from the body stresses the back even if there is little being lifted because the upper body must be supported. When loads are being lifted, bending over at the waist increases the distance the load is held away from the body and increases the stress placed on the back.
- **Forceful gripping**—employees lift multiple birds at one time, usually by the legs. Lifting two to three birds in each hand is not uncommon. Birds

are cold and slick, and employees usually wear rubber gloves that are also slick and may not fit well. All these factors increase the finger force that must be exerted. Exerting significant finger force can stretch and fray the tendons of the hand and can create a contact trauma to the tendon and sheath where they come in contact with bone or tendon. These types of actions increase the risk of tendonitis and carpal tunnel syndrome.

Task 2: Tail Cutter—The tail is cut from the bird before the bird is placed on the cone. Standard scissors are generally used to perform the operation. Hazards of this task may include:

- **Ergonomic hazards from use of scissors**—using traditional scissors forces the fingers to repeatedly open and close the blade, which can stress tendons, increasing the risk of tenosynovitis and carpal tunnel. Contact trauma to sides of fingers can damage nerves, which can cause numbness and tingling in the tips of the fingers and thumb.
- **Standing for a long time**—standing for a long time reduces blood flow to the legs, forces isolated muscles to work for an extended time, and increases risk of fatigue and varicose veins.

Task 3: Saw Operator—Employees may use a saw with a manual feed to cut legs/thighs or wings away from the main carcass, or may load a machine that automatically performs cuts. Manual feed saws can be used to remove legs from the back, divide the legs, cut wings away from the breast, and split the breast in two. After being loaded, automated machines perform the same cuts as described above. Hazards of this task may include:

- **Reaching to access product, saws, or machine-load areas**—employee reaches repeatedly to conveyor or shelf to obtain birds for processing. Reaches are also necessary to place birds into the automatic saw feed mechanism and perform manual cuts. Repetitive reaching stresses the shoulder and upper back.
- **Cuts and lacerations**—the nature of this task involves employees working with unguarded saws. Cuts, lacerations, and amputations are possible.
- **Standing for a long time**—standing for a long time reduces blood flow to the legs and forces isolated muscles to work for extended periods of time. This increases the risk of fatigue and varicose veins.

Task 4: Rehang—Rehang is generally not necessary since most cutting is performed on a cone line. If the cutting is to be performed from a shackle conveyor, the bird must be rehung. Some automated cutters, such as a "multi cut" machine, must be loaded, and this is technically a rehang type of activity. Hazards of this task may include:

- **Reaching up, forward, or to the side to access the shackle**—employees may bend to lift chickens from the supply conveyor and then reach out and away, sometimes above shoulder height, to place them on multi-cut machines or shackle conveyors. Injuries to the shoulder, back, and neck are common due to awkward postures and high repetition. Employees at the beginning of the line often work faster than those near the end of the line because there is always a full supply of birds and all positions are open.
- **Standing for a long time**—standing for a long time reduces blood flow to the legs, forces isolated muscles to work for an extended time, and increases risk of fatigue and varicose veins.

Task 5: Cone Line Feeder—Most plants use a cone line as the main staging area for removing appendages and meat from the body of the bird. The feeder places the eviscerated carcass onto the cone, which is integrated into a conveyor line. This line moves the bird past employees who remove parts from the carcass.

In some plants, parts are removed from birds hanging from a shackle conveyor. The process may be automated using multi-cut machinery. In these cases the cone line feeder is replaced by a rehang worker. Hazards of this task may include:

- **Reaching**—employees repeatedly reach to a conveyor or shelf to obtain birds for processing and reach to place birds on the cone. Repetitive reaching stresses the shoulder and upper back.
- **Standing for a long time**—standing for a long time reduces blood flow to the legs, forces isolated muscles to work for an extended time, and increases risk of fatigue and varicose veins.

Task 6: Wing Cutter—Wing cutters use knives to cut the wings from the bird. This may be a multistep process where several workers along the line each perform one of the necessary cuts, or all cuts can be done by a single operator. Hazards of this task may include:

- **Ergonomic hazards for use of knives**—workers use a knife to cut the wings away from the rest of the carcass. The cutting motion may entail some bending of the wrist. Factors such as poorly fitting gloves, slick handles, inappropriately sized handles, or dull knives increase the force that must be used. Finger force and bending of the wrist are recognized risk factors for the development of many hand injuries. Minimize these factors when performing cutting tasks.

- **Cuts and lacerations**—employees are performing highly repetitive tasks using knives close to other employees. Cuts and lacerations are possible to the employee and those standing nearby because employees are exposed to sharp knife blades. Any cut not treated at once will normally become infected as a result of working with poultry.
- **Reaching**—employees repeatedly reach to the bird on the cone to perform cutting tasks and may need to reach to a bin or a tub to deposit removed item. Repetitive reaching stresses the shoulder and upper back.
- **Standing for a long time**—standing for a long time reduces blood flow to the legs, forces isolated muscles to work for an extended time, and increases risk of fatigue and varicose veins.

Task 7: Leg/Thigh Cutter—cutters use knives to cut the legs/thigh unit from the bird. This may be a multistep process where several workers along the line each perform one of the necessary cuts, or all cuts can be done by a single operator. Hazards of this task may include:

- **Ergonomic hazards from use of knives**—workers use a knife to cut the wings away from the rest of the carcass. The cutting motion may entail some bending of the wrist. Factors such as poorly fitting gloves, slick handles, inappropriately sized handles, or dull knives increase the force that must be used. Finger force and bending of the wrist are recognized risk factors for the development of many hand injuries. Minimize these factors when performing cutting tasks.
- **Cuts and lacerations**—employees are performing highly repetitive tasks using knives close to other employees. Cuts and lacerations are possible to the employee and those standing nearby because employees are exposed to sharp knife blades. Any cut not treated at once will normally become infected as a result of working with poultry.
- **Reaching**—employees repeatedly reach to the bird on the cone to perform cutting tasks and may need to reach to a bin or a tub to deposit removed item. Repetitive reaching stresses the shoulder and upper back.

Task 8: Back/Breast Separator—Employees may use a saw with a manual feed to separate the breast section form the back. This manual feed technique can be used to remove legs from the back, divide the legs, cut wings away from the breast, and split the breast in two. After being loaded, automated multi-cut machines perform the same cuts as described above. Hazards of this task may include:

- **Reaching to access product, saws, or machine-load areas**—employee repeatedly reaches to conveyor or shelf to obtain birds for processing. Repetitive reaching stresses the shoulder and upper back.
- **Cuts and lacerations**—the nature of this task involves employees working with an unguarded saw. Cuts, lacerations, and amputations are possible.
- **Standing for a long time**—standing for a long time reduces blood flow to the legs, forces isolated muscles to work for an extended time, and increases risk of fatigue and varicose veins.

Task 9: Trimmer/Cleanup—Employee obtains separated pieces of poultry from conveyor and uses scissors to trim excess skin, fat, and pieces of bone. Hazards of this task may include:

- **Ergonomic hazards for use of scissors**—use of traditional scissors forces the fingers to repeatedly open and close the blade, which can stress tendons, increasing the risk of tenosynovitis and carpal tunnel. Contact trauma to sides of fingers can damage nerves, which can cause numbness and tingling in the tips of the fingers and thumb.
- **Standing for a long time period**—standing for a long time reduces blood flow to the legs, forces isolated muscles to work for extended periods of time, and increases risk of fatigue and varicose veins.
- **Reaching hazard**—employees repeatedly reach to conveyor or shelf to obtain parts to be trimmed and reach to place finished parts in tubs or baskets. Repetitive reaching stresses the shoulder and upper back.

Support Task: Knife Person—A knife person collects dull knives from employees along the processing lines and replaces them with sharp ones. This employee may also sharpen knives that have been collected. Hazards of the task may include:

- **Slips, trips, and falls**—workers walk all over the facility on wet floors that may have bird skin, bird parts, and ice on them, creating a slipping hazard. Metal drain covers on the floor are also very slippery and pose a hazard. A falling worker may contact dangerous equipment or cut him/herself on a knife blade.
- **Hazards from use of grinders**—employees may suffer cuts, lacerations, skin abrasions, contusions, or eye damage during use of grinders to sharpen knives. Grinding wheels may break up or explode. Bits and pieces of knife blades may be thrown off during sharpening.

Process 2: Deboning—All parts of the chicken that have sufficient meat are candidates for a bone-out process. The deboning process generally follows the cutting operation when deboning legs and thighs. The breast meat, however, may be removed while the carcass is still on the cone at the end of the cutting process.

Did You Know?

In some facilities, the birds are aged in a cooler before the meat is separated from the bone.

While most deboning cuts are performed with a knife, some of the processes, such as removal of the meat from the bone of the leg, can be more easily and safely accomplished by using other cutting tools such as a Whizzard knife. Trimming tasks in this process are generally performed with scissors.

After the meat has been removed from the bone, it may be quick-frozen and bagged with little additional trimming or processing, or it may be moved to a separate specialty trim line where it is trimmed and cut according to customer requirements.

Task 1: Skin Puller—The employee uses pliers or a similar tool to pull skin from breasts, thighs, and legs. Hazards of this task may include:

- **Standing for a long time**—standing for a long time reduces blood flow to the legs, forces isolated muscles to work for an extended time, and increases risk of fatigue and varicose veins.
- **Forceful hand exertions**—employees exert high finger force with both hands to open and close the tool and to hold the product while the skin is being pulled. Using gloves and handling cold product increases the amount of finger force that must be exerted. Repeatedly exerting high finger force can stretch and fray the tendon if there are not sufficient periods of rest. Repeatedly stretching the tendon can lead to tendinitis or tenosynovitis. Using standard in-line tools can cause employees to bend the wrist, which, in combination with high finger force, can cause contact trauma between the tendon and the bones and ligaments of the wrist. Contact between these entities can cause irritation and inflammation, leading potentially to tendinitis, tenosynovitis, and carpal tunnel syndrome.

Task 2: Line Loader—Parts are transported from the cutting stations to the deboning stations in hand-carried tubs. Tubs may be lifted from cart shelves or lifted from the floor and carried to deboning stations. Tubs are emptied onto a conveyor or staging shelf, which is generally at waist- to shoulder-height. This task may also be repeated at the trimming lines where product that has been deboned may again be placed in a tub and transported to the trimmers. Hazards of the task may include **bending at the waist to lift tubs of product**—repeatedly bending forward and reaching out away from the body stresses the back even if there is a little weight being lifted because the upper body must be supported. When loads are being lifted, bending over at the waist increases the distance the load is held away from the body and increases the stress placed on the back. Bending and lifting heavy loads such as those encountered at these stations greatly increases the risk of injury to the low back.

Task 3: Deboner—Employees remove bones from various poultry parts, including breasts, thighs, and legs. The task may be performed with a standard knife on a flat cutting surface or a tilted cutting surface. Legs may be deboned on a specialized conveyor line using a Whizzard knife. Breast meat may be removed directly from the carcass while it is still on the cone line. Hazards of this task may include:

- **Ergonomic hazards from use of knives**—workers use a knife to cut the meat away from the bone. Most knives have a straight, in-line design. Using this type of knife on a horizontal cutting surface forces employees to bend their wrists to perform the cut. Bending the wrist while exerting finger force is stressful to the tendons and muscles of the hand and forearm. Factors such as poorly fitting gloves, slick handles, inappropriately sized handles, frozen meat, or dull knives increase the force that must be used. Minimize finger force and bending of the wrist when performing the cutting task.
- **Cuts and lacerations**—employees are performing highly repetitive tasks using knives close to other employees. Cuts and lacerations are possible to the employee and those standing nearby because employees are exposed to sharp knife blades. Any cut not treated at once will normally become infected as a result of working with poultry.
- **Reaching**—employees repeatedly reach to a conveyor or shelf to obtain parts for deboning and reach to place finished product in tubs or receptacles. Repetitive reaching stresses the shoulders and upper back.
- **Standing for a long time**—standing for a long time reduces blood flow to the legs, forces isolated muscles to work for an extended time, and increases risk of fatigue and varicose veins.

Task 4: Tender Puller—Tender pullers use their fingers to pull tenders away from the breastbone after the main section of breast has been removed. This task may be performed while the carcass is on the cone line or from a cut full breast placed on flat work surface. Hazards of this task may include:

- **Reaching**—employees repeatedly reach to conveyor or shelf to obtains breasts, reach to cones to pull tenders, or reach to place finished product in tubs or receptacles. Repetitive reaching stresses the shoulders and upper back.
- **Standing for a long time**—standing for a long time reduces blood flow to the legs, forces isolated muscles to work for an extended time, and increases risk of fatigue and varicose veins.

Task 5: Trimmer—Trimming is usually the last processing step before packaging or quick freezing. Trim lines often produce specialty products according to customer specification. Trimmers remove pieces of bone, fat, tendons, gristle, or blemishes in the meat as well as perform specialty cutting to produce tenders and nuggets. Many of the items, such as bone and fat, that are easily grasped can be pulled away from the meat using only the fingers. Although a knife can be used, for the majority of these operations the tool of choice is usually scissors. Hazards of this task may include:

- **Ergonomic hazards from use of scissors**—use of traditional scissors forces the fingers to repeatedly open and close the jaws, which can stress tendons, increasing the risk of tenosynovitis and carpal tunnel. Contact trauma to the sides to fingers can damage nerves, which can cause numbness and tingling in the tips of the fingers and thumb.
- **Standing for a long time**—standing for a long time reduces blood flow to the legs, forces isolated muscles to work for an extended time, and increases risk of fatigue and varicose veins.
- **Reaching**—employees repeatedly reach to a conveyor or shelf to obtain parts for trimming and reach to place finished product in tubs or receptacles. Repetitive reaching stresses the shoulders and upper back.

Task 6: Quality Control Inspector—The employee pulls selected processed parts from the conveyor line and visually inspects them for compliance with quality standards. Hazards of the task may include:

- **Reaching**—employees repeatedly reach to a conveyor or shelf to obtain parts for inspection. Reaches are also necessary to place inspected parts in tubs or back on the line. Repetitive reaching stresses the shoulders and upper back.
- **Standing for a long time**—standing for a long time reduces blood flow to the legs, forces isolated muscles to work for an extended time, and increases risk of fatigue and varicose veins.

Support Task: Knife Person—A knife person collects dull knives from employees along the processing lines and replaces them with sharp ones. This employee may also sharpen knives that have been collected. Hazards of this task may include:

- **Slips, trips, and falls**—workers walk all over the facility on wet floors that may have bird skin, bird parts, and ice on them, creating a slipping hazard. Metal drain covers on the floor are also very slippery and pose a hazard. A falling worker may contact dangerous equipment, or cut him/herself on a knife blade.
- **Hazards from use of grinders**—employees may suffer cuts, lacerations, skin abrasions, contusions, or eye damage during use of grinders to sharpen knives. Grinding wheels may break up or explode. Bits and pieces and knife blades may be thrown off during sharpening.

Packout—Packaging is necessary to get the processed product from the plant to the consumer. It is generally a two-part procedure. First, the bird or bird parts are placed in a bag or package; and second, the package is placed in a shipping box. Poultry can be packaged in a wide variety of formats that range from minimal processing to maximum processing. Packaging options are:

- **Whole Bird Bulk Packaging**: The whole bird can be bulk-boxed and sent to large users such as broiler restaurants or secondary processors.
- **Whole Bird Individual Packaging**: The whole bird is individually bagged and boxed for supermarket sale.
- **Bone-In Product**: Parts are packaged and sold as consumer product or as bulk sale for large commercial users.
- **Bone-Out Product**: Parts are packaged and sold as consumer product or as bulk sale for large commercial users.

No matter how a bird is packaged, it is almost always placed in a large cardboard box for shipping. The steps in building, filling, weighing, and stacking these boxes is almost always the same.

Task 1: Box Maker—The box maker is responsible for building boxes out of flat box stock and providing them to the stations where they will be filled with product. This task may be automated. Hazards of this task may include:

- **Congested work area**—problems can occur at this task because it is often an unplanned workstation, originally intended to be temporary. Placing such a workstation in a process line without adequate planning and space allocation can result in a highly congested work area, creating the potential for injury from working in awkward postures and coming in contact with other moving machinery.

- **Hot glue burns, splashes**—problems can occur at this task station from exposure to hot glue as boxes are made. There is potential for employee burns and splashing hot glue when adding glue sticks to the pot, or when cleaning the machine.
- **Poor access to emergency exits**—box-making operations are often placed in remote and hard-to-get-to locations, which makes access to emergency exits difficult. Quick and easy access to emergency exits from any room or workstation is very important to the safety of the employee. Without it, employees could be trapped or killed if unable to reach emergency exits during an emergency.
- **Bending and reaching**—employees are required to repeat the same motion over and over, (i.e., reaching and or bending to obtain box stock and placing finished boxes on conveyor), which can result in work-related musculoskeletal disorders.

Task 2: Box Packer—Packing boxes involves taking packed trays, bags, or whole birds, from the conveyor and depositing them in a box. Employees generally work at a boxing station located next to a conveyor, which supplies product to be boxed. Hazards of this task may include:

- **Repetitive reaching and lifting**—employees repeatedly reach to a conveyor to obtain product for processing and may reach to place product in the box. Repetitive reaching stresses the shoulders and upper back.
- **Standing for a long time**—standing for a long time reduces blood flow to the legs, forces isolated muscles to work for an extended time, and increases risk of fatigue and varicose veins.

Task 3: Scale Operator—A scale operator pulls filled boxes from a conveyor and places them onto scales. The operator then adds or removes product until the desired weight is achieved. This employee usually affixes a label showing the box weight before placing the weighed box back on the conveyor. Hazards of this task may include:

- **Lifting heavy loads**—employees may experience back strain from reaching, twisting, and bending if the task requires them to lift boxes, which may weigh forty to eighty pounds, from the conveyor to the scales.
- **Standing for a long time**—standing for a long time reduces blood flow to the legs, forces isolated muscles to work for an extended time, and increases risk of fatigue and varicose veins.

Task 4: Box Sealer—After weighing takes place, the box sealer adds ice as required and places a lid on the box. The sealer pushes the box down the conveyor to the stack off employee. Hazards of this task may include **standing for a long time**—standing for a long time reduces blood flow to the legs, forces isolated muscles to work for an extended time, and increases risk of fatigue and varicose veins.

Task 5: Stack Off—The stack off employee is responsible for removing boxes, which may weigh forty to eighty pounds, from the conveyor and stacking on pallets. Palletized loads are then stored or loaded onto trucks. Hazards of this task may include:

- **Bending and twisting while lifting heavy loads**—stacking full boxes forces employees to bend the torso forward to place the box on the pallet, or to lift the box up to or above head height to place it on the top of the stack. This can result in back injuries, since boxes can weigh up to eighty pounds.
- **Slips, trips, and falls**—employees must often move pallets of product with a hand jack while stepping on wet, slippery surfaces, resulting in possible slips and falls.

Warehousing—Once the chicken is packed in its shipping container, it is moved from the processing floor. Options are moving to a truck for immediate shipment or placement in a warehouse for storage. Lifting and moving heavy loads using awkward body postures is common practice. This operation includes the following tasks:

Task 1: Forklift/Pallet Jack Operators: Operators move loaded pallet around the worksite and onto trucks. Hazard of this task may include:

- **Collision**—forklifts are driven in heavily congested areas where many other operations are being performed. Visibility and sightlines can be limited, increasing the chance of collisions between machinery and employees.
- **Defective forklift**—over long periods of time the solid rubber wheel on the front of the forklift may develop a flat spot, resulting in unstable loads and poor handling.
- **Being struck by falling loads**—there is a potential for pallets to collapse or tip on the driver; uneven floor areas increase the danger.
- **Fumes from battery charging**—forklift batteries need to be removed by hoist and routinely charged. Exposure to fumes or vapors from battery-charging process can be dangerous.
- **Hazards from working with pallet jacks**—while using hand or electric jacks, employees may have the problems of backing into walls, running

over own feet, or running into other employees as they move through freezer doors. It is difficult to see who or what is on the other side of the doors, because the windows may be high and covered with frost.

Task 2: Freezer/Cooler Worker: The worker manually loads/unloads boxes of frozen product on warehouse shelves. Occasional palletizing of loads may be necessary. Hazards of this task may include:

- **Bending and twisting while lifting heavy loads**—stacking boxes over head height forces employees to bend the torso forward to place the box on the pallet, or to lift the box up to or above head height to place it on the top of the stack. Even when boxes are moved by conveyors directly to refrigerated trucks, employees remove the boxes from the conveyor and then stack them in the truck above head height or at floor level. This can result in back and shoulder injuries.
- **Exposure to cold environments**—employees working in cold environments must wear additional clothing, which can restrict movement and increase the force they must exert when performing lifting operations. Additionally, employees burn more energy in these environments to keep warm, so fatigue may occur more rapidly, which increases the risk of injury. Cold areas may also have ice forming on work surfaces where slipping can cause injury of strain, especially if a load is being carried when the slip occurs.

Preserved Fruits and Vegetables Industry

The preserved fruits and vegetables (or canned products) industry includes not only canned and frozen fruits and vegetables but also canned and frozen specialties, dried and dehydrated fruits and vegetables, pickles, and salad dressings. It is important to point out that the canned products of this industry are distinguished by their processing rather than by the container.

OSHA (2007) points out that the food processing industry occupies a powerful position within the food and fiber system. The industry has been likened to the center of an hourglass: raw agricultural commodities from more than two million farms and ranches flow through roughly 20,000 processors, which in turn sell their array of processed products to more than half a million food wholesalers and retailers. Over a hundred million domestic households consume the meat and dairy products, canned and frozen fruit and vegetables, milled grains, bakery products, beverages, and seafood.

The importance of food processing lies in its various economic functions. Foremost, processors convert food materials into finished, consumer-ready products using labor, machinery, energy, and management. They employ handling, manufacturing, and packaging techniques to add economic value to raw commodities harvested from the farm or the sea. Virtually all agricultural products are processed to some degree before reaching consumers. The value added varies by commodity: steers become meat, potatoes are turned into french fries, wheat is made into flour, apples become juice or sauce, and fresh salmon emerges as canned salmon. The farm value of fruit and vegetable products at the retail level—frozen peas, for instance—is about 20 percent. Thus, 80 percent of the retail value is "added" to the raw product during processing distribution.

Processors serve as middlemen within the food system. Consumer demand and agricultural supply information come together at the food processing center. For instance, a tight supply of frozen corn at the retail level is eventually translated into higher processor prices, a greater willingness to pay for key inputs, and a price signal to farmers to expand production or sell off their stored crop. In contrast, an unexpectedly short crop induces processors to raise their prices to retailers and distributors, which subsequently prompts a decrease in consumer demand.

The canning of fruits and vegetables is a growing, competitive industry, especially the international export portion. The industry is made up of establishments primarily engaged in canning fruits, vegetables, and fruit and vegetable juices; processing ketchup and other tomato sauces; and producing natural and imitation preserves, jams, and jellies.

Along with industry growth, technology and automation have had a large impact on the fruit and vegetable processing business. For instance, much of the manual, labor-intensive work of the past has been replaced by machines and computer-operated product processing programs that control machine operation. This does not mean that all humans in the industry have been replaced by machines and computers. Instead, it means that the total number of humans employed today has been reduced to perform tasks that in the past required a much larger workforce.

Again, it is important to point out that food manufacturing is an industry marked by a high rate of on-the-job injury and/or illness. This is the case even though, as mentioned, significant technological upgrades and computerized operation have been implemented. Even with the reduced number of workers per plant, the fruit and vegetable processing industry continues to contribute to this high on-the-job injury/illness rate. Many of the potential hazards and their sources in the fruit and vegetable processing industry are the same as many of the other sectors making up the food processing industry as a whole. For example, typical hazards include being struck by falling objects; being caught at point of operation in conveyors; slips, trips, and falls; being struck by flying objects from box staple machines; contact with toxic or noxious substances released from lift trucks and other sources; and noise generated from conveyors and other machinery.

Important Note: To be able to recognize and understand the extent of potential hazard exposure in the fruit and vegetable processing industry, it is important to be familiar with the actual fruit and vegetable preservation (production) process. This is the approach USEPA (1995) takes in their canned fruits and vegetable process description that we describe below. Keep in mind that USEPA primarily points out potential emission hazards affecting the environment and workers involved in the processing procedures. While certain process emissions certainly can be hazardous both to the environment and to workers' safety and well-being, it is our purpose in this text to provide an overview of several different exposures, that is, potential safety and health hazards that could be present during the processing of fruit and vegetables.

Did You Know?

According to OSHA (2007), more than 46 percent of injuries and illnesses reported annually in the fruit and vegetable preservation industry consist of sprains, strains, and bruises.

Fruit and Vegetable Preservation Process Description

The primary objective of food processing is the preservation of perishable foods in a stable form that can be stored and shipped to distant markets during all months of the year. Processing also can change foods into new or more usable forms and make foods more convenient to prepare (USEPA, 1995).

The goal of the canning process is to destroy any microorganisms in the food and prevent recontamination by microorganisms. Heat is the most common agent used to destroy microorganisms. Removal of oxygen can be used in conjunction with other methods to prevent the growth of oxygen-requiring microorganisms.

In the conventional canning of fruits and vegetables, there are basic process steps that are similar for both types of products. However, there is a great diversity among all plants and even those plants processing the same commodity. The differences include the inclusion of certain operations for some fruits or vegetables, the sequence of the process steps used in the operations, and the cooking or blanching steps. Production of fruit or vegetable juice occurs by a different sequence of operations, and there is a wide diversity among these plants. Typical canned products include beans (cut and whole), beets, carrots, corn, peas, spinach, tomatoes, apples, peaches, pineapple, pears, apricots, and cranberries. Typical juices are orange, pineapple, grapefruit, tomato, and cranberry. A typical commercial canning operation may employ the following general processes: washing, sorting/grading, preparation, container filling, exhausting, container sealing, heat sterilization, cooling, labeling/casing, and storage for shipment. In these diagrams, no attempt has been made to be product-specific and include all process steps that would be used for all products. One of the major differences in the sequence of operations between fruit and vegetable canning is the blanching operation. Most of the fruits are not blanched prior to can filling, whereas many of the vegetables undergo this step. Canned vegetables generally require more severe processing than do fruits because the vegetables have much lower acidity and contain more heat-resistant soil organisms. Many vegetables also require more cooking than fruits to develop their most desirable flavor and texture. The methods used in the cooking step vary widely among factories. With many fruits, preliminary treatment steps (e.g., peeling, coring, halving, pitting) occur prior to any heating or cooking step, but with vegetables, these treatment steps often occur after the vegetable has been blanched. For both fruits and vegetables, peeling is done either by a mechanical peeler, steam peeling, or lye peeling. The choice depends upon the type of fruit or vegetables.

Some citrus fruit processors produce dry citrus peel, citrus molasses, and d-Limonene from the peels and pulp residue collected from the canning and juice operations. Other juice processing facilities use concentrates, and raw commodity processing does not occur at the facility. The peels and residue are collected and ground in a hammermill, lime is added to neutralize the acids, and the product is pressed to remove excess moisture. The liquid from the press is screened to remove large particles, which are recycled back to the press, and the liquid is concentrated to molasses in an evaporator. The pressed peel is sent to a direct-fired hot-air drier. After passing through a condenser to remove the d-Limonene (the major component of the oil extracted from citrus rind), the exhaust gases from the drier are used as the heat source for the molasses evaporator.

Equipment for conventional canning has been converting from batch to continuous units. In continuous retorts, the cans are fed through an airlock, then rotated through the pressurized heating chamber, and subsequently cooled through a second section of the retort in a separate cold-water cooler. Commercial methods or sterilization of canned foods with a pH of 4.5 or lower include use of static retorts, which are similar to large pressure cookers. A newer unit is the agitating retort, which mechanically moves the can and the food, providing quicker heat penetration. In the aseptic packaging process, the problem with slow heat penetration in the in-container process is avoided by sterilizing and cooling the food separate from the container.

Presterilized containers are then filled with the sterilized and cooled product and are sealed in a sterile atmosphere.

Did You Know?

Blanching is a cooking term that describes a process of food preparation wherein the food substance, usually a fruit or vegetable, is plunged into boiling water, removed after a brief, timed interval, and finally plunged into iced water or placed under cold running water (shocked) to halt the cooking process.

Fruit and Vegetable Processing: An Environmental Health Concern

Based on the fruit and vegetable processes and dehydration operations described above, the reader should be able to discern and/or recognize many of the potential hazards workers are (or can be) exposed to during each work shift. During each of the processes described, workers can be exposed to, to name a few, machines or mechanical transmission apparatus power systems; many forms of energy and stored energy (kinetic and potential energy sources); electrical wiring and components; process chemicals (e.g., lime) and off-gases generated from the processes; wet, hot, cold, humid work areas; slip, trip, and fall hazards; and a wide assortment of powered industrial trucks.

Because of the potential for exposure to these hazards, it is not surprising that OSHA keeps a close eye on the fruit and vegetable processing industry. In

TABLE 5.6
Top Ten 29 CFR Violations Cited, FY 2005 (IMIS Database, 2005)

Standard	*# Cited*	*# Inspected*	*Description*
1910.219	53	25	Mechanical power—transmission apparatus
1910.147	52	29	The control of hazardous energy, lockout/tagout
1910.212	44	32	Machines, general requirements
1910.305	35	23	Electrical, wiring methods, components & equipment
1910.1200	32	19	Hazard communication
1910.23	31	23	Guarding floor & wall openings & holes
1910.146	31	11	Permit—required confined spaces
1910.303	30	22	Electrical systems design, general requirements
1910.178	29	21	Powered industrial trucks
1910.132	22	15	Personal protective equipment, general requirements

Source: IMIS Database (2005).

addition to monitoring the industry, OSHA typically cites processing plants for various violations of standards. Consider, for example, the following top ten violations cited by OSHA: FY 2005 Fruit and Vegetable Industry.

In addition to typical physical and chemical hazards that fruit and vegetable process workers may be exposed to during a shift, air emissions (requiring ventilation protection) may also arise from a variety of sources in the canning/dehydration of fruits and vegetables. Along with being potential hazardous exposures for workers, these emissions may also contaminate the environment (e.g., odor problems) (Jones et al., 1979; Woodroof and Luh, 1986; Van Langenhove et al., 1991; Buttery et al., 1990; Rafson, 1977).

Particulate matter (PM) emissions result mainly from solids handling, solids size reduction, and drying (e.g., citrus peel driers). Some of the particles are dusts, but others (particularly those from thermal processing operations) are produced by condensation of vapors and may be in the low-micrometer or submicrometer particle-size range.

The VOCs (volatile organic compounds) emissions may potentially occur at almost any stage of processing, but most usually are associated with thermal processing steps, such as cooking, and evaporative concentration. The cooking technologies in canning processes are very high moisture processes so that predominant emissions will be steam or water vapor. The waste gases from these operations may contain PM or, perhaps, condensable vapors, as well as malodorous VOC. Particulate matter, condensable materials, and the high moisture content of the emissions many interfere with the collection or destruction of these VOC. The condensable materials also may be malodorous.

Did You Know?

VOCs are organic chemicals that have a high vapor pressure and easily form vapors at normal temperature and pressure.

Popcorn Workers Lung (Flavorings Industry)

Popcorn workers lung, the term given to a debilitating condition diagnosed in some workers employed at flavor manufacturing jobsites that use the chemical diacetyl, is an emerging spotlight issue—generating a political tug-of-war in Washington over how best to mitigate the hazard. Approved as a food additive in the early 1980s, diacetyl is a naturally occurring substance in a wide variety of foods, including butter, milk, cheese, fruit, wine, and beer. It provides a buttery flavor when added to foods like microwave popcorn. The chemical has been linked to a severe, irreversible lung disease known as bronchiolitis obliterans, which has sickened and killed a number of workers nationwide. (Marvin V. Greene, 2007)

One of America's favorite snack foods—popcorn—is at the center of a national health controversy. The chemical diacetyl, used to make artificial butter flavoring, has been linked to a respiratory disease called "popcorn lung" in hundreds of people. Labor unions and prominent occupational health scientists are calling on federal authorities to set an emergency standard for the chemical in the workplace. (*Living on Earth*, 2006)

WARNING!

Breathing certain flavoring chemicals in the workplace may lead to severe lung disease (CDC, 2005).

The National Institute for Occupational Safety and Health (NIOSH) points out that the occurrence of severe lung disease in workers who make flavorings or use them to produce microwave popcorn has revealed an unrecognized occupational health risk. Flavorings are often complex mixtures of many chemicals (Conning, 2000). The safety of these chemicals is usually established for humans consuming small amounts in foods (Pollitt, 2000), not for food industry workers inhaling them. Production workers employed by flavoring manufacturers (or those who use flavorings in the production process) often handle a large number of chemicals, many of which can be highly irritating to breathe in high concentrations.

NIOSH investigated the occurrence of severe lung disease in workers at a microwave popcorn packaging plant. Eight former workers at this plant developed illness characterized by fixed airways obstruction on lung function tests (Akpinar-Elci et al., 2002). An evaluation of the current workforce at this plant showed an association between exposure in vapors from flavorings used in the production process and decreased lung function (Kreiss et al., 2002a). Similar fixed obstructive lung disease has also occurred in workers at other plants that use or manufacture flavorings (NIOSHA, 1986; Lockey et al., 2002). In animal tests, inhaling vapors from a heated butter flavoring used in microwave popcorn production caused severe injury to airways (Hubbs et al., 2002a).

Medical test results in affected workers (including some lung biopsy results) are consistent with bronchiolitis obliterans, an uncommon lung disease characterized by fixed airways obstruction (Akpinar-Elci et al., 2002). In bronchiolitis obliterans, inflammation and scarring occur in the smallest airways of the lung and can lead to severe and disabling shortness of breath. The disease has many known causes, such as inhalation of certain chemicals, certain bacterial and viral infections, organ transplantation, and reactions to certain medications (King, 2000). Known causes of bronchiolitis obliterans due to occupational or other environmental exposures include gases such as nitrogen oxides (e.g., silo gas), sulfur dioxide, chlorine, ammonia, phosgene, and other irritant gases (King, 1998). Recent NIOSH investigations strongly suggest that some flavoring chemicals can also cause bronchiolitis obliterans in the workplace. (Some workers exposed to flavorings in one of these plants were also found to have occupational asthma.)

Bronchiolitis obliterans (Popcorn Workers Lung Disease)

Nicknamed "popcorn lung" or "popcorn workers' lung" due to onset of this disease from inhalation of airborne diacetyl—a butter flavoring used in popcorn and in many other food flavorings such as those in candy and even potato chips.

The main respiratory symptoms experienced by workers affected by fixed airways obstruction include cough (usually without phlegm) and shortness of breath on exertion. These symptoms typically do not improve when the worker goes home at the end of the workday or on weekends or vacations. The severity of the lung symptoms can range from only a mild cough to severe cough and shortness of breath on exertion. Usually these symptoms are gradual in onset and progressive, but severe symptoms can occur suddenly. Some workers may experience fever, night sweats, and weight loss. Before arriving at a final diagnosis, doctors of affected workers initially thought that the symptoms might be due to asthma, chronic bronchitis, emphysema, pneumonia, or smoking. Severe cases may not respond to medical treatment. Affected workers generally notice a gradual reduction or cessation of cough years after they are no longer exposed to flavoring vapors, but shortness of breath on exertion persists. Several with very severe disease were placed on lung transplant waiting lists. Workers exposed to flavorings may also experience eye, nose, throat, and skin irritation. In some cases, chemical eye burns have required medical treatment. Medical testing may reveal several of the following findings:

- Spirometry, a type of breathing test, most often shows fixed airways obstruction (i.e., difficulty blowing air out fast and no improvement with asthma medications), and sometimes shows restriction (i.e., decreased ability to fully expand the lungs).
- Lung volume may show hyperinflation (i.e., too much air in the lungs due to air trapping beyond obstructed airways).

- Diffusing capacity of the lung (DLCO) is generally normal, especially early in the disease.
- Chest x-rays are usually normal but may show some hyperinflation.
- High-resolution computerized tomography scans of the chest at full inspiration and expiration may reveal heterogeneous air trapping on the expiratory view as well as haziness and thickened airway walls.
- Lung biopsies may reveal evidence of constrictive bronchiolitis obliterans (i.e., severe narrowing or complete obstruction of the small airways). An open lung biopsy, such as by thoracoscopy, is more likely to be diagnostic than a transbronchial biopsy. Special processing, staining, and review of multiple tissue sections may be necessary for a diagnosis.

Current Exposure Limits

Flavorings are composed of various natural and man-made substances. They may consist of a single substance, but more often they are complex mixtures of several substances. The Flavor and Extract Manufacturers Association (FEMA) evaluates flavoring ingredients to determine whether they are "generally recognized as safe" (GRAS) under the conditions of intended use through food consumption. Though considered safe to eat, ingredients may be harmful to breathe in the forms and concentrations to which food and chemical industry workers may be exposed.

Occupational exposure guidelines have been developed for only a small number of the thousands of ingredients used in flavorings. For example, OSHA permissible exposure limits (PELs) and/or NIOSHA recommended exposure limits (RELs) have been established for only 46 (<5%) of the 1,037 flavoring ingredients considered by the flavorings industry to represent potential respiratory hazards due to possible volatility and irritants properties (alpha, beta-unsaturated aldehydes and ketones, aliphatic aldehydes, aliphatic carboxylic acids, aliphatic amines, and aliphatic aromatic thiols and sulfides) (Hallagan, 2002). Material safety data sheets (MSDSs) contain information about known occupational hazards of specific chemicals, but they may not be based on the most up-to-date information in the case of newly recognized occupational health risks.

Consumer Risk Alert—Lung Disease

Marcus Kabel (AP) reports that consumers, not just factory workers, may be in danger from fumes from butter flavoring in microwave popcorn, according to a warning letter to federal regulators from a doctor at a leading lung research hospital. A pulmonary specialist at Denver's National Jewish Medical and Research Center has written to federal agencies to say doctors there think they have the first case of a consumer who developed lung disease from the fumes of microwaving popcorn several times a day for years. (*The Virginian-Pilot,* Norfolk, VA, 09/05/07)

Popcorn Lung: An Environmental Health Concern

Case Study 1. In September 2003, a man aged twenty-nine years with no history of smoking, lung disease, or respiratory symptoms developed progressive shortness of breath on exertion, decreased exercise tolerance, intermittent wheezing, left-side chest pain, and a productive cough two years after beginning employment as a flavor compounder. His job involved measuring diacetyl and other ingredients to prepare batches of powder flavorings. The workplace did not have effective methods for controlling exposure to the flavoring chemicals, such as local exhaust ventilation or adequate use of respirators to reduce exposure to organic compounds and powders. The worker reported wearing a paper dust mask and occasionally a cartridge respirator for organic vapors. However, he never received a fit test for the respirator. He had a beard at the time, which precluded a proper fit, and he was not adequately protected from both volatile organic chemicals and particulates.

In November 2003, the man went to his primary-care physician and was treated with antibiotics and bronchodilators for suspected bronchitis and allergic rhinitis. In January 2004, he stopped working because of his respiratory symptoms. His shortness of breath became more severe, with dyspnea after walking ten to fifteen feet. A high-resolution computed tomography (HRCT) scan of his chest showed cylindrical bronchiectasis in the lower lobes, with scattered peribronchial ground-glass opacities. In April 2004, spirometry showed severe obstructive lung disease, with a forced expiratory volume in 1 second (FEV_1) of 28 percent of the predicted normal value, without bronchodilator response. Static lung volumes by body plethysmography were consistent with severe air trapping. Diffusing capacity was normal.

In October 2004, the patient was referred for an occupational pulmonary consultation. Paired inspiratory and expiratory HRCT scans showed central peribronchial thickening with central airway dilation and subtle areas of mosaic attenuation scattered throughout the lungs, predominately in the right lower lobe. The diagnosis of work-related bronchiolitis obliterans was made on the basis of history, fixed airway obstruction with normal diffusing capacity, and typical HRCT findings (CDC, 2007). Diacetyl is considered the cause of this patient's disease on the basis of its known toxic effects; however, exposure to other less well-characterized flavoring chemicals might also have contributed.

Case Study 2. During 2002, a nonsmoking woman aged forty years, who had no history of lung disease of respiratory symptoms when she began working as a flavor compounder, experienced nasal congestion and cough after five years on the job, which involved mixing dry powders with diacetyl and other ingredients to make artificial butter flavoring. The workplace did not have exposure-control measures such as local exhaust ventilation, and employees did not use respirators appropriately. The worker reported wearing a paper dust mask that had not been fit tested and did not provide adequate protection from either volatile organic compounds or particulates. The woman was treated with antibiotics and antihistamines by her primary-care physician. She experienced progressively worsening shortness of breath on exertion, decreasing exercise tolerance, and a nonproductive cough. In November 2005, she visited a pulmonary specialist who suspected work-related asthma and treated her with bronchodilators and oral corticosteroids, producing minimal improvement. An HRCT of the chest showed several small areas of patchy ground-glass opacities throughout the lungs.

In December 2005, the patient stopped working because of her respiratory symptoms. Spirometry revealed severe obstructive lung disease, with an FEV_1 of 18 percent of the predicted normal value, without bronchodilator response. Static lung volumes by body plethysmography were consistent with severe air trapping. Diffusing capacity was normal. Left thoracotomy with wedge resection of he left lower lobe did not indicate bronchiolitis obliterans in this area of the lung. However, other findings of peribronchial inflammation, interstitial fibrosis, and non-caseating-type granulomas suggested an inflammatory process. The diagnosis of work-related bronchiolitis obliterans was made on the basis of history, fixed airway obstruction with normal diffusing capacity, and typical HRCT findings (CDC, 2007).

Follow-up Report on Consumer Popcorn Lung Incident

Josh Funk (AP) reports that microwave popcorn fans worried about the potential for lung disease from butter flavoring fumes should know this: the sole reported case of the disease in a non-factory worker involves a man who popped the corn every day and inhaled from the bag. "He really liked microwave popcorn. He made two or three bags every day for ten years," said William Allstetter, a spokesman for National Jewish Medical and Research Center in Denver, where the man's respiratory illness was diagnosed. "He told us he liked the smell of popcorn, so he would open and inhale from freshly popped bags," Allstetter said. And the patient said he did this for a decade. (*The Virginian-Pilot*, 09/06/07, Norfolk, VA)

Discussion Questions

1. How can foodborne pathogens spread among the animals themselves be preventive?
2. What do you think is the microbial cause of outbreaks in which no pathogen can be identified by current methods?
3. How can food and water that animals consume be made safer?
4. How can we dispose of animal manure successfully, without threatening the food supply and the environment?
5. How can basic food safety principles be most effectively taught to schoolchildren?

References and Recommended Reading

ACGIH, 2001. *Industrial ventilation: a manual of recommended practice*, 24th ed. Cincinnati, OH: American Conference of Governmental Industrial Hygienists.

Akpinar-Elci M., R. Kanwal, and K. Kreiss, 2002. Bronchiolitis obliterans syndrome in popcorn plant workers. *Am. J. Respir Crit Care Med* 165:A526.

Analysis of Workers' Compensation Laws. The Chamber of Commerce of the United States, Washington, D.C., Annual.

ATS, 2005. Standardization of Spirometry. *European Respiratory Journal.* Vol. 26(2).

Balch, P., 2010. *Prescription for Nutritional Healing,* 5th edition. New York: Avery Trade.

Best Food Nation, 2007. *America's Poultry Industry.* Accessed 08/04/07 at http://best-food nation.com/poultry.asp.

BLS, 2005. *Food Manufacturing.* Accessed 7/28/07 at http://www.bls.gov/oco/cg/cgs011.htm.

BLS (Bureau of Labor Statistics), 1985. *Supplementary Data System.* Washington, D.C.: Department of Labor.

Brauer, R.L., 1994. *Safety and Health for Engineers.* New York: Van Nostrand Reinhold.

Buttery, R.G. et al., 1990. Identification of Additional Tomato Paste Volatiles, *Journal of Agricultural and Food Chemistry,* 38(3):792–795.

CALMIS, 2007. Food Manufacturing. Accessed 7/30/07 at www. Caljobs.ca.gov.

CDC, 2005. NIOSH: *Preventing Lung Disease in Workers Who Use or Make Flavorings.* Accessed 08/18/07 at http://www.cdc.gov/niosh/docs/2004-110/.

CDC, 2007. Fixed obstructive lung disease among workers in the flavor-manufacturing industry—California, 2004–2007. *MMWR* 56(16):389–393.

Conning, D.M., 2000. Toxicology of food and food additives. In Ballantyne B., T.C. Marrs, and T. Syversen, eds. *General and applied toxicology.* 2nd ed. London: Macmillan Reference Ltd., 1977–1992.

CoVan, J., 1995. *Safety Engineering.* New York: Wiley.

Deroiser, N.W., 1970. *The Technology of Food Preservation*, 3rd edition. Westport, CT: The Avi Publishing Company.

FEMA, 2004. *Respiratory Health and Safety in the Flavor Manufacturing Workplace*. Washington, D.C.: The Flavor and Extract Manufacturers Association of the United States.

Ferry, T., 1990. *Safety and Health Management Planning*. New York: Van Nostrand Reinhold.

Florida Chemical Company, 2007. *Citrus oils and d-Limonene*. Winter Haven, FL.

Greene, M.V., 2007. Popcorn workers' lung spurs tug-of-war over regulation. *Safety + Health*. Vol. 176, No. 2:32–33.

Hallagan, J.B., 2002. Letter of November 26, 2002, from J.B. Hallagan, Flavor and Extract Manufacturers Association of the United States, to R. Kanwai, Division of Respiratory Disease.

IMIS Database, 2005. *Top 10 Violations Cited*. Accessed 08/16/07 at www.osha.gov.

Hubbs, A.F., L.S. Battelli, W.T. Goldsmith, D.W. Porter, D. Frazer, S. Friend, et al., 2002a. Necrosis of nasal and airway epithelium in rates inhaling vapors of artificial butter flavoring. *Toxicol Appl Pharmacol* 185:128–135.

Hubbs, A., V. Castranova, W. Jones, D. Porter, W. Goldsmith, B. Kullman, et al., 2002b. Workplace safety and food ingredients: the example of butter flavoring. *In*: Abstracts of papers, 24th ACS National Meeting, Boston, MA. August 18–22. Washington, D.C.: American Chemical Society, AGFD-148.

Jones, J.L. et al., 1979. *Overview of Environmental Control Measure and Problems in the Food Processing Industries*. Industrial Environmental Research Laboratory, Cincinnati, OH, Kenneth Dostal, Food and Wood Products Branch. Grant No. R804642-01.

Kanwai, R., 2002a. Letter of April 19, 2002, from R. Kanwai and S. Martin, Division of Respiratory Disease Studies, National Institute for Occupational Safety and Health, Centers for Disease Control and Prevention, Department of Health and Human Services, to Keith Heuermann, B.K. Heuermann, Popcorn, Inc., Phillips, Nebraska.

Kanwai, R., 2002b. Letter of November 18, 2002, from R. Kanwai, Division of Respiratory Disease Studies, National Institute for Occupational Safety and Health, Centers for Disease Control and Prevention, Department of Health and Human Services, to Greg Hoffman, America Popcorn Company, Sioux City, Iowa.

King, T.E., 1998. Bronchiolitis. In: Fishman, A.E., ed. *Pulmonary diseases and disorders*. New York: McGraw-Hill, 825–847.

King, T.E., 2000. Bronchiolitis. *Eur Respir Mon* 14:244–266.

Kreiss, K., A. Gomaa, G. Kullman, K. Fedan, E.J. Simoes, and P.L. Enright, 2002a. Clinical bronchiolitis obliterans in workers at a microwave-popcorn plant. *N Engl J Med* 347:330–338.

Kreiss, K., A. Hubbs, and G. Kullman, 2002b. Correspondence: bronchiolitis in popcorn-factory workers. *N Engl J Med* 347:1981–1982.

Living on Earth, 2006. *Popcorn Production Harms Workers*. Accessed 08/17/07 at http://www.Loe.org/shows/segments.htm?.

Lockey, J., R. McKay, E. Barth, J. Dahisten, and R. Baughman, 2002. Bronchiolitis obliterans in the food flavoring manufacturing industry. *Am J Respir Crit Care Med* 165:A461.

Luh, B.S., and J.G. Woodroof, eds. 1988. *Commercial Vegetable Processing*, 2nd edition. New York: Van Nostrand Reinhold.

National Safety Council, 2007. *Farm Facts.* Accessed 7/28/07 at http.www.nsc.org/farmsafe/facts.htm.

NCDOL, 2005. *A guide to safe work practices in the poultry processing industry.* Raleigh, NC: North Carolina Department of Labor.

NFPA, 2004. *NFPA 70E Standard for Electrical Safety in the Workplace.* Quincy, MA: National Fire Protection Association.

NIOSH, 1986. Hazard evaluation and technical assistance report: International Bakers Services, Inc., South Bend, IN. Cincinnati, OH: U.S. Department of Health and Human Services, Public Health Service, Centers for Disease Control, National Institute for Occupational Safety and Health, NIOSHA Report No. HETA 85-171-1710.

NSC, 1979. *Meat Industry Safety Guidelines.* Chicago: National Safety Council.

OSHA, 2007a. *OSHA Fact Sheet: Farm Safety.* Accessed 7/27/07 at www.osha.gov.

OSHA, 2007b. *Poultry Processing.* Accessed 08/04/07 at http://www.osha.gov/SLTC/poultryprocessing/index.html.

OSHA, 2007c. *Preserved Fruits and Vegetables.* Washington, D.C.: U.S. Department of Labor. Accessed 08/13/07 at www.osha.gov.

OSHA eTool, 2007. *Poultry Processing Industry eTool.* Accessed 08/05/07 at www.osha.gov/SLTC/etools/poultry/general_hazards.html.

Parmet, A.J., and S. Von Essen, 2002. Rapidly progressive, fixed airway obstructive disease in popcorn workers: a new occupational pulmonary illness? *J. Occup Environ Med* 44:216–218.

Pollitt, F.D., 2000. Regulation of food additives and food contract materials. In: Ballantyne, B., T.C. Marrs, and T. Syversen, eds. *General and applied toxicology*: 2nd ed. London: Macmillan Reference Ltd., 1653–1660.

Rafson, H.J., 1977. Odor Emission Control for the Food Industry. *Food Technology*, June.

Rasmussen, C., 2003. Muckraker's Own Life as Compelling as His Writing. *Los Angeles Times*, May 11, 2003, Metro 4.

Right Off the Docket, 1986. Cleveland, OH: Penton Educational Division, Penton Publishing.

Sahakian, N., 2003. Letter of January 13, 2003, from N. Sahakian, Division of Respiratory Disease Studies, National Institute for Occupational Safety and Health, Centers for Disease Control and Prevention, Department of Health and Human Services, to Gary Sanders, Agrilink Foods, Ridgeway, Illinois.

Smith, R.L., et al., 2003. GRAS flavoring substances 21. *Food Technology* 57(5).

Somogyi, L.P., and B.S. Luh, 1986. *Dehydration of Fruits, Commercial Fruit Processing*, 2nd ed. Woodroof J.G., and B.S. Luh, eds. AVI Publishing Company.

Somogyi, L.P., and B.S. Luh, 1988. *Vegetable Dehydration, Commercial Vegetable Processing*. 2nd ed., Luh B.S., and J.G. Woodroof, eds. An AVI Book Published by Van Nostrand Reinhold.

Spellman, F.R., 1998. *Surviving an OSHA Audit: A Manager's Guide.* Lancaster, PA: Technomic Publishing Company.

USATODAY.com, 2012. Pink Slime eliminated from fast food, but not school lunches. Accessed 03/09/12 at http://yourlife.usatoday.com/health/story/2012-03-09/Pink-slime-elimianted-from-fast-food.

USDOC, 1990. *1987 Census of Manufacturers: Preserved Fruits and Vegetables.* Washington, D.C.: U.S. Department of Commerce, Bureau of Census.

USDOL, 2007. *The Unique Hazards of Packing.* Accessed 08/02/07 at www.osha.gov/Publications/OSHA3108/osha3108.html.

U.S. Department of Labor, 1988. *Safety and Health Guide for the Meatpacking Industry.* Washington, D.C.: OSHA 3108.

USEPA, 1995. *Canned Fruits and Vegetables.* Accessed 08/14/07 at http://www.epa.gov; *Dehydrated fruits and vegetables*—emission factor documentation for AP-42, Section 9.8.2. Washington, D.C.: U.S. Environmental Protection Agency.

Van Langenhove, H.J., et al., 1991. Identification of Volatiles Emitted During the Blanching Process of Brussels Sprouts and Cauliflower. *Journal of the Science of Food and Agriculture* 55:483–487.

Woodroof, J.G., and B.S. Luh, 1986. *Commercial Fruit Processing.* Westport, CT: The Avi Publishing Company.

Young, J.H., 1989. *Pure food: securing the Federal Food and Drugs Act of 1906.* Princeton: Princeton University Press.

6

Vector-Borne Disease

Knowing not enough; we must apply. Willing is not enough; we must do.

—Goethe

We call them dumb animals, and so they are, for they cannot tell us how they feel, but they do not suffer less because they have no words.

—Anna Sewell, author of *Black Beauty*

Pests—An Environmental Health Concern

(This section from CDC Tox Town—*Pests*. Accessed 03.13/12 at http://toxtown.nhlm,nih.gov/text_version/locations.php? id=43.) Pests can be a health hazard to humans and animals, as well as destructive to homes and other buildings. Pests that sting or bite include ants, bees, fleas, bedbugs, flies, lice, mosquitoes, spiders, and ticks. Pests that ingest food or fabric include centipedes, cockroaches, moths, and silverfish. Wood-destroying pests include carpenter ants, termites, and wood-boring beetles. Animal pests include lizards, mice, moles, rabbits, raccoons, rats, snakes, skunks, squirrels, and voles.

Many pests pose human health threats. Cockroaches can carry and transmit diseases, including salmonella, and cause allergies. Fire ant stings can cause severe allergic reactions. Fleas can bite, causing allergic skin problems. Flies can carry bacteria, viruses, and several diseases. Mosquitoes in the United States can carry serious diseases such as encephalitis, dengue fever, and West Nile Virus; in some other countries, mosquitoes can carry malaria and yellow fever. Mosquito bites can also result in infections, allergic reactions, pain, and aching.

Rats can carry and transmit diseases, including salmonella; rat-bite fever, a bacterial illness; hantavirus, an often deadly respiratory disease; and occasionally, plague. Rats can infest buildings with mites and fleas, which can also spread disease. Their dander can cause asthma and nasal inflammation.

Ticks can carry and transmit serious diseases such as Lyme disease and Rocky Mountain spotted fever. Wasp stings can cause pain, itching, swelling, and allergic reactions that can even cause death.

Pests can threaten food, buildings, and structures. Cockroaches can contaminate food and transmit bacteria that cause food poisoning. Termites can cause severe destruction to wooden structures, including homes, buildings, and utility poles.

The 411 on Vector-Borne Disease

(Based on CDC's Division of Vector-Borne Diseases. Accessed 03/12/12 at http://www.cod.gove/ncidod/dvbid.) The chapter covers vector-borne diseases that have substantial environmental components and contribute greatly to society's burdens of morbidity and mortality. As a matter of fact and record, some of the world's most destructive diseases are vector-borne—that is, they are transmitted to humans by ticks, mosquitoes, or fleas. Moreover, vector-borne diseases are among the most complex of all infectious diseases to prevent and control. Not only is it difficult to predict the habits of the vectors, but most vector-borne agents can infect animals, as well. Consider, for example, the rapid, unstoppable spread of West Nile Virus (WNV) across the United States. Lyme disease, another emerging infection, has resulted in 20,000 annual reported human cases in the United States in recent years. As mentioned earlier, dengue fever causes thousands of cases of illness in U.S. territories and in U.S. travelers, and millions of cases worldwide. Then there is *Yersinia pestis*, the ancient disease known as *plague*, which continues to cause focal outbreaks in the United States, and is a significant health threat in Africa and Asia.

In this book, we focus on CDC's four branches of vector-borne disease: bacterial, arboviral (arthropod-borne), dengue, and rickettsial zoonosis.

Vector-Borne Bacterial Diseases

Lyme disease is caused by the bacterium *Borrelia burgdorferi* and is transmitted to humans through the bite of infected backlegged ticks. Typical symptoms include fever, headache, fatigue, and a characteristic skin rash called erythema migrans. If left untreated, infection can spread to joints, the heart, and the nervous system. Lyme disease is diagnosed based on symptoms, physical findings (e.g., rash), and the possibility of exposure to infected ticks; laboratory testing is helpful if used correctly and performed with validated methods. Most cases of Lyme disease can be treated successfully with a few weeks of antibiotics. Steps to prevent Lyme disease include using insect repellent, removing ticks promptly, applying pesticides, and reducing tick habitat. The ticks that transmit Lyme disease can occasionally transmit other tick-borne diseases, as well (CDC, 2011a).

Plague is an infectious disease of animals and humans caused by a bacterium named *Yersinia pestis*. People usually get plague from being bitten by a rodent flea that is carrying this bacterium or by handling an infected animal. Millions of people in Europe died from plague in the Middle Ages, when human homes and places of work were inhabited by flea-infested rats. Today, modern antibiotics are effective against plague, but if an infected person is not treated promptly, the disease is likely to cause illness or death.

Wild rodents in certain areas around the world are infected with plague. Outbreaks in people still occur in rural communities or in cities. They are usually associated with infected rats and rat fleas that live in the home. In the United States, the last urban plague epidemic occurred in Los Angeles in 1924–25. Since then, human plague in the United States has occurred as mostly scattered cases in rural areas (an average of ten to fifteen people each year). Globally, the World Health Organization reports 1,000 to 3,000 cases of plague every year. In North America, plague is found in certain animals and their fleas from the Pacific Coast to the Great Plains, and from southwestern Canada to Mexico. Most human cases occur in southern Colorado, then in California, southern Oregon, and far western Nevada. Plague also exists in Africa, Asia, and South America.

Did You Know?

You can get plague from another person. When the other person has plague pneumonia and coughs, droplets containing plague bacteria can be spread into the air. If a non-infected person inhales these infectious droplets, they also can become infected (CDC, 2005).

Relapsing fever is a bacterial infection characterized by recurring episodes of fever, headache, muscle and joint aches, and nausea. There are two types of relapsing fever:

- **Tick-borne relapsing fever** (TBRF) is caused by certain species of *Borrelia* spirochetes, a gram negative bacteria 0.2 to 0.5 microns in width

and five to twenty microns in length. They are visible with light microscopy and have the corkscrew shape typical of spirochetes. Relapsing fever spirochetes have a unique process of DNA rearrangement that allows them to periodically change the molecules on their outer surface. This process, called antigenic variation, allows the spirochete to evade the host immune system and cause relapsing episodes of fever and other symptoms. Three species cause TBRF in the United States: *Borrelia hermssi*, *B. parkerii*, and *B. turicatae*. The most common cause is *B. hermssi*. It occurs in the western United States in fourteen western states: Arizona, California, Colorado, Idaho, Kansas, Montana, Nevada, New Mexico, Oklahoma, Oregon, Texas, Utah, Washington, and Wyoming, and is usually linked to sleeping in rustic, rodent-infested cabins in mountainous areas (CDC, 2012a).
- **Louse-borne relapsing fever** (LBRF) is transmitted by the human body louse and is generally restricted to refugee settings in developing regions of the world where conditions of overcrowding and social disruption prevail. For example, LBRF causes sporadic illness and outbreaks in sub-Saharan Africa, particularly in regions affected by war and in refugee camps. LBRF is commonly found in Ethiopia, Sudan, Eritrea, and Somalia. Illness can be severe, with a mortality of 30 to 70 percent in outbreaks (CDC, 2012a).

Tularemia is a disease of animals and humans caused by the bacterium *Francisella tularensis*. Rabbits, hares, and rodents are especially susceptible and often die in large numbers during outbreaks. Humans can become infected through several routes, including:

- Tick and deer fly bites
- Skin contact with infected animals
- Ingestion of contaminated water
- Laboratory exposure
- Inhalation of contaminated dusts or aerosols

Symptoms vary depending upon the route of infection. Although tularemia can be life-threatening, most infections can be treated successfully with antibiotics (CDC, 2011b).

Vector-Borne Arboviral Diseases

Chikungunya fever is a viral disease transmitted to humans by the bite of infected mosquitoes. Chikungunya virus is a member of the genus *Alpha-*

virus, in the family *Togaviridae*. Chikungunya fever is diagnosed based on symptoms, physical findings (e.g., joint swelling), laboratory testing, and the possibility of exposure to infected mosquitoes. There is no specific treatment for Chikungunya fever; care is based on symptoms. Chikungunya infection is not usually fatal. Steps to prevent infection with Chikungunya virus include use of insect repellent, protective clothing, and staying in areas with screens. Chikungunya virus was first isolated from the blood of a febrile patient in Tanzania in 1953, and has since been cited as the cause of numerous human epidemics in many areas of Africa and Asia and most recently in limited areas of Europe.

Did You Know?

Mosquitoes become infected with Chikungunya virus when they feed on an infected person. Infected mosquitoes can then spread the virus to other humans when they bite them. Monkeys, and possibly other wild animals, may also serve as reservoirs of the virus. *Aedes aegypti*, a household container breeder and aggressive daytime biter that is attracted to humans, is the primary vector of Chikungunya virus to humans. *Aedes albopictus* (the Asia tiger mosquito) has also played a role in human transmission in Asia, Africa, and Europe. Various forest-dwelling mosquito species in Africa have been found to be infected with the virus (CDC, 2008a).

Eastern equine encephalitis virus (EEEV) is transmitted to humans by the bite of an infected mosquito. Eastern equine encephalitis (EEE) is a rare illness in humans, and only a few cases are reported in the United States each year. Most cases occur in the Atlantic and Gulf Coast states. Most persons infected with EEEV have no apparent illness. Severe cases of EEE (involving encephalitis, an inflammation of the brain) begin with the sudden onset of headache, high fever, chills, and vomiting. The illness may then progress into disorientation, seizures, or coma. EEE is one of the most severe mosquito-transmitted diseases in the United States with approximately 33 percent mortality and significant brain damage in most survivors. There is no specific treatment for EEE; case is based on symptoms. You can reduce your risk of being infected with EEEV by using insect repellent, wearing protective clothing, and staying indoors while mosquitoes are most active (CDC, 2010a).

Did You Know?

All residents of and visitors to areas where virus activity has been identified are at risk of infection with EEEV, particularly people who engage in outdoor work and recreational activities in these areas. People over age fifty and younger than age fifteen are at greatest risk of severe disease (encephalitis) following infection. EEEV infection is thought to confer lifelong immunity against re-infection (CDC, 2010b).

Japanese encephalitis is a flavivirus antigenically related to St. Louis encephalitis virus. In regard to frequency of occurrence, there is fewer than one case per year in U.S. civilians and U.S. military personnel (CDC, 2003).

La Crosse encephalitis virus (LACV) is transmitted to humans by the bite of an infected mosquito. Most cases of LACV disease occur in the upper Midwestern and mid-Atlantic and southeastern states. Many people infected with LACV have no apparent symptoms. Among people who become ill, initial symptoms include fever, headache, nausea, vomiting, and tiredness. Some of those who become ill develop severe neuroinvasive disease (disease that affects the nervous system). Severe LACV disease often involves encephalitis (an inflammation of the brain) and can include seizures, coma, and paralysis. Severe disease occurs most often in children under the age of sixteen. In rare cases, long-term disability or death can result from La Crosse encephalitis. There is no specific treatment for LACV infection—care is based on symptoms. If you or a family member have symptoms of severe LACV disease or any symptoms causing you concern, consult a health-care provider for proper diagnosis (CDC, 2009a). Less than 1 percent of LACV encephalitis cases are fatal.

St. Louis encephalitis virus (SLEV) is transmitted to humans by the bite of an infected mosquito. Most cases of SLEV disease have occurred in eastern and central states. Most persons infected with SLEV have no apparent illness. Initial symptoms of those who become ill include fever, headache, nausea, vomiting, and tiredness. Severe neuroinvasive disease (often involving encephalitis, an inflammation of the brain) occurs more commonly in older adults. In rare cases, long-term disability or death can result. There is no specific treatment for SLEV infection; care is based on symptoms. You can reduce your risk of being infected with SLEV by using insect repellent, wearing protective clothing, and staying indoors while mosquitoes are most active (CDC, 2009c).

Did You Know?

Many people infected with SLEV have no apparent illness. People with mild illness often have only a headache and fever. More severe disease is marked by fever, headache, neck stiffness, stupor, disorientation, coma, tremors, occasional convulsions (especially in infants), and spastic (but rarely flaccid) paralysis. The risk of severe disease generally increases with age (CDC, 2009d).

West Nile encephalitis (WNE) is a potentially serious illness. Experts believe WNE is established as a seasonal epidemic in North America that flares up in the summer and continues into the fall. People typically develop symptoms between three and fourteen days after they are bitten by the infected mosquito.

Western equine encephalitis (WEE) is the causative agent of a relatively uncommon viral disease. WEE is a recombinant virus between two other closely related yet distinct encephalitis viruses. There have been under 700 confirmed cases in the United States since 1964 (Ryan and Ray, 2004).

Yellow fever virus is found in tropical and subtropical areas in South America and Africa. The virus is transmitted to humans by the bite of an infected mosquito. Yellow fever is a very rare cause of illness in U.S. travelers. Illness ranges in severity from a self-limited febrile illness to severe liver disease with bleeding. Yellow fever disease is diagnosed based on symptoms, physical findings, laboratory testing, and travel history, including the possibility of exposure to infected mosquitoes. There is no specific treatment for yellow fever; care is based on symptoms. Steps to prevent yellow fever virus infection include using insect repellent, wearing protective clothing, and getting vaccinated (CDC, 2011c).

Dengue Infection

As mentioned earlier, dengue infection is a leading cause of illness and death in the tropics and subtropics. As many as 100 million people are infected yearly. Dengue is caused by any one of four related viruses transmitted by mosquitoes. There are not yet any vaccines to prevent infection with dengue virus (DENV), and the most effective protective measures are those that avoid mosquito bites. When infected, early recognition and prompt supportive treatment can substantially lower the risk of developing severe disease.

Dengue has emerged as a worldwide problem only since the 1950s. Although dengue rarely occurs in the continental United States, it is endemic in Puerto Rico, and in many popular tourist destinations in Latin America and Southeast Asia; periodic outbreaks occur in Samoa and Guam (CDC, 2011d).

Did You Know?

Aedes aegypti, the principal mosquito vector of dengue viruses, is an insect closely associated with humans and their dwellings. People not only provide the mosquitoes with blood meals but also water-holding containers in and around the home needed to complete their development. The mosquito lays her eggs on the sides of containers with water, and eggs hatch into larvae after a rain or flooding. A larva changes into a pupa in about a week and into a mosquito in two days (CDC, 2012c).

Rocky Mountain spotted fever (RMSF) is a tickborne disease caused by the bacterium *Rickettsia rickettisii*. This organism is a cause of potentially fatal human illness in North and South America, and is transmitted to humans by the bite of an infected tick species. In the United States, these include the American dog tick (*Dermacentor variabilis*), Rocky Mountain wood tick (*Dermacentor andersoni*), and brown dog tick (*Rhipicephalus sanguineus*). Typical symptoms include: fever, headache, abdominal pain, vomiting, and muscle pain. A rash may also develop, but is often absent in the first few days, and in some patients, never develops. Rocky Mountain spotted fever can be a severe or even fatal illness if not treated in the first few days of symptoms. Doxycycline is the first-line treatment for adults and children of all ages, and is most effective if started before the fifth day of symptoms. The initial diagnosis is made based on clinical signs and symptoms, and medical history, and can later be confirmed by using specialized laboratory tests (CDC, 2011e).

Anaplasmosis is a tickborne disease caused by the bacterium *Anaplasma phagocytophilum*. It was previously known as human granulocytic ehrlichiosis (HGE) and has more recently been called human granulocytic anaplasmosis (HGA). Anaplasmosis is transmitted to humans by tick bites primarily from the black-legged tick (*Ixodes scapularis*) and the western black-legged tick (*Ixodes pacificus*). Of the four distinct phases in the tick life-cycle (egg, larvae, nymph, adult), nymphal and adult ticks are most frequently associated with transmission of anaplasmosis to humans. Typical symptoms include: fever, headache, chills, and muscle aches. Usually, these symptoms occur within one to two weeks of a tick bite. Anaplasmosis is initially diagnosed based on symptoms and clinical presentation, and later confirmed by the use of specialized laboratory tests (CDC, 2012d). Note that ticks are arachnids, relatives of spiders.

Ehrlichiosis is the general name used to describe several bacterial diseases that affect animals and humans. Human ehrlichiosis is a disease caused by at least three different ehrlichial species in the United States: *Ehrlichia chaffeensis, Ehrlichia ewingii*, and a third *Ehrlichia* species provisionally called *Ehrlichia muris-like* (EML). Ehrlichiae are transmitted to humans by the bite of an infected tick. The lone star tick (*Amblyomma americanum*) is the primary vector of both *Ehrlichia chaffeensis* and *Ehrlichia ewingii* in the United States. Typical symptoms include fever, headache, fatigue, and muscle aches. Usually, these symptoms occur within one to two weeks following a tick bite. Ehrlichiosis is diagnosed based on symptoms, clinical presentation, and later confirmed with specialized laboratory tests. The first-line treatment for adults and children of all ages is doxycycline.

Discussion Questions

1. Discuss how vectors play a role in transmission.
2. Discuss the impact of malaria on environmental health.
3. Describe methods that environmental health professionals might use to control vector-borne disease.

References and Recommended Reading

Bahmanyar, M., and D.C. Cavanaugh, 1976. *Plague Manual.* Geneva: World Health Organization.

Butler, T., 1983. *Plague and other Yersinia infections.* New York: Plenum Press.

Campbell, G., and D.T. Dennis, 1998. Plague and other *Yersinia* infections. In: Kasper D., et al., eds. Harrison's principles of internal medicine. 14th ed. New York: McGraw Hill.

CDC, 2003. *Japanese Encephalitis Fact Sheet.* Accessed 03/14/12 at http://www.cod.gov/ncidod/dvibid/jencephalitis/facts.htm.

CDC, 2005. *Plague Fact Sheet.* Accessed 03/13/12 at www.codc.gov/incidod/dvbid/plague.

CDC, 2006. *West Nile Virus Fact Sheet.* Accessed 03/14/12 at http://www.cdc.gov/ncidod/dvbid/westnile/wnv_factsheet.htm.

CDC, 2008. *Chikungunya Fact Sheet.* Accessed 03/14/12 at http://www.cdc.gov/ncidod/dvbid/Chikungunya?CH_FactSheet.html.

CDC, 2009a. *La Crosse Encephalitis.* Accessed 03/14/12 at http://www.cdc.gov/LAC/.

CDC, 2009b. *La Cross Encephalitis Fact Sheet.* Accessed 03/14/12 at http://www.cdc.gov/Lac/tech/fact.html.

CDC, 2009c. *Saint Louis Encephalitis.* Accessed 03/14/12 at http://www.cdc.gov/sle/.

CDC, 2009d. *Saint Louis Encephalitis Fact Sheet.* Accessed 03/14/12 at http://www.cfidc.gov/sle/technical/fact.html.

CDC, 2010a. *Eastern Equine Encephalitis.* Accessed 03/14/12 at http:// www.cdc.gov/EasternEquineEncephalitis/.

CDC, 2010b. *Technical Fact Sheet: Eastern Equine Encephalitis.* Accessed 03/14/12 at http://www.codc.gov/Eastern Equine Encephalitis/tech/fastSheet.html.

CDC, 2011a. *Lyme Disease.* Accessed 03/13/12 at http://www.cdc.gov/lyme/.

CDC, 2011b. *Tularemia.* Accessed 03/14/12 at http://www.cdc.gov/Tularemia.

CDC, 2011c. *Yellow Fever.* Accessed 03/15/12 at http://www.cdc.yellowfever/.

CDC, 2011d. *Dengue.* Accessed 03/15/12 at http://www.cdc.gov/dngue/.

CDC, 2011e. *Rocky Mountain Spotted Fever.* Accessed 03/15/12 at http://www.codc.gov/rmsf/.

CDC, 2012a. *Tick-borne Relapsing Fever (LBRF)*. Accessed 03/13/12 at http://www.cdc.gov/Relapsing-fever.

CDC, 2012b. *Chikungunya Virus*. Accessed 03/14/12 at http://www.cdc.gov/ncidod/dvbid/Chikungunya/index.html.

CDC, 2012c. *Dengue Entomology and Ecology*. Accessed 03/15/12 at http://www.cdc.dengue/entomologyEcology/index.html.

CDC, 2012d. *Anaplasmosis*. Accessed 03/15/12 at http://www.cdc.gov/anaplasmosis/.

Cutler, S.J., A. Abdissa, and J.F. Trape, 2009. New concepts for the old challenges of African relapsing fever borreliosis. *Clin Microbiol Infect* 15:400-406.

Gage, K.L., 1998. Plague. In: Colliers, L., A. Balows, M. Sussman, and W.J. Hausies, eds. Topley and Wilson's microbiology and microbiological infections, vol. 3. London: Edward Arnold Press.

Hayes, E.B., and D.T. Dennis, 2004. Relapsing Fever. In: Kasper, D.L., E. Braunwalk, A.S. Fauci, S.L. Hauser, D.L. Longer, and J.L. Jameson, eds. *Harrison's Principles of Internal Medicine*, 16th edition.

Perry, R.D., and J.D. Fetherston, 1997. *Yersinia pestis*—etiologic agent of plague. *Clin Microbiol Rev*. 10:35–36.

Rahlenbeck, S.I., and A. Gebre-Yohannes, 1995. Louse-borne relapsing fever and its treatment. *Tropical & Geographical Medicine* 47(2):49–52.

Raoult D., and V. Roux, 1999. The body louse as a vector of reemerging human diseases. *Clin Infect Dis*. 29(4):888–911.

Ryan, K.J., and C.G. Ray, eds., 2004. Eds. *Sherris Medical Microbiology*, 4th ed. New York: McGraw Hill.

7

Outdoor/Indoor Air Quality

Take a course in good water and air; and in the eternal youth of Nature you may review your own. Go quietly, alone; no harm will befall you.

—John Muir

All of us face a variety of risks to our health as we go about our day-to-day lives. . . . Indoor air pollution is one risk that you can do something about.

—EPA

Sun and Air

The air staggers under the sun, and heat morasses
Flutter the birds down; wind barely climbs the hills,
Saws thin and splinters among the roots of the grasses;

All air sickens, and falls into barn shadows, spills
    Into hot hay and heat-hammered road dust, makes no sound,
Waiting the sun's siege out to collect its wills.

As a hound stretched sleeping will all of a sudden bound,
Air will rise all sinews, crack-crying, tear tether,
Plow sheets of powder high, heave sky from milling ground.

So sun and air, when these two goods war together,
Who else can tune day's face to a softest laugh,
Being sweet beat the world with a most wild weather,

Trample with light or blow all heaven blind with chaff.

(R. Wilber, 1947, 365)

(INFORMATION IN THIS INTRODUCTORY SECTION is based on F. R. Spellman, 2009, *The Science of Environmental Pollution,* 2nd ed. Boca Raton, FL: CRC Press.) Do you ever think about atmospheric air—the air we breathe? Unless you are an air scientist, an air technology specialist, or a practitioner of air science, probably not. The rest of us? No, not really. We usually only think about air, again, the air we breathe, when there is not enough of it to breathe or when it offends us. Air, like water and soil, is simply one of those "things" that are all around us. Air, water, and soil are everywhere and seemingly limitless. We breathe air, we drink water, we plant our crops in soil (and are sometimes interred in it) . . . they are just out there, all three of them—especially air, which literally enshrouds the planet we live on. On the average, we each need at least thirty pounds of air every day to live, but we need only about three pounds of water and less of food. A person can live about five weeks or so without food and about five days without water, but only about five minutes without air. Why don't we give air a second or even a first thought? Normally, we don't—and this is somewhat surprising when you consider that if we take that air away, life ends. Simply put, air is necessary for the survival of most life forms on earth.

As one of the three vital environmental mediums (air, water, soil), air is important—initially, the most important—without air, the other two are meaningless. Maintaining air quality is even more important—important not only to our environmental health but also to our very existence. In this chapter, we give air a lot of room, a lot of attention, and a lot of respect. We talk about the environmental problems and more—about air itself, about air pollution, and about indoor air quality.

All About Air

Any discussion about air has to begin with the earth's atmosphere. Since intelligent human life appeared on earth, we have described the atmosphere in many different ways—often in ways that glorify its mysteriousness. However, more often it is quite simply described as that mixture of gases that surrounds the earth, or as that envelope of gases that encapsulates the earth.

To put earth's atmosphere into a visual perspective, we use the following popular analogy: if the earth were the size of an apple, the atmosphere would be no thicker than the apple's skin. Beyond the atmosphere (the skin) is an infinite expanse of space.

Our lives depend on a relatively thin layer of gases—gases that are prevented from escaping earth by the pull of gravity. A few quick facts:

- Three-quarters of the mass of earth's atmosphere lies below a height of 35,000 feet.
- Atmospheric pressure decreases with height in the atmosphere.
- The air at the top of Mount Everest is only one-third as thick as at sea level.
- The weight of atmosphere is 5.7 × 1015 tons, about one-millionth of the weight of the earth.

In its lowest layer, the atmosphere consists of nitrogen (78 percent) and oxygen (21 percent). The other 1 percent is largely argon, with very small quantities of other gases, including water vapor (on average about 0.7 percent by volume) and carbon dioxide (see table 7.1).

The atmosphere plays a major part in the various cycles of nature (biogeochemical cycles), including the water cycle, carbon cycle, and nitrogen cycle. It is stratified with altitude with respect to temperature. With increasing altitude, air pressure decreases exponentially. The density of air decreases, but not drastically. The principal industrial source of nitrogen, oxygen, and argon is atmospheric, obtained by fractional distillation of liquid air.

Stratification of the Atmosphere

The lowest level of the atmosphere, the troposphere, is heated by the earth, which is warmed by infrared and visible radiation from the sun. It extends from sea level to a height of about 10 to 17 kilometers (6 to 11 miles), and contains approximately 75 percent of the total air mass on earth, and virtually all the water vapor. Warm air cools as it rises in the troposphere,

TABLE 7.1
Composition of Air

Gas	*% By Volume*	*Molecular Weight*
Nitrogen (N2)	78.08	28.02
Oxygen (O2)	20.98	32.00
Argon (Ar)	0.9	39.88
Carbon dioxide	0.04	44.00
Helium		trace
Hydrogen		trace
Methane		trace
Ozone		trace
Krypton		trace
Neon		trace
Xenon		trace

causing rain and most other weather phenomena; all cloud formations, precipitation, and seasonal changes occur in the troposphere. Prevalent turbulence (mainly winds) in the troposphere causes diffusive (Fickian) transport (i.e., the movement of molecules in air from higher to lower concentration regions) of atmospheric chemicals, along with advective transport (i.e., the transport of a pollutant as it is carried along by the mass movement of air from higher to lower regions of pressure). In the troposphere, chemicals mix in a few weeks, whereas it may take years to mix in the upper layer. There is a general decrease of temperature with height throughout this layer at a mean rate of 6.4° C per km. Typically, the temperature at the top of the layer is –55° C.

The standard atmospheric pressure is 1 atm or 14.7 lb/in2 at sea level. At an altitude of fifteen kilometers, this decreases to 1.6 psi. Between the layers of the atmosphere, pauses (or regions of transition) exist. The transition between the troposphere and the next stratification is called the tropopause, a thin layer of relatively stable temperature. Above the tropopause, the air does not hold enough oxygen to support life.

The layer above the troposphere (the stratosphere) extends up to fifty kilometers, and in terms of air circulation patterns is a stable layer. It is deeper than the troposphere, but contains only a small part of the total air mass because of its lower density. However, along with the troposphere, almost all of the remaining air mass is contained here; 99 percent of air occurs within the two layers. Most of the atmospheric natural ozone (O3) is contained in the stratosphere. This stratospheric ozone plays a significant environmental role as a barrier to harmful ultraviolet (UV) radiation from the sun.

Layers of the atmosphere above the stratosphere include the mesosphere, the ionosphere, and the thermosphere. These portions of the atmosphere are essentially unaffected by air pollution and thus are not discussed further in this text.

Did You Know?

Only when we lack for air or the air we breathe is fouled do we give it much thought. Once in a while, air and air quality problems come to our attention through the media. When this occurs, we usually read or hear about recurring horror stories relating to the damage we are doing to our atmosphere (the source of our air). We hear about what atrocities the industries and our internal combustion engines have committed against air, how the chemicals used in many modern products have choked the atmosphere with pollutants that have the power to affect our health, our well-being, our environment, our lives. Smog can be offensive and deadly. The burning of fossil fuels has given us acid rain. We have air-quality alerts in smaller urban areas—not simply in big cities anymore. Many fear a dangerous climate change—a warming trend known as the greenhouse effect. Man-made chemical compounds are shredding the ozone layer, increasingly exposing us to the sun's ultraviolet radiation.

Air Pollution

Air pollution—what is it? This question might be too broad; let's try again. Atmospheric air pollution—what is it? The standard answer might be, "Atmospheric air pollution is the contamination of the atmosphere with the harmful by-products of human activity." Another answer might be "contamination of the atmosphere by the discharge (accidental or deliberate) of one or more of a wide range of toxic airborne substances." Are you confused?

Well, we understand. Actually we like the second definition better than the first, because the second definition's qualifications about "accidental or deliberate" and its inclusion of "a wide range of toxic airborne substances" are truer to the actual situation.

Let's see if we can clear up the confusion and come up with our own definition of air pollution, of what air pollution really is, at least in terms of how it is used in this text.

First, we need to agree that air pollutants are those substances in the air that are not "normal" constituents of the atmosphere (oxygen, nitrogen, water vapor, etc.); these "abnormal" substances are, of course, air pollutants.

Secondly, another reason we like the second definition is because it is not implicit in its assumption that air pollution is anthropogenic (caused by human activities alone). This is certainly not the case, because air pollution can result from natural sources, and the resulting pollutants can be much more severe and long-lasting than air pollution from human activities. For example, an eruption from a volcano can spew vast quantities of dust and gases into the atmosphere in a relatively short period of time. Pollutants from such an eruption can reach the upper atmosphere and act as sunlight reflectors, causing global temperatures to fall slightly for a few years or longer. Other natural air pollutants include smoke from forest fires, salt sea spray, dust blown from desert areas, and pollen grains.

What does all this mean? What it means is that we need to come up with a more definitive definition of air pollution. And we have. Our definition: Air pollution is the natural or man-induced presence of certain substances in the air in high enough concentrations and for long enough duration to cause undesirable effects.

Air pollutants can be classified as either primary or secondary. Primary air pollutants are made up of constituents emitted directly into the air from a stationary source (a source contributor), such as a factory smokestack. Secondary air pollutants, however, are made up of elements emitted directly from sources, which then form new substances in the atmosphere as a result of complex chemical reactions involving the primary pollutants and sunlight, such as non-methane VOC conversion to ozone (O3).

Primary air pollutants may also be categorized by being either stationary (e.g., power plants) or mobile (e.g., automobiles) sources.

Indoor and outdoor (or ambient) air pollutants are other frequently used classifications. In this text, we focus on ambient air pollutants but point out that indoor air pollution is beginning to receive the attention it merits.

At the present, two types of ambient air pollutants are regulated under the Clean Air Act (CAA): criteria and hazardous air pollutants. Under the National Ambient Air Quality Standards (NAAQS) promulgated under CAA, five primary pollutants and one secondary pollutant are characterized. These six pollutants are hazardous to human health or welfare and are generated in relatively large quantities.

Transportation is the largest contributor of pollutants at approximately 46 percent. Fuel combustion at stationary facilities is the next highest contributor at almost 30 percent. Industrial processes and miscellaneous sources contribute about 6 percent each, while solid waste disposal accounts for about 2 percent.

Criteria Air Pollutants

The criteria air pollutants consist of the five primary criteria pollutants of sulfur dioxide (SO2), nitrogen oxides (NOx), carbon monoxide (CO), particulates, and particulate lead. Ozone (O3) is the secondary criteria pollutant regulated under the NAAQS.

Sulfur Dioxide—the element sulfur may be contained in certain fossil fuels such as coal. The sulfur is oxidized during burning of these fuels, producing sulfur dioxide gas. Sulfur dioxide is a colorless gas with a sharp, choking odor; it irritates the eyes, nose, and throat, and causes chronic bronchitis. It is a primary pollutant since it is emitted directly in the form of SO2. Approximately 25 million tons of SO2 are discharged into the atmosphere in the United States each year, mostly from fossil fuel combustion used in electric utility power plants. SO2 is a constituent of acid rain.

Nitrogen Oxides—characterized collectively as NOx, nitrogen oxides come in many forms. Most emissions are initially in the form of nitrogen oxide (NO), which by itself is not harmful at the concentrations usually found in the atmosphere. However, when NO oxidizes to NO2 in the presence of sunlight, it can react with hydrocarbons to form photochemical smog. NO2 also reacts with the hydroxyl radical (OH-) to form nitric acid (HNO3), a contributor to acid rain. NO is colorless, but when converted to NO2, it tends to give smog a reddish-brown color. The largest source of nitrogen oxides is from the oxidation of nitrogen compounds during the combustion of certain fossil fuels, such as coal and gasoline.

Carbon Monoxide—incomplete combustion of fossil fuels produces carbon monoxide (CO). Completely invisible, carbon monoxide is colorless, odorless, and tasteless. CO can be a serious health hazard, causing effects that range from slight headaches to nausea to death, depending on concentration and conditions.

Particulate Matter—particulates are extremely small fragments of solids or liquid droplets suspended in air. Particulates (lead being the exception) are distinguished on the basis of particle size and source, rather than by chemical composition. Most particulates range in size from 0.1 to 100 microns. The particulate materials of most concern for adverse effects on human health (ones that can cause upper respiratory infection, cardiac disorder, bronchitis, asthma, pneumonia, and emphysema) are generally less than 10 microns in size and are referred to as PM10. Particulates also reduce visibility and intensify certain chemical reactions in the atmosphere.

Lead Particulates—in fume form (less than 0.5 microns in size) lead particulates are toxic and can lead to anemia, destructive behavior, learning disabilities, seizures, brain damage, and death. Major sources of lead (Pb) in the past were motor vehicles that burned gasoline containing lead-based antiknock additives. The USEPA now requires the use of unleaded gasoline, but lead is still emitted from petroleum refining and smelting operations and other industrial activities.

Ozone (O3)—is the key component of photochemical smog (smog = smoke + fog); it is a secondary air pollutant in the troposphere formed by a complex chemical reaction between nitrogen dioxide (NO2) and volatile organic compounds (VOCs). VOCs are hydrocarbons that are gaseous under normal atmospheric conditions. The reactions are initiated by ultraviolet energy in sunlight. Along with producing smog, ozone irritates the eyes, nose, and throat, reduces lung function, damages plants, and causes cracking of paints, rubber, and textiles.

Deposition of Pollutants in the Atmosphere

When pollutants are released from their sources into the atmosphere, their fate may be affected by three different mechanisms: (1) physical deposition,

(2) transport and dispersal downwind, and (3) photochemical and oxidizing atmospheric chemical reactions.

Physical deposition is the process whereby gravitational settling, impaction, absorption, and wet deposition with precipitation occurs, effectively removing many pollutants from the atmosphere near the source.

Dispersion of pollutants in the atmosphere occurs when pollutants emitted from smokestacks or automobile exhausts are dispersed and transported away from the source following the motion of the gas in which they are borne.

Atmospheric reactions are the result of photochemical reactions driven by energy from the sun, and chemical transformations as a result of oxidation reactions, both of which contribute to many atmospheric pollution problems. For example, oxides of sulfur and nitrogen are oxidized in the atmosphere to form acids, contributing to acid rain.

Atmospheric Pollution Problems

Many air pollution problems have been recognized, ranging from small areas affected by a single industry, to citywide problems contributed by multiple contaminants, to the global-scale contamination by universal pollutants. Specifically, the problems pertain to:

- Acid deposition
- Smog formation
- Stratospheric ozone depletion
- Climate change

Acid Deposition—During the 1970s, the press began reporting about a new environmental hazard: acid deposition. They described it in grim terms as "acid rain" and in even grimmer terms as "death from the sky."

So, what is acid deposition? Are the reports of its effects on earth as grim as the reporters first claimed in the 1970s?

Acid deposition (acid rain) results when gaseous emissions of sulfur oxides (SOx) and nitrogen oxides (NOx) interact with water vapor and sunlight and are chemically converted to strong acidic compounds such as sulfuric acid (H2SO4) and nitric acid (HNO3). Along with other organic and inorganic chemicals, these compounds are deposited on the earth as aerosols and particulates (dry deposition), or are carried to the earth by raindrops, snowflakes, fog, or dew (wet deposition).

Keep in mind that most rainfall is naturally slightly acidic. Decomposing organic matter, the movement of the sea, and volcanic eruptions all contribute to natural acidity levels in precipitation. However, the principal contributor to the acid rain phenomenon is atmospheric carbon dioxide, which causes carbonic acid to form. Acid rain (in the pollution sense) with a pH <5.6 is produced by the conversion of the primary pollutants sulfur dioxide and nitrogen oxides to secondary pollutants—sulfuric acid and nitric acid, respectively. These complex processes depend upon physical dispersion and chemical conversion rates.

Although we think of acidic precipitation as a relatively new problem, it is not a modern phenomenon (a 1970s and onward event), nor does it result solely from industrial pollution. However, the rise in manufacturing that began with the Industrial Revolution literally dwarfs all other contributions to the problem.

The chief source of sulfur dioxide emissions is the burning of fossil fuels (including oil and coal). Nitrogen oxide, formed mostly from internal combustion engine emissions, is readily transformed into nitrogen dioxide. They mix in the atmosphere to form sulfuric acid and nitric acid.

Acid deposition is usually seen in or near industrialized and highly populated areas. For example, in the United States, most of the blame for the problem can be traced to the many coal-burning electric utility plants that are found in the Midwest and the East. Approximately 68 percent of their smokestacks account for the sulfur dioxide emitted annually. Approximately 11 percent comes from industrial burning. Other emissions—from automobile exhausts, for example—account for the other 21 percent.

Acid deposition is felt most severely in the eastern United States, because of wind patterns. Most of the United States lies within 30° to 60° North Latitude, where the prevailing wind comes from the west. The wind picks up much of the sulfur and nitrogen emitted in the Midwest, carries it eastward to mingle with the emissions already there, and then fans the entire mass out in a broad north-to-south pattern. The fact is, no matter what part of the world is affected, the hardest hit regions always lie in the path of the winds approaching from industrial and highly populated areas.

Note that in dealing with atmospheric acid deposition, the earth's ecosystems are not completely defenseless; through natural alkaline substances in soil or rocks that buffer and neutralize acids, they can deal with a certain amount of acid. The American Midwest and southern England possess highly alkaline soil (limestone and sandstone), which provides some natural neutralization. Areas with thin soil and those laid on granite bedrock, however, suffer most from acid rain. Many of our environments have little ability to neutralize acid deposition.

We have explained what acid deposition is, but we still need to answer the other question: are the reports of its effects on earth as grim as reported?

Scientists continue to study the acid deposition problem. Over the years, acid deposition has been accused of doing much damage. For example, it has been blamed for damaging freshwater life, forests, food crops, buildings, and even our physical health. What is the truth? Again, scientists are studying these potential effects and how living beings are damaged and/or killed by acid deposition. This complex subject has many variables, many of them related to the fact that pollution can travel over very long distances. For example, bodies of water in Canada and New York suffer the effects of coal-burning in the Ohio Valley. Because they are so seriously affected by it, and for other reasons, most of the scientific studies have taken place in the world's lakes. In lakes, acid precipitation quickly affects population. Smaller organisms often die off first, leaving the larger animals to die by starvation.

In the United States, some locations report that rainfall acidity levels have fallen well below pH 5.6. In the northeastern United States, for example, the average pH of rainfall is 4.6, and rainfall with a pH of 4.0, which is 100 times more acidic than distilled water, is not uncommon.

So what is the answer? Is acidic deposition as bad as they say it is ("death from the sky")? Despite intensive, ongoing research into most aspects of acid deposition, the jury is still out. Scientists still have many areas of uncertainty and disagreement—actually, scientific opinion is divided on the possible answer to this question. However, let us point out that most scientists do not believe acid deposition is solely at fault for all the damage being done. Nor do most think it completely innocent of such damage. Rather, they feel that much of the havoc attributed to acid deposition is actually the result of a combination of factors, some of which—such as the weather—are fashioned by nature, while others are anthropogenic pollutants. Acid deposition has simply joined these forces to worsen its impact.

And so, where do we stand today in our attempt to solve the acid deposition puzzle? Let's attempt to answer this question with a question: have you ever attempted to complete a jigsaw puzzle when a few of the pieces were missing? If you have, then you can appreciate where we are in solving the acid deposition puzzle; we know a few of the puzzle pieces are still missing. The parts of the puzzle that are complete show us a picture of a wide range of acid deposition damage. This damage includes destabilizing of ecosystems such as loss of forests. Lakes, streams, and watersheds, and aquatic life within, are victims of acid deposition. Man-made monuments (statues, for example) have been disfigured by the corrosive behavior of acid deposition. We also see (in the part of the puzzle that is complete) that even human health is at risk from trace metals leached into the water supply by acidified groundwater. These things we can see in the puzzle; these things we know. The problem is, we do not know what we do not know—we don't know what information is contained on the missing pieces of the puzzle. The "puzzle picture" is incomplete at this time. And because of this—because we are not exactly sure of the "total" impact of acid deposition, but nonetheless have suspicions about its consequences, the USEPA in the 1990 CA Act strengthened its requirements to permanently reduce levels of SO2 and nitrogen oxides. We feel that this was a prudent move.

Smog Formation—Smog, a term coined more than forty years ago from the words **sm**oke and f**og**, is a dirty, yellow-brown, cloudy formation (created in the lower portion of the troposphere near the ground level) resulting from the photochemical reaction of sunlight on the oxides of nitrogen and hydrocarbons emitted from automobiles. When coined forty years ago, the term *smog* actually referred to the clouds of sulfur dioxide, sulfuric acid droplets, and heavy suspended particles discharged by industries. Since then, this type of smog (known as gray smog) had been identified for what it really is: a rare type of smog. The smog we are most familiar with (Los Angeles–type smog) is principally a gas formed through the photochemical process on warm days when nitrogen oxides and hydrocarbons in the atmosphere react to sunlight and turn themselves into ozone, a major constituent of smog—the type of smog that causes irritation of eyes and throat, impairs pulmonary function, damages plants and crops, and makes rubber-based products crack and leak.

An ally to smog that is particularly annoying (and dangerous) is a condition known variously as a thermal inversion, or temperature inversion. To understand how a thermal inversion works, we have to begin with the fact that as the altitude increases, the atmosphere becomes thinner and the pressure drops. This characteristic action causes a parcel of exhausted gas moving upward to undergo an expansion that reduces its temperature. This natural phenomenon of reduction or lapse in temperature is known as the adiabatic lapse rate (the rate at which the temperature of the gas parcel drops due to elevation change). As we said earlier, under the assumption that the process is adiabatic, no exchange of heat occurs between the gas parcel and the surroundings.

As Dolan puts it, in simple terms, consider that "when warm air at ground level rises, becoming cooler at it travels upward, pollutants are carried along on the upward rise and then can be better dispersed in the atmosphere. A thermal inversion, which often arrives in hot weather, sees a layer of warm air settle in over an area. It is warmer than the air below it, and thus keeps the lower air—and the pollution in it—from rising. Trapped as it is, the lower air becomes increasingly filthy as more and more pollutants are added from car exhaust systems and industrial burnings" (p. 22).

Typically, such thermal inversions bring about results that induce a wide variety of physical discomforts, including watery eyes, coughing, sneezing, choking, nausea, and shortness of breath. However, records indicate that such events have also caused deaths. For example, in 1952 in London, such an event caused an ultimate death toll of 12,000. Another attack, this one in Donora, Pennsylvania, in 1948, caused twenty deaths. Still another attack in New York City in 1966 caused the deaths of 168 people.

Stratospheric Ozone Depletion—Try to imagine, if you can, a hole in the sky as large as the United States—a hole extending some 29,000 feet up through the layer. The hole we are referring to is the giant hole first reported in 1985, above Antarctica. The hole is better known as the "ozone hole."

Ozone hole? "So what?" you say. "What's it to me?" To answer this question, we need to understand exactly what the ozone layer is and its importance for life on earth.

Ozone is the Jekyll and Hyde of the earth's atmosphere; it has a good side and a bad side. Ozone in the troposphere, created by interaction of hydrocarbons, nitrogen oxides, and sunlight, leads to smog-related problems. This is the "Hyde," or bad side, of ozone. The stratosphere also contains ozone, at altitudes of ten to thirty kilometers—ozone that is responsible for absorbing (and shielding us from) harmful ultraviolet rays of the sun before they hit the earth. This is the "Jekyll," or good, side of ozone.

A large percentage of the ozone in the stratosphere is created during electrical storms near the equator—then the winds pick it up and shift it toward the polar regions. At the same time, the newly created ozone rises until it reaches the stratosphere where it spreads like a thin veil over the world and performs its vital function of blocking the flow of harmful UV radiation. The stratospheric ozone layer is analogous to a layer of gaseous "sun block" that protects us from UV radiation.

The obvious question is, What causes the hole in the ozone layer? Scientists theorize that increasing concentrations of the synthetic chemicals known as CFCs (and others such as Halon) are breaking down the ozone layer. CFCs are organic chemicals called chlorofluorocarbons. Sources of CFC gases include aerosol spray cans, refrigerants, industrial solvents, and foam insulation; they do not degrade easily. As a result, when released in the lower atmosphere (troposphere), they rise into the stratosphere, where they are broken down by UV light. The chlorine atoms from CFCs react with ozone to convert it into two molecules of oxygen. Each chlorine atom can destroy many ozone molecules; it is estimated that one chlorine atom is capable of destroying over 10,000 ozone molecules. The result—the ozone hole.

Climate Change—Is earth's climate changing? Are warmer times or colder times on the way? Is the greenhouse effect going to affect our climate, and if so, do we need to worry about it? Will the tides rise and flood New York? Does the ozone hole portend disaster right around the corner?

These and many other questions related to climate change have come to the attention of us all. A constant barrage of newspaper headlines, magazine articles, and television news reports on these topics have, over the years, inundated us. Recently, we've seen constant reports on El Niño and its devastation of the West Coast of the United States and Peru and Ecuador—and its reduction of the number, magnitude, and devastation of hurricanes that annually blast the East Coast of the United States.

Scientists have been warning us of the catastrophic harm that can be done to the world by atmospheric warming. One view states that the effect could bring record droughts, record heat waves, record smog levels, and an increasing number of forest fires.

Another caution put forward warns that the increasing atmospheric heat could melt the world's icecaps and glaciers, causing ocean levels to rise to the point where some low-lying island countries would disappear, while the coastlines of other nations would be drastically altered for ages—or perhaps for all time.

What's going on? We hear plenty of theories put forward by doomsayers, but are they correct? If they are correct, what does it all mean? Does anyone really know the answers? Should we be concerned? Should we invest in waterfront property in Antarctica? Should we panic?

No. While no one really knows the answers, and while we should be concerned, no real cause for panic exists.

Should we take some type of decisive action—should we come up with quick answers and put together a plan to fix these problems? What really needs to be done? Is there anything we can do?

The key question to answer here is, What really needs to be done? We can study the facts, the issues, the possible consequences—but the key to successfully combating these issues is to stop and seriously evaluate the problems. We need to let scientific fact, common sense, and cool-headedness prevail. Shooting from the hip is not called for, makes little sense—and could have titanic consequences for us all.

The other question that has merit here is, Will we take the correct actions before it is too late? The key words here are *correct actions*. Eventually, we will have to take action. But we do not yet know what those actions should be.

Then there is another side of the dilemma or argument to contemplate: if, as the experts say, we are experiencing global warming, is it really a problem? In this day and age of the high cost of limited energy supplies, is global warming all that bad? Is the alternative to global warming—another global ice age—a better option? Which do you prefer?

Indoor Air Pollution

Consider the quote "the air is the air." However, in regard to the air we breathe, according to the USEPA (2003), few of us realize that we all face a variety of risks to our health as we go about our day-to-day lives. Driving our cars, flying in planes, engaging in recreational activities, and being exposed to environmental pollutants all pose varying degrees of risk. Some risks are simply unavoidable. Some we choose to accept because to do otherwise would restrict our ability to lead our lives the way we want. And some are risks we might decide to avoid if we had the opportunity to make informed choices. Indoor air pollution is one risk that we can do something about.

In the last several years, a growing body of scientific evidence has indicated that the air within homes and other buildings can be more seriously polluted than the outdoor air in even the largest and most industrialized cities. Other research indicates that people spend approximately 90 percent of their time indoors. Thus, for many people, the risks to health may be greater due to exposure to air pollution indoors than outdoors (USEPA, 2003).

In addition, people who may be exposed to indoor air pollutants for the longest periods of time are often those most susceptible to the effects of indoor air pollution. Such groups include the young, the elderly, and the chronically ill, especially those suffering from respiratory or cardiovascular disease.

The impact of energy conservation on inside environments may be substantial, particularly with respect to decreases in ventilation rates (Hollowell et al., 1979a) and "tight" buildings constructed to minimize infiltration of outdoor air (Woods, 1980; Hollowell et al., 1979b). The purpose of constructing "tight buildings" is to save energy—to keep the heat or air-conditioning inside the structure. The problem is that indoor air contaminants within these tight structures are not only trapped within but also can be concentrated, exposing inhabitants to even more exposure.

What about the air quality problems in the workplace? In this section, we discuss this pervasive but often overlooked problem. In this regard, we discuss the basics of indoor air quality as related to the workplace environment and the major contaminants that currently contribute to this problem. Moreover, mold and mold remediation, although not new to the workplace, are the new buzzwords attracting the environmental health practitioner's attention these days.

What Is Indoor Air Quality (IAQ)?

According to Byrd (2003), Indoor Air Quality (IAQ) refers to the effect, good or bad, of the contents of the air inside a structure on its occupants. Stated differently, IAQ, in this text, refers to the quality of the air inside workplaces as represented by concentrations of pollutants and thermal (temperature and relative humidity) conditions that affect the health, comfort, and performance of employees. Usually, temperature (too hot or too cold), humidity (too dry or too damp), and air velocity (draftiness or motionlessness) are considered "comfort" rather than indoor air quality issues. Unless they are extreme, they may make someone uncomfortable, but they won't make a person ill. Other factors affecting employees, such as light and noise, are important indoor environmental quality considerations, but are not treated as core elements of indoor air quality problems. Nevertheless, most environmental health practitioners must take these factors into account in investigating environmental quality situations.

Byrd (2003) further points out that good IAQ is the quality of air that has no unwanted gases or particles in it at concentrations that will adversely affect someone. Poor IAQ occurs when gases or particles are present at an excessive concentration so as to affect the satisfaction of health of occupants.

In the workplace, poor IAQ may only be annoying to one person; however, at the extreme it could be fatal to all the occupants in the workplace.

The concentration of the contaminant is crucial. Potentially infectious, toxic, allergenic, or irritating substances are always present in the air. Note that there is nearly always a threshold level below which no effect occurs.

Why Is IAQ Important to Workplace Owners?

Workplace structures (buildings) exist to protect workers and equipment from the elements and to otherwise support worker activity. Workplace buildings should not make workers sick, cause them discomfort, or otherwise inhibit their ability to perform. How effectively a workplace building functions to support its workers and how efficiently the workplace building operates to keep costs manageable is a measure of the workplace building's performance.

The growing proliferation of chemical pollutants in industrial and consumer products, the tendency toward tighter building envelopes and reduced ventilation to save energy, and pressures to defer maintenance and other building services to reduce costs have fostered IAQ problems in many workplace buildings. Employee complaints of odors, stale and stuffy air, and symptoms of illness or discomfort breed undesirable conflicts between workplace occupants and workplace managers. Lawsuits sometimes follow.

If IAQ is not well managed on a daily basis, remediation of ensuing problems and/or resolution in court can be extremely costly. Moreover, air quality problems in the workplace can lead to reduced worker performance. So it helps to understand the causes and consequences of IAQ and to manage your workplace buildings to avoid these problems.

Worker Response to Poor Air Quality

Worker responses to pollutants, climatic factors, and other stressors such as noise are generally categorized according to the type and degree of responses and the time frame in which they occur. Workplace managers should be generally familiar with these categories, leaving detailed knowledge to environmental health professionals such as industrial hygienists.

Acute Effects—are those that occur immediately (e.g., within twenty-four hours) after exposure. Chemicals released from building materials may cause headaches, or mold spores may result in itchy eyes and runny nose in sensitive individuals shortly after exposure. Generally, these effects are not long lasting and disappear shortly after exposure ends. However, exposure to some biocontaminants (fungi, bacteria, viruses) resulting from moisture problems, poor maintenance, or inadequate ventilation have been known to cause serious, sometimes life-threatening respiratory diseases that themselves can lead to chronic respiratory conditions.

Chronic Effects—are long-lasting responses to long-term or frequently repeated exposure. Long-term exposure to even low concentrations of some chemicals may induce chronic effects. Cancer is the most commonly associated long-term health consequence of exposure to indoor air contaminants. For example, long-term exposure to environmental tobacco smoke, radon, asbestos, and benzene increase cancer risk.

Discomfort—is typically associated with climatic conditions, but workplace building contaminants may also be implicated. Workers complain of being too hot or too cold or experience eye, nose, or throat irritation because of low humidity. However, reported symptoms can be difficult to interpret. Complaints that the air is "too dry" may result from irritation from particles on the mucous membranes rather than low humidity, or "stuffy air" may mean that the temperature is too warm or there is lack of air movement, or "stale air" may mean that there is a mild but difficult to identify odor. These conditions may be unpleasant and cause discomfort among workers, but there is usually no serious health implication involved. Absenteeism, work performance, and employee morale, however, can be seriously affected when building managers fail to resolve these complaints.

Performance Effects—are significant measurable changes in a worker's ability to concentrate or perform mental or physical tasks and have been shown to result from modest changes in temperature and relative humidity. In addition, recent studies suggest that the similar effects are associated with indoor pollution due to lack of ventilation or the presence of pollution sources. Estimates of performance losses from poor indoor air quality for all buildings suggest a 2 to 4 percent loss on average. Future research should further document and quantify these effects.

Workplace Building Associated Illnesses

The rapid emergence of IAQ problems and associated occupant complaints have led to terms that describe illnesses or effects particularly associated with buildings. These include sick building syndrome, building-related illness, and multiple chemical sensitivity.

Sick Building Syndrome (SBS)—is a catch-all term that refers to a series of acute complaints for which there is no obvious cause and where medical tests reveal no particular abnormalities. It describes situations in which more than 20 percent of the building occupants experience acute health and comfort effects that appear to be linked to time spent in a building because all other probable causes have been ruled out. The 20 percent figure is arbitrarily set, as there will always be some workers complaining about adverse health effects associated with occupancy of a building. However, if the figure is 20 percent or more, it is considered that there must be some determinable cause, which can be remedied. Symptoms include headaches; eye, nose, and throat irritation; dry cough; dry or itchy skin; dizziness and nausea; difficulty in concentration; fatigue; and sensitivity to odors (USEPA, 2001).

SBS is attributed to inadequate ventilation, chemical contaminants from indoor and outdoor sources, and biological contaminants such as molds, bacteria, pollens, and viruses. Passon et al. (1996), in their paper, *Sick-Building Syndrome and Building Related Illnesses* explain how increased air tightness of buildings in the 1970s, as a means of reducing energy consumption, has created environmental conditions conducive to the "proliferation of microorganisms (including mold) in indoor environments." Once growth has occurred, harmful organisms can be spread by improperly designed and maintained ventilation systems (USEPA, 2001).

A single causative agent (e.g., contaminant) is seldom identified and complaints may be resolved when building operational problems and/or occupant activities identified by investigators are corrected.

Increased absenteeism, reduced work efficiency, and deteriorating employee morale are the likely outcomes of SBS problems that are not quickly resolved.

Building-Related Illness (BRI)—refers to a defined illness with a known causative agent resulting from exposure to the building air. While the causative agent can be chemical (e.g., formaldehyde), it is often biological. Typical sources of biological contaminants are humidification systems, cooling towers, drain pans or filters, other wet surfaces, or water-damaged building material. Symptoms may be specific or mimic symptoms commonly associated with the flu, including fever, chills, and cough. Serious lung and respiratory conditions can occur. Legionnaires' disease, hypersensitivity pneumonitis, and humidifier fever are common examples of building-related illness.

Multiple Chemical Sensitivity (MCS)—it is generally recognized that some workers can be sensitive to particular agents at levels that do not have an observable effect in the general population. In addition, it is recognized that certain chemicals can be sensitizers in that exposure to the chemical at high levels can result in sensitivity to that chemical at much lower levels.

Some evidence suggests that a subset of the worker population may be especially sensitive to low levels of a broad range of chemicals at levels common in today's home and working environments. This apparent condition has come to be known as multiple chemical sensitivity (MCS).

Workers reported to have MCS apparently have difficulty being in most buildings. There is significant professional disagreement concerning whether MCS actually exists and what the underlying mechanism might be. Building managers may encounter occupants who have been diagnosed with MCS. Resolution of complaints in such circumstances may or may not be possible. Responsibility to accommodate such workers is subject to negotiation and may involve arrangements to work at home or in a different location.

Building (Structure) Factors Affecting IAQ

Building factors affecting IAQ can be grouped into two factors: factors affecting indoor climate and factors affecting indoor air pollution.

Factors Affecting Indoor Climate—the thermal environment (temperature, relative humidity, and airflow) includes important dimensions of indoor air quality for several reasons. First, many complaints of poor indoor air may be resolved by simply altering the temperature or relative humidity. Second, people who are thermally uncomfortable will have a lower tolerance to other building discomforts. Third, the rate at which chemicals are released from building material is usually higher at higher building temperatures. Thus, if occupants are too warm, it is also likely that they are being exposed to higher pollutant levels.

Factors Affecting Indoor Air Pollution—much of the building fabric, its furnishings and equipment, its occupants and their activities produce pollution. In a well-functioning building, some of these pollutants will be directly exhausted to the outdoors and some will be removed as outdoor air enters that building and replaces the air inside. The air outside may also contain contaminants that will be brought inside in this process. This air exchange is brought about by the mechanical introduction of outdoor air (outdoor air ventilation rate), the mechanical exhaust of indoor air, and the air exchanged through the building envelope (infiltration and exfiltration).

Pollutants inside can travel through the building as air flows from areas of higher atmospheric pressure to areas of lower atmospheric pressure. Some of these pathways are planned and deliberate so as to draw pollutants away from occupants, but problems arise when unintended flows draw contaminants into occupied areas. In addition, some contaminants may be removed from the air through natural processes, as with the absorption of chemicals by surfaces or the settling of particles onto surfaces. Removal processes may also be deliberately incorporated into the building systems. Air filtration devices, for example, are commonly incorporated into building ventilation systems.

Thus, the factors most important to understanding indoor pollution are a) indoor sources of pollution, b) outdoor sources of pollution, c) ventilation parameters, d) airflow patterns and pressure relationships, and e) air filtration systems.

Sources of Indoor Air Pollutants

Air quality is affected by the presence of various types of contaminants in the air. Some are in the form of gases. These would be generally classified as toxic chemicals. The types of interest are combustion products (carbon monoxide, nitrogen dioxide), volatile organic compounds (formaldehyde, solvents, perfumes and fragrances, etc.), and semi-volatile organic compounds (pesticides). Other pollutants are in the form of animal dander, etc.; soot; particles from buildings, furnishings, and occupants such as fiberglass, gypsum powder, paper dust, lint from clothing, carpet fibers, etc.; dirt (sandy and earthy material), etc.

Burge and Hoyer (1998) point out many specific sources for contaminants that result in adverse health effects in the workplace, including the workers (contagious diseases, carriage of allergens, and other agents on clothing); building compounds (VOCs, particles, fibers); contamination of building components (allergens, microbial agents, pesticides); and outdoor air (microorganisms, allergens, and chemical air pollutants).

When workers complain of IAQ problems, the environmental health practitioner is called upon to determine if the problem really is an IAQ problem. If he or she determines that some form of contaminant is present in the workplace, proper remedial action is required. This usually includes removing the source of the contamination.

Indoor Contaminant Transport

Contaminants reach worker breathing-zones by traveling from the source to the work by various pathways. Normally, the contaminants travel with the flow of air.

As mentioned, air moves from areas of high pressure to areas of low pressure. That is why controlling workplace air pressure is an integral part of controlling pollution and enhancing building IAQ performance.

Air movements should be from occupants, toward a source, and out of the building rather than from the source to the occupants and out the building. Pressure differences will control the direction of air motion and the extent of occupant exposure.

Driving forces change pressure relationships and create airflow. Major driving forces include: wind, stack effect, HVAC/fans, flues and exhaust, and elevators.

Common Airflow Pathways

Contaminants travel along pathways—sometimes over great distances. Pathways may lead from an indoor source to an indoor location or from an outdoor source to an indoor location. The location experiencing a pollution problem may be close by, in the same or an adjacent area, but it also may be a great distance from, and/or on a different floor from a contaminant source.

Knowledge of common pathways helps to track down the source and/or prevent contaminants from reaching building occupants. Ventilation can be used to either exhaust contaminants from a fixed source, or dilute contaminants from all sources within a space.

Discussion Questions

1. Write your own comprehensive definition and description of air pollution. Include the elements you consider most important and why.
2. Discuss the problems associated with air pollution migration.
3. What are the chief types of air pollutants? What are their common sources and most detrimental effects?
4. What mechanisms affect the fate of air pollutants? How does each work?
5. Describe and discuss the four chief problem areas related to air pollutants.
6. Discuss air pollution and its possible relationship to climate change.

7. What are the possible and probable ways global warming could affect human populations?
8. Define air, both chemically and from a human use perspective, as well.
9. Describe the physical makeup of the atmosphere.
10. What physical properties are critical to how air behaves? Discuss and describe them.

References and Recommended Reading

Associated Press, 1997. In the *Virginian-Pilot* (Norfolk, VA). "Does warming feed El Niño?" A-15, December 7.

Associated Press, 1998. In the *Lancaster New Era* (Lancaster, PA). "Ozone hole over Antarctica at record size," September 28.

Associated Press, 1998. In the *Lancaster New Era* (Lancaster, PA). "Tougher air pollution standards too costly, Midwestern states say." September 25.

Dolan, E.F., 1991. *Our Poisoned Sky*. New York: Cobblehill Books.

Hansen, J.E. et al., 1986. "Climate Sensitivity to Increasing Greenhouse Gases," In: *Greenhouse Effect and Sea Level Rise: A Challenge for this Generation*, Barth, M.C., and J.G. Titus, eds. New York: Van Nostrand Reinhold.

Hansen, J.E. et al., 1989. Greenhouse Effect of Chlorofluorocarbons and Other Trace Gases, *Journal of Geophysical Research* 94 November, 16, 416–17, 421.

Hesketh, H.E., 1991. *Air Pollution Control: Traditional and Hazardous Pollutants*, Lancaster, PA: Technomic Publishing Company.

Godish, T., 1997. *Air Quality*, 3rd ed., Boca Raton, FL: Lewis.

Masters, G.M., 1991. *Introduction to Environmental Engineering and Science*. Englewood Cliffs, NJ: Prentice-Hall.

Peavy, H.S., D.R. Rowe, and G. Tchobanglous, 1985. *Environmental Engineering*, New York: McGraw-Hill.

Time magazine, 1998. Global Warming: It's Here . . . And Almost Certain to Get Worse. August 24.

USA Today, 1997. Global Warming: Politics and economics further complicate the issue, A-1, 2, December 1.

Wilber, R., 1988, 1947. *New and Collected Poems*, New York: Harcourt, Brace, Jovanovich.

8
Water Quality

> Water and air, the two essential fluids on which all life depends, have become global garbage cans.
>
> —Jacques Cousteau

> Every eight months, nearly 11 million gallons of oil run off our streets and driveways into our waters—the equivalent of the Exxon Valdez oil spill.
>
> —*America's Living Oceans* (Pew Oceans Report, 2003)

Drinking Water—An Environmental Health Concern

(INFORMATION IN THIS SECTION comes from CDC 2011 ToxTown, *Drinking Water.* Accessed 03/19/12 at http://toxtown.nlm.nih.gov/text_version/Locations.php?id=18.) Drinking water comes from groundwater, wells, rivers, lakes, streams, and reservoirs. People in cities usually drink water from lakes, rivers, and reservoirs that is filtered and cleaned through water treatment plants. People in rural areas frequently drink water pumped from a private well.

Every drinking water supply is affected by activities close by and many miles away. All water contains some impurities. The U.S. Environmental Protection Agency (EPA) sets acceptable limits for more than eighty contaminants that may be in drinking water and pose a risk to human health.

Drinking water can be contaminated by natural conditions and recreational activities, including boating. It can be contaminated by pesticides, fertilizers, animal waste, damaged septic systems, leaking underground storage tanks, landfills, mining, and industrial releases to air and water. Drinking water can also be contaminated by runoff from farms, storms, urban areas, and industrial or construction sites. Endocrine disruptors can also be found in drinking water.

Waterborne diseases are a major human health concern because many people can be affected if a source of drinking water is contaminated. Bacteria and viruses can cause acute, or short-term, human health effects if they are found in drinking water at high levels.

More serious, chronic health effects can occur if people consume a contaminant in drinking water at levels above EPA standards for many years. Contaminants that can have long-term health effects include solvents, radioactive elements, and minerals such as arsenic. Contaminated drinking water can cause chronic health effects such as cancer, liver or kidney problems, and reproductive problems.

What's in the Water?

(Information in this section is adapted from USGS 2006. *A primer on water quality*. Accessed 03/19/12 at http://pubs.usgs.gov/fs.fs-027-01/.) Is the water safe to drink? Can fish and other aquatic life thrive in streams and lakes that are affected by human activities? What is the water quality? To answer these questions, it is helpful to understand what "water quality" means, how it is determined, and the natural processes and human activities that affect water quality.

Water quality can be thought of as a measure of the suitability of water for a particular use based on selected physical, chemical, and biological characteristics. To determine water quality, scientists first measure and analyze characteristics of the water such as temperature, dissolved mineral content, and number of bacteria. Selected characteristics are then compared to numeric standards and guidelines to decide if the water is suitable for a particular use.

Some aspects of water quality can be measured right in the stream or at the well. These include temperature, acidity (pH), dissolved oxygen (DO), and electrical conductance (an indirect indicator of dissolved minerals in the water). Analyses of individual chemicals generally are done at a laboratory.

Standards and guidelines are established to protect water for designated uses such as drinking, recreation, agricultural irrigation, or protection and maintenance of aquatic life. Standards for drinking water quality ensure that public drinking-water supplies are as safe as possible. The U.S. Environmental Protection Agency (USEPA) and the states are responsible for establishing the standards for constituents in water that have been shown to pose a risk to human health. Other standards protect aquatic life, including fish, and fish-eating wildlife such as birds.

Natural water quality varies from place to place, with the seasons, with climate, and with the types of soils and rocks through which water moves. When water from rain or snow moves over the land and through the ground, the water may dissolve minerals in rocks and soil, percolate through organic material such as roots and leaves, and react with algae, bacteria, and other microscopic organisms. Water may also carry plant debris and sand, silt, and clay to rivers and streams, making the water appear "muddy" or *turbid*. When water evaporates from lakes and streams, dissolved minerals are more concentrated in the water that remains. Each of these natural processes changes the water quality and potentially the water use.

The most common dissolved substances in water are minerals or salts that, as a group, are referred to as *dissolved solids*. Dissolved solids include *common constituents* such as calcium, sodium, bicarbonate, and chloride; plant *nutrients* such as nitrogen and phosphorus; and *trace elements* such as selenium, chromium, and arsenic.

In general, the common constituents are not considered harmful to human health, although some constituents can affect the taste, smell, or clarity of water. Plant nutrients and trace elements in water can be harmful to human health and aquatic life if they exceed standards or guidelines.

Dissolved gases such as oxygen and radon are common in natural waters. Adequate oxygen levels in water are a necessity for fish and other aquatic life. Radon gas can be a threat to human health when it exceeds drinking-water standards.

Human activities, such as urban and industrial development, farming, mining, combustion of fossil fuels, stream-channel alteration, or animal-feeding operations, can change the quality of natural waters. As an example of the effects of human activities on water quality, consider nitrogen and phosphorus fertilizers that are applied to crops and lawns. These plant nutrients can be dissolved easily in rainwater or snowmelt runoff. Excess nutrients carried to streams and lakes encourage abundant growth of algae, which leads to low oxygen in the water and the possibility of fish kills.

Chemicals such as pharmaceutical drugs, dry-cleaning solvents, and gasoline that are used in urban and industrial activities have been found in streams and groundwater. After decades of use, pesticides are now widespread in streams and groundwater, though they rarely exceed the existing standards and guidelines established to protect human health. Some pesticides have not been used for twenty to thirty years, but they are still detected in fish and streambed sediment at levels that pose a potential risk to human health, aquatic life, and fish-eating wildlife. There are so many chemicals in use today that determining the risk to human health and aquatic life is a complex task. In addition, mixtures of chemicals typically are found in water, but health-based standards and guidelines have not been established for chemical mixtures.

Note that the quality of water for drinking cannot be assured by chemical analyses alone. The presence of bacteria in water, which are normally found in the intestinal tracts of humans and animals, signal that disease-causing pathogens may be present. Giardia and cryptosporidium are pathogens that have been found occasionally in public-water supplies and have caused illness in a large number of people in a few locations. Pathogens can enter our water from leaking septic tanks, wastewater-treatment discharge, and animal wastes.

Did You Know?

Endocrine disruptors are chemicals that may interfere with the production or activity of hormones in the human endocrine system. These chemicals may occur naturally or be manufactured. The term *endocrine disruptors* describes a diverse group of chemicals that are suspected or known to affect human hormones. Effects on human hormones can range from minor to serious depending on the specific endocrine receptor and the amount of exposure. Because these chemicals are found in products you use every day and because you are exposed to many endocrine receptors at the same time, it is difficult to determine the public health effects of these chemicals.

Water Pollution

This country's waterways have been transformed by *omission*. Without beavers, water makes its way too quickly to the sea; without prairie dogs, water runs over the surface instead of sinking into the aquifer; without bison, there are no ground-water-recharge ponds in the grasslands and the riparian zone is trampled; without alligators, the edge between the water and land is simplified. Without forests, the water runs unfiltered to the waterways, and there is less deadwood in the channel, reducing stream productivity. Without floodplains and meanders, the water moves more swiftly, and silt carried in the water is more likely to be swept to sea.

The beaver, the prairie dog, the bison, and the alligator have been scarce for so long that we have forgotten how plentiful they once were. Beaver populations are controlled, because they flood fields and forests, while wetlands acreage decrease annually. Prairie dogs are poisoned, because they compete with cattle for grass, while the grasslands grow more barren year by year. Buffalo are generally seen as photogenic anachronisms, and alligators are too reptilian to be very welcome. But all of these animals once shaped the land in ways that improve water quality. (Alice Outwater, *Water: A Natural History*, 1996, 175-76)

Is water contamination really a problem—a serious problem? In answer to the first part of the question we can say it depends upon where your water comes from. As to the second part of the question, we refer you to a book (or the film based upon the book) that describes a case of toxic contamination—one you might be familiar with—*A Civil Action*, written by Jonathan Harr. The book and film portray the legal repercussions connected with polluted water supplies in Woburn, Massachusetts. Two wells became polluted with industrial solvents, in all apparent likelihood causing twenty-four of the town's children, who lived in neighborhoods supplied by those wells, to contract leukemia and die.

Many who have read the book or have seen the movie may mistakenly get the notion that Woburn, a toxic "hot spot," is a rare occurrence. Nothing could be further from the truth. Toxic "hot spots" abound. Most striking are areas of cancer clusters—a short list includes:

- Woburn, where about two dozen children were stricken with leukemia over twelve years, a rate several times the national average for a community of its size.
- Storrs, Connecticut, where wells polluted by a landfill are suspected of sickening and killing residents in nearby homes.
- Bellingham, Washington, where pesticide-contaminated drinking water is thought to be linked to a sixfold increase in childhood cancers.

As *USA Today* (1999) points out, these are only a few examples of an underlying pathology that threatens many other communities. Meanwhile, cancer is now the primary cause of childhood death from disease.

Water contamination is a problem—a very serious problem. In this chapter a wide range of water contaminants and the contaminant sources—and their impact on drinking water supplies from both surface water and groundwater sources—are discussed. In addition, the point is made that when it comes to fresh water pollution, nature is not defenseless in mitigating the situation. The point is made that nature, through its self-purification process in running water systems, is able to fight back against pollution—to a point, at least.

Did You Know?

Keep in mind that when we specify "water pollutants" we are in most cases speaking about pollutants that somehow get into the water (by whatever means) from the interactions of the other two environmental mediums: air and soil. Probably the best example of this is the acid rain phenomenon—pollutants originally emitted only into the atmosphere land on earth and affect both soil and water. Consider that 69 percent of the anthropogenic (human-generated) lead and 73 percent of the mercury in Lake Superior reach it by atmospheric deposition (Hill, 1997).

Sources of Contaminants

If we were to list all the sources of contaminants and the contaminants themselves (the ones that can and do foul our water supply systems), along with a brief description of each contaminant, we could easily fill a book—probably several volumes. To give you some idea of the magnitude of the problem, a condensed list of selected sources and contaminants (a "short list") includes:

Subsurface Percolation—hydrocarbons, metals, nitrates, phosphates, microorganisms, and cleaning agents (TCE).

Injection Wells—hydrocarbons, metals, non-metals, organics, organic and inorganic acids, microorganisms, and radionuclides.

Land Application—Nitrogen, phosphorous, heavy metals, hydrocarbons, microorganisms, and radionuclides.

Landfills—organics, inorganics, microorganisms, and radionuclides.

Open Dumps—organics, inorganics, and microorganisms.

Residential (Local) Disposal—organic chemicals, metals, non-metal inorganics, inorganic acids, and microorganisms.

Surface Impoundments—organic chemicals, metals, non-metal inorganics, inorganic acids, microorganisms, and radionuclides.

Waste Tailings—arsenic, sulfuric acid, copper, selenium, molybdenum, uranium, thorium, radium, lead, manganese, and vanadium.

Waste Piles—arsenic, sulfuric acid, copper, selenium, molybdenum, uranium, thorium, radium, lead, manganese, and vanadium.

Materials Stockpiles—coal pile: aluminum, iron, calcium, manganese, sulfur, and traces of arsenic, cadmium, mercury, lead, zinc, uranium, and copper. Other materials piles: metals/non-metals and microorganisms.

Graveyards—metals, non-metals, and microorganisms.

Animal Burial—contamination is site-specific—depending on disposal practices, surface and subsurface, hydrology, proximity of the site to water sources, type and amount of disposed material, and cause of death.

Above-Ground Storage Tanks—organics, metal/non-metal inorganics, inorganic acids, microorganisms, and radionuclides.

Underground Storage Tanks—organics, metal, inorganic acids, microorganisms, and radionuclides.

Containers—organics, metal/non-metal inorganics, inorganic acids, microorganisms, and radionuclides.

Open Burning and Detonating Sites—inorganics, including heavy metals; organics, including TNT.

Radioactive Disposal Sites—radioactive cesium, plutonium, strontium, cobalt, radium, thorium, and uranium.

Pipelines—organics, metals, inorganic acids, and microorganisms.

Material Transport and Transfer Operations—organics, metals, inorganic acids, microorganisms, and radionuclides.

Irrigation Practices—fertilizers, pesticides, and naturally occurring contamination and sediments.

Pesticide Applications—1200 to 1400 active ingredients. Contamination already detected: alachlor, aldicarb, atrazine, bromacil, carbofuran, cyanazine, DBCP, DCPA, 1,2-dichloropropane, dyfonate, EDB, metolachlor, metribyzen, oxalyl, siazine, and 1,2,3-trichloropropane. The extent of groundwater contamination cannot be determined with current data.

Animal Feeding Operations—nitrogen, bacteria, viruses, and phosphates.

De-Icing Salts Applications—chromate, phosphate, ferric ferocyanide, na-ferrocyan, and chlorine.

Urban Runoff—suspended solids and toxic substances, especially heavy metals and hydrocarbons, bacteria, nutrients, and petroleum residues.

Percolation of Atmospheric Pollutants—sulfur and nitrogen compounds, asbestos, and heavy metals.

Mining and Mine Drainage—coal (acids), toxic inorganics (heavy metals), and nutrients. Phosphates: radium, uranium, and fluorides, and metallic ores (sulfuric acid, lead, cadmium, arsenic, sulfur, cyan).

Production Wells—Oil (1.2 million abandoned production wells), irrigation (farms), and all potential to contaminate: installation, operation, and plugging techniques.

Construction Excavation—pesticides, diesel fuel, oil, salt, and variety of others.

Important Note: Before discussing specific water pollutants, it is important to examine several terms important to the understanding of water pollution. One of these is point source. The USEPA defines a *point source* as "any single identifiable source of pollution from which pollutants are discharged, e.g., a pipe, ditch, ship, or factory smokestack." For example, the outlet pipes of an industrial facility or a municipal wastewater treatment plant are point sources. In contrast, *non-point sources* are widely dispersed sources and are a major cause of stream pollution. An example of a non-point source of pollution is rainwater carrying topsoil and chemical contaminants into a river or stream. Some of the major sources of non-point pollution include water runoff from farming, urban areas, forestry, and construction activities.

The word *runoff* signals a non-point source that originated on land. Runoff may carry a variety of toxic substances and nutrients, as well as bacteria and viruses with it. Non-point sources now comprise the largest source of water pollution, contributing approximately 65 percent of the contamination in quality-impaired streams and lakes.

Radionuclides

Note that in chapter 9, we address in greater detail the impact radiation has on environmental health; however, because of their potential impact on our water supply, we also discuss radionuclides in this section.

When radioactive elements decay, they emit alpha, beta, or gamma radiations caused by transformation of the nuclei to lower energy states. In drinking water, radioactivity can be from natural or artificial radionuclides (the radioactive metals and minerals that cause contamination). These radioactive substances in water are of two types: radioactive minerals and radioactive gas. The USEPA reports that some 50 million Americans face increased cancer risk because of radioactive contamination of their drinking water.

Because of their occurrence in drinking water and their effects on human health, the natural radionuclides of chief concern are radium-226, radium-228, radon-222, and uranium. The source of some of these naturally occurring radioactive minerals is typically associated with certain regions of the country where mining is active or was active in the past. Mining activities expose rock strata, most of which contains some amount of radioactive ore. Uranium mining, for example, produces runoff. Another source of natural radioactive contamination occurs when underground streams flow through various rockbed and geologic formations containing radioactive materials. Other natural occurring sources where radioactive minerals may enter water supplies are smelters and coal-fired electrical generating plants. Sources of man-made radioactive minerals in water are nuclear power plants, nuclear weapons facilities, radioactive materials disposal sites, and mooring sites for nuclear-powered ships. Hospitals also contribute radioactive pollution when they dump low-level radioactive wastes into sewers. Some of these radioactive wastes eventually find their way into water supply systems.

While radioactive minerals such as uranium and radium in water may present a health hazard in these particular areas, a far more dangerous threat exists in the form of radon. *Radon* is a colorless, odorless gas created by (a by-product of) the natural decay of minerals in the soil. Normally present in all water in minute amounts, radon is especially concentrated in water that has passed through rock strata of granite, uranium, or shale.

Radon enters homes from the soil beneath the house, through cracks in the foundation, through crawl spaces and unfinished basements, and in tainted water, and is considered the second leading cause of lung cancer in the United States (about 20,000 cases each year), following cigarette smoking. Contrary to popular belief, radon is not a threat from surface water (lake, river, or above-ground reservoir), because radon dissipates rapidly when water is exposed to air. Even if the water source is groundwater, radon is still not a threat if the water is exposed to air (aerated) or if it is processed through an open tank during treatment. Though studies show that where high concentrations of radon occur within the air in a house, most of the radon comes through the foundation and from the water. However, radon in the tap water, showers, baths, and cooking (with hot water) will cause high concentrations of radon in the air.

Note, however, that radon is a threat from groundwater taken directly from an underground source—either a private well or from a public water supply and whose treatment of the water does not include exposure to air. Because radon in water evaporates quickly into air, the primary danger is from inhaling it from the air in a house, not from drinking it.

The Chemical Cocktail

When we hold a full glass of tap water in our hand (though we usually don't hold it; instead we fill the glass and drink the water, and that is that—just a routine, boring, and repetitive action), if we were to inspect the contents, a few possibilities might present themselves. The contents might appear cloudy or colored (making us think that the water is not fit to drink). The contents might look fine, but carry the prevalent odor of chlorine. Most often, when we take the time to look at water drawn from the tap, it simply looks like water and we drink it or use it to cook with, or whatever.

The fact is, typically a glass of treated water is a chemical cocktail (Kay, 1996). While water utilities in communities seek to protect the public health by treating raw water with certain chemicals, what they are in essence doing is providing us a drinking water product that is a mixture of various treatment chemicals and their by-products. For example, the water treatment facility typically adds chlorine to disinfect—chlorine can produce contaminants. Another concoction is formed when ammonia is added to disinfect. Alum and polymers are added to the water to settle out various contaminants. The water distribution system and appurtenances also need to be protected to prevent pipe corrosion or soft water, so the water treatment facility adds caustic soda, ferric chloride, and lime, which in turn works to increase the aluminum, sulfates, and salts in the water. Thus, when we hold that glass of water before us, and we perceive what appears to be a full glass of crystal-clear, refreshing water—what we really see is a concoction of many chemicals mixed with water, forming the chemical cocktail.

The most common chemical additives used in water treatment are chlorine, fluorides, and flocculants. Because we have already discussed fluorides, our discussion in the following sections is on the by-products of chlorine and flocculant additives.

By-Products of Chlorine

To lessen the potential impact of that chemical cocktail, the biggest challenge today is to make sure the old standby—chlorine—won't produce as many new contaminants as it destroys. At the present time, weighing the balance of the argument, arguing against chlorine and the chlorination process is difficult. Since 1908, chlorine has been used in the United States to kill off microorganisms that spread cholera, typhoid fever, and other waterborne diseases. However, in the 1970s, scientists discovered that while chlorine does not seem to cause cancer in lab animals, it can—in the water treatment process—create a whole list of by-products that do. The by-products of chlorine—organic hydrocarbons called *trihalomethanes* (usually discussed as total trihalomethanes, or TTHMs)—present the biggest health concern.

The USEPA classifies three of these trihalomethanes by-products—chloroform, bromoform, and bromodichloromethane—as probable human carcinogens. The fourth, dibromochloromethane, is classified as a possible human carcinogen.

The USEPA set the first trihalomethane limits in 1979. Most water companies met these standards initially, but the standards were tightened after the 1996 SDWA Amendments. The USEPA is continuously studying the need to regulate other cancer-causing contaminants, including haloacetic acids (HAAs) also produced by chlorination.

Most people concerned with protecting public health applaud the USEPA's efforts in regulating water additives and disinfection by-products. However, some of those in the water treatment and supply business express concern. A common concern often heard from water utilities having a tough time balancing the use of chlorine without going over the regulated limits revolves around the necessity of meeting regulatory requirements by lowering chlorine amounts to meet by-product standards, and at the same ensuring that all the pathogenic microorganisms are killed off. Many make the strong argument that while no proven case exists that disinfection by-products cause cancer in humans, many cases—a whole history of cases—show that if we don't chlorinate water, people get sick and sometimes die from waterborne disease.

Because chlorine and chlorination is now prompting regulatory pressure and compliance with new, demanding regulations, many water treatment facilities are looking for options. Choosing alternative disinfection chemical processes is feeding a growing business enterprise. One alternative that is currently being given widespread consideration in the United States is ozonation, which uses ozone gas to kill microorganisms. Ozonation is Europe's favorite method—and it doesn't produce trihalomethanes. But the USEPA doesn't yet recommend wholesale switchover to ozone to replace chlorine or chlorination systems (sodium hypochlorite or calcium hypochlorite versus elemental chlorine). The USEPA points out that ozone also has problems: (1) it does not produce a residual disinfectant in the water distribution system, (2) it is much more expensive, and (3) in salty water, it can produce another carcinogen, bromate.

At the present time, drinking water practitioners (in the real world) are attempting to fine-tune water treatment. It all boils down to a delicate balancing act. The drinking water professional doesn't want to cut back disinfection—if anything, they'd prefer to strengthen it.

The compound question is, How do we bring into parity the microbial risks versus the chemical risks? How do you reduce them both to an acceptable level? The answer? —no one is quite sure how to do this. The problem really revolves around the enigma associated with a "we don't know what we don't know" scenario.

The disinfection by-products problem stems from the fact that most U.S. water systems produce the unwanted by-products when the chlorine reacts to decayed organics: vegetation and other carbon-containing materials in water. Communities that take drinking water from lakes and rivers have a tougher time keeping the chlorine by-products out of the tap than those that use clean groundwater.

In some communities, when a lot of debris is in the reservoir, the water utility switches to alternate sources—wells, for example. In other facilities, chlorine is combined with ammonia in a disinfection method called *chloramination*. This method is not as potent as pure chlorination, but stops the production of unwanted trihalomethanes.

In communities where rains wash leaves, trees, and grasses into the local water source (lake or river), hot summer days trigger algae blooms, upping the organic matter that can produce trihalomethanes. Spring runoff in many communities exacerbates the problem. With increased runoff comes agricultural waste, pesticides, and quantities of growth (leaves, branches, and some live plants) falling into the water that must be dealt with.

Nature's conditions in summer diminish some precursors for trihalomethanes—the bromides in salty water.

The irony is that under such conditions, nothing unusual is visible in the drinking water. However, cloudy water from silt (dissolved organics from decayed plants) is enough to create trihalomethanes.

With the advent of the new century, most cities will strain out (filter) the organics from their water supplies before chlorinating to prevent the formation of trihalomethanes and haloacetic acids.

In other communities, the move is already on to switch from chlorine to ozone and other disinfectant methods. The National Resources Defense Council states that in fifteen to twenty years, most U.S. systems will likely catch up with Europe and use ozone to kill resistant microbes like cryptosporidium. Note that when this method is employed, the finishing touch is usually accomplished by filtering through granulated activated carbon, which increases the cost slightly (estimated at about $100 or more per year per hookup) that customers must pay.

Total Trihalomethanes (TTHM)

Total trihalomethanes are a by-product of chlorinating water that contains natural organics (USEPA, 2012). A U.S. Environmental Protection Agency survey discovered that trihalomethanes are present in virtually all chlorinated water supplies. Many years ago the USEPA required large towns and cities to reduce TTHM levels in potable water. However, recent changes in national drinking water quality standards now require that water treatment systems of smaller towns begin to reduce TTHM. It is important to note that TTHM do not pose a high health risk compared to waterborne diseases, but they are among the most important water quality issues to be addressed in the U.S. water supply.

A major challenge for drinking water practitioners is how to balance the risks from microbial pathogens and disinfection by-products. Providing protection from these microbial pathogens while simultaneously ensuring decreasing health risks to the population from disinfection by-products (DBPs) is important. The Safe Drinking Water Act (SDWA) Amendments, signed by the president in August 1996, required the USEPA to develop rules to achieve these goals. The new Stage 1 Disinfectant and Disinfection By-product Rule and Interim Enhanced Surface Water Treatment Rule are the first of a set of rules under the Amendments.

Public Health Concerns

Most Americans drink tap water that meets all existing health standards all the time. These new rules will further strengthen existing drinking water standards and thus increase protection for many water systems.

The USEPA's Science Advisory Board concluded in 1990 that exposure to microbial contaminants such as bacteria, viruses, and protozoa (e.g., *Giardia lamblia* and *Cryptosporidium*) was likely the greatest remaining health risk management challenge for drinking water suppliers. Acute health effects from exposure to microbial pathogens is documented, and associated illness can range from mild to moderate cases lasting only a few days, to more severe infections that can last several weeks and may result in death for those with weakened immune systems.

While disinfectants are effective in controlling many microorganisms, they react with natural organic and inorganic matter in source water and distribution systems to form potential DBPs. Many of these DBPs have been shown to cause cancer and reproductive and developmental effects in laboratory animals. More than 200 million people consume water that has been disinfected. Because of the large population exposed, health risks associated with DBPs, even if small, need to be taken seriously.

Existing Regulations

Microbial Contaminants: The Surface Water Treatment Rule, promulgated in 1989, applies to all public water systems using surface water sources or groundwater sources under the direct influence of surface water. It establishes maximum contaminant level goals (MCLGs) for viruses, bacteria, and *Giardia lamblia*. It also includes treatment technique requirements for filtered and unfiltered systems specifically designed to protect against the adverse health effects of exposure to these microbial pathogens. The Total Coliform Rule, revised in 1989, applies to all PWSs and establishes a maximum contaminant level (MCL) for total coliforms.

Disinfection By-Products: In 1979, the USEPA set an interim MCL for total trihalomethanes of 0.10 mg/l as an annual average. This applies to any community water system serving at least 10,000 people that adds a disinfectant to the drinking water during any part of the treatment process.

Information Collection Rule

To support the M-DBP rule-making process, the Information Collection Rule establishes monitoring and data reporting requirements for large public water systems (PWSs) serving at least 100,000 people. This rule is intended to provide the USEPA with information on the occurrence in drinking water of microbial pathogens and DBPs. The USEPA is collecting engineering data on how PWSs currently control such contaminants as part of the Information Collection Rule.

Groundwater Rule

The USEPA developed a groundwater rule that specifies the appropriate use of disinfection, and equally importantly, addresses other components of groundwater systems to ensure public health protection. More than 158,000 public or community systems serve almost 89 million people through groundwater systems. Ninety-nine percent (157,000) of groundwater systems serve fewer than 10,000 people. However, systems serving more than 10,000 people serve 55 percent—more than 60 million—of all people who get their drinking water from public groundwater systems.

Filter Backwash Recycling

The 1996 SDWA Amendments require that the USEPA sets a standard on recycling filter backwash within the treatment process of public water systems. The regulation applies to all public water systems, regardless of size.

Opportunities for Public Involvement

The USEPA encourages public input into regulation development. Public meetings and opportunities for public comment on M-DBP rules are announced in the Federal Register. EPA's Office of Groundwater and Drinking Water also provides this information for the M-DBP rule and other programs in its online Calendar of Events (www.epa.gov).

Flocculants

In addition to chlorine and sometimes fluoride, water treatment plants often add several other chemicals, including flocculants, to improve the efficiency of the treatment process—and they all add to the cocktail mix. *Flocculants* are chemical substances added to water to make particles clump together, which improves the effectiveness of filtration. Some of the most common flocculants are polyelectrolytes (polymers)—chemicals with constituents that cause cancer and birth defects and are banned for use by several countries. Although the USEPA classifies them as "probable human carcinogens," it still allows their continued use. Acrylamide and epichlorohydrin are two flocculants used in the United States that are known to be associated with probable cancer risk (Lewis, 1996; Spellman, 2007).

Groundwater Contamination

Note that groundwater under the direct influence of surface water comes under the same monitoring regulations as does surface water (i.e., all water open to the atmosphere and subject to surface runoff. The legal definition of *groundwater under the direct influence of surface water* is any water beneath the surface of the ground with (1) significant occurrence of insects or microorganisms, algae, or large diameter pathogens such as *Giardia lamblia*, or (2) significant and relatively rapid shifts in water characteristics such as turbidity, temperature, conductivity, or pH, which closely correlate to climatological or surface water conditions. Direct influence must be determined for individual sources in accordance with criteria established by the state. The state determines for individual sources in accordance with criteria established by the state, and that determination may be based on site-specific measurements of water quality and/or documentation of well construction characteristics and geology with field evaluation.

Generally, most groundwater supplies in the United States are of good quality and produce essential quantities. The full magnitude of groundwater contamination in the United States is, however, not fully documented, and federal, state, and local efforts continue to assess and address the problems (Rail, 1985).

Groundwater supplies about 25 percent of the fresh water used for all purposes in the United States, including irrigation, industrial uses, and drinking water (about 50 percent of the U.S. population relies on groundwater for drinking water). John Chilton (1998) points out that the groundwater aquifers beneath or close to Mexico City provide the areas with more than 3.2 billion liters per day. But, as groundwater pumping increases to meet water demand, it can exceed the aquifers' rates of replenishment, and in many urban aquifers, water levels show long-term decline. With excessive extraction comes a variety of other undesirable effects, including:

- increased pumping costs
- changes in hydraulic pressure and underground flow directions (in coastal areas, this results in seawater intrusion)
- saline water drawn up from deeper geological formations
- poor-quality water from polluted shallow aquifers leaking downward

Severe depletion of groundwater resources is often compounded by a serious deterioration in its quality. Thus, without a doubt, the contamination of a groundwater supply should be a concern of those drinking water practitioners responsible for supplying a community with potable water provided by groundwater.

Despite our strong reliance on groundwater, groundwater has for many years been one of the most neglected and ignored natural resources. Groundwater has been ignored because it is less visible than other environmental resources—rivers or lakes, for example. What the public cannot see or observe, the public doesn't worry about—or even think about. However, recent publicity about events concerning groundwater contamination is making the public more aware of the problem, and the regulators have also taken notice.

Are natural contaminants a threat to human health—harbingers of serious groundwater pollution events? No, not really. The main problem with respect to serious groundwater pollution has been human activities. When we (all of us) improperly dispose of wastes, or spill hazardous substances onto/into the ground, we threaten groundwater, and in turn, the threat passes on to public health.

Algae

You do not have to be a scientist to understand that algae can be a nuisance. Many ponds and lakes in the United States are currently undergoing *Eutrophication*, the enrichment of an environment with inorganic substances (e.g., phosphorus and nitrogen), causing excessive algae growth and premature aging of the water body. The average person may not know what eutrophication means—however, when eutrophication occurs and especially when filamentous algae like *Caldophora* breaks loose in a pond or lake and washes ashore, algae makes its stinking, noxious presence known.

Algae are a form of aquatic plants and are classified by color (e.g., green algae, blue-green algae, golden-brown algae, etc.). Algae come in many shapes and sizes. Although they are not pathogenic, algae do cause problems with water/wastewater treatment plant operations. They grow easily on the walls of troughs and basins, and heavy growth can plug intakes and screens. Additionally, some algae release chemicals that give off undesirable tastes and odors.

As mentioned, algae are usually classified by their color. However, they are also commonly classified based on their cellular properties or characteristics. Several characteristics are used to classify algae including (1) cellular organization and cell wall structure; (2) the nature of the chlorophyll(s); (3) the type of motility, if any; (4) the carbon polymers that are produced and stored; and (5) the reproductive structures and methods.

Many algae (in mass) are easily seen by the naked eye—others are microscopic. They occur in fresh and polluted water, as well as in saltwater. Since they are plants, they are capable of using energy from the sun in photosynthesis. They usually grow near the surface of the water because light cannot penetrate very far through the water.

Algae are controlled in raw waters with chlorine and potassium permanganate. Algae blooms in raw water reservoirs are often controlled with copper sulfate.

Important Note: By producing oxygen, which is utilized by other organisms, including animals, algae play an important role in the balance of nature.

Underground Storage Tanks (USTs)

If we looked at a map of the United States marked with the exact location of every underground storage tank (UST) shown (we wish such a map existed!), with the exception of isolated areas, most of us would be surprised at the large number of tanks buried underground. With so many buried tanks, that structural failures arising from a wide variety of causes have occurred over the years should come as no surprise. Subsequent leaking has become a huge source of contamination that affects the quality of local groundwaters.

Important Note: A UST is any tank, including any underground piping connected to the tank that has at least 10 percent of its volume below ground (USEPA, 1987).

The fact is, leakage of petroleum and its products from USTs occurs more often than we generally realize. This widespread problem has been a major concern and priority in the United States for well over a decade. In 1987, the USEPA promulgated regulations for many of the nation's USTs, and much progress has been made in mitigating this problem to date.

When a UST leak or past leak is discovered, the contaminants released to the soil and thus to groundwater, for the average person, would seem rather straightforward to identify: fuel oil, diesel, and gasoline. However, even though it is true that these are the most common contaminants released from leaking USTs, others also present problems. For example, in the following section, we discuss one such contaminate, a by-product of gasoline—one that is not well-known—to help illustrate the magnitude of leaking USTs.

MTBE

In December 1997, the USEPA issued a drinking water advisory titled *Consumer Acceptability Advice and Health Effects Analysis on Methyl Tertiary-Butyl Ether (MTBE)*. The purpose of the advisory was to provide guidance to communities exposed to drinking water contaminated with MTBE.

Important Note: A USEPA Advisory is usually initiated to provide information and guidance to individuals or agencies concerned with potential risk from drinking water contaminants for which no national regulations currently exist. Advisories are not mandatory standards for action, and are used only for guidance. They are not legally enforceable, and are subject to revision as new information becomes available. The USEPA's Health Advisory Program is recognized in the Safe Drinking Water Act Amendments of 1996, which state in section 102(b)(1)(F): "The Administrator may publish health advisories (which are not regulations) or take other appropriate actions for contaminants not subject to any national primary drinking water regulation." As its title indicates, this Advisory includes consumer acceptability advice as "appropriate" under this statutory provision, as well as a health effects analysis.

What Is MTBE?

MTBE is a volatile, organic chemical. Since the late 1970s, MTBE has been used as an octane enhancer in gasoline. Because it promotes more complete burning of gasoline (thereby reducing carbon monoxide and ozone levels), it is commonly used as a gasoline additive in localities that do not meet the National Ambient Air Quality Standards.

In the Clean Air Act of 1990, Congress mandated the use of reformulated gasoline (RFG) in areas of the country with the worst ozone or smog problems. RFG must meet certain technical specifications set forth in the Act, including a specific oxygen content. Ethanol and MTBE are the primary oxygenates used to meet the oxygen content requirement. MTBE is used in about 84 percent of RFG supplies. Currently, thirty-two areas in a total of eighteen states are participating in the RFG program, and RFG accounts for about 30 percent of gasoline nationwide.

Studies identify significant air quality and public health benefits that directly result from the use of fuels oxygenated with MTBE, ethanol, or other chemicals. The refiners' 1995/96 fuel data submitted to the USEPA indicates that the national emissions benefits exceeded those required. The 1996 Air Quality Trends Report shows that toxic air pollutants declined significantly between 1994 and 1995. Early analysis indicates this progress may be attributable to the use of RFG. Starting in the year 2000, required emission reductions are substantially greater, at about 27 percent for volatile organic compounds, 22 percent for toxic air pollutants, and 7 percent for nitrogen oxides.

Important Note: When gasoline that has been oxygenated with MTBE comes in contact with water, large amounts of MTBE dissolves. At 25°C, the water solubility of MTBE is about 5,000 milligrams per liter for a gasoline that is 10 percent MTBE by weight. In contrast, for a non-oxygenated gasoline, the total hydrocarbon solubility in water is typically about 120 milligrams per liter. MTBE sorbs only weakly to soil and aquifer material; therefore, sorption will not significantly retard MTBE's transport by groundwater. In addition, the compound generally resists degradation in groundwater (Squillace et al., 1998).

Why Is MTBE a Drinking Water Concern?

A limited number of instances of significant contamination of drinking water with MTBE have occurred because of leaks from underground and above ground petroleum storage tank systems and pipelines. Due to its small molecular size and solubility in water, MTBE moves rapidly into groundwater, faster than do other constituents of gasoline. Public and private wells have been contaminated in this manner. Non-point sources (such as recreational watercraft) are most likely to be the cause of small amounts of contamination in a large number of shallow aquifers and surface waters. Air deposition through precipitation of industrial or vehicular emissions may also contribute to surface water contamination. The extent of any potential for buildup in the environment from such deposition is uncertain.

Is MTBE in Drinking Water Harmful?

Based on the limited sampling data currently available, most concentrations at which MTBE has been found in drinking water sources are unlikely to cause adverse health effects. However, the USEPA is continuing to evaluate the available information and is doing additional research to seek more definitive estimates of potential risks to humans from drinking water.

There isn't data on the effects on humans of drinking MTBE-contaminated water. In laboratory tests on animals, cancer and noncancer effects occur at high levels of exposure. These tests were conducted by inhalation exposure or by introducing the chemical in oil directly to the stomach. The tests support a concern for potential human hazard. Because the animals were not exposed through drinking water, significant uncertainties exist concerning the degree of risk associated with human exposure to low concentrations typically found in drinking water.

How Can People Protect Themselves?

MTBE has a very unpleasant taste and odor, and these properties make contaminated drinking water unacceptable to the public. The Advisory recommends control levels for taste and odor acceptability that will also protect against potential health effects.

Studies conducted on the concentrations of MTBE in drinking water determined the level at which individuals can detect the odor or taste of the chemical. Humans vary widely in the concentrations they are able to detect. Some who are sensitive can detect very low concentrations. Others do not taste or smell the chemical, even at much higher concentrations. The presence or absence of other natural or water treatment chemicals sometimes masks or reveals the taste or odor effects.

Studies to date have not been extensive enough to completely describe the extent of this variability, or to establish a population response threshold. Nevertheless, we conclude from the available studies that keeping concentrations in the range of 20 to 40 micrograms per liter (μg/L) of water or below will likely avert unpleasant taste and odor effects, recognizing that some people may detect the chemical below this.

Concentrations in the range of 20 to 40 μg/L are about 20,000 to 100,000 (or more) times lower than the range of exposure levels in which cancer or noncancer effects were observed in rodent tests. This margin of exposure lies within the range of margins of exposure typically provided to protect against cancer effects by the National Primary Drinking Water Standards under the Federal Safe Drinking Water Act—a margin greater than such standards typically provided to protect against noncancer effects. Protection of the water source from unpleasant taste and odor as recommended also protects consumers from potential health effects.

The USEPA also notes that occurrences of groundwater contamination observed at or above this 20-40 μg/l taste and odor threshold—that is, contamination at levels that may create consumer acceptability problems for water supplies—have, to date, resulted from leaks in petroleum storage tanks or pipelines, not from other sources.

Recommendations for State or Public Water Suppliers

Public water systems that conduct routine monitoring for volatile organic chemicals can test for MTBE at little additional cost, and some states are already moving in this direction.

Public water systems detecting MTBE in their source water at problematic concentrations can remove MTBE from water using the same conventional treatment techniques that are used to clean up other contaminants originating from gasoline releases—air stripping and granular activated carbon (GAC), for example. However, because MTBE is more soluble in water and more resistant to biodegradation than other chemical constituents in gasoline, air stripping and GAC treatment requires additional optimization, and must often be used together to effectively remove MTBE from water. The costs of removing MTBE are higher than when treating for gasoline releases that do not contain MTBE. Oxidation of MTBE using UV/peroxide/ozone treatment may also be feasible, but typically has higher capital and operating costs than air stripping and GAC.

The bottom line: Because MTBE has been found in sources of drinking water, many states are phasing out the sale of gasoline with MTBE.

Important Note: Of the sixty volatile organic compounds (VOCs) analyzed in samples of shallow ambient groundwater collected from eight urban areas during 1993–1994 as part of the U.S. Geological Survey's National Water Quality Assessment Program, MTBE was the second most frequently detected compound (after trichloromethane [chloroform]) (Squillace et al., 1998).

Industrial Wastes

Since industrial waste represents a significant source of groundwater contamination, water practitioners and others expend an increasing amount of time in abating or mitigating pollution events that damage groundwater supplies.

Groundwater contamination from industrial wastes usually begins with the practice of disposing of industrial chemical wastes in surface impoundments—unlined landfills or lagoons, for example. Fortunately, these practices, for the most part, are part of our past. Today, we know better. For example, we now know that what is most expedient or least expensive does not work for industrial waste disposal practices. We have found through actual experience that the long run has proven just the opposite—for society as a whole (with respect to health hazards and the costs of cleanup activities) to ensure clean or unpolluted groundwater supplies is very expensive—and utterly necessary.

Septic Tanks

Septage from septic tanks is a biodegradable waste capable of affecting the environment through water and air pollution. The potential environmental problems associated with the use of septic tanks are magnified when you consider that subsurface sewage disposal systems (septic tanks) are used by almost one-third of the United States population.

Briefly, a septic tank and leaching field system traps and stores solids while the liquid effluent from the tank flows into a leaching or absorption field, where it slowly seeps into the soil and degrades naturally.

The problem with subsurface sewage disposal systems such as septic tanks is that most of the billions of gallons of sewage that enter the ground each year are not properly treated. Because of faulty construction or lack of maintenance, not all of these systems work properly.

Experience has shown that septic disposal systems are frequently sources of fecal bacteria and virus contamination of water supplies taken from private wells. Many septic tank owners dispose of detergents, nitrates, chlorides, and solvents in their septic systems, or use solvents to treat their sewage waste. A septic tank cleaning fluid that is commonly used contains organic solvents (trichloroethylene, or TCE)—potential human carcinogens that in turn pollute the groundwater in areas served by septic systems.

Landfills

Humans have been disposing of waste by burying it in the ground since time immemorial. In the past, this practice was largely uncontrolled, and the disposal sites (i.e., garbage dumps) were places where municipal solid wastes were simply dumped on and into the ground without much thought or concern. Even in this modern age, landfills have been used to dispose of trash and waste products at controlled locations that are then sealed and buried under earth. Now such practices are increasingly seen as a less than satisfactory disposal method, because of the long-term environmental impact of waste materials in the ground and groundwater.

Unfortunately, many of the older (and even some of the newer) sites were located in low-lying areas with high groundwater tables. *Leachate* (seepage of liquid through the waste), high in BOD, chloride, organics, heavy metals, nitrate, and other contaminants, has little difficulty reaching the groundwater in such disposal sites. In the United States, literally thousands of inactive or abandoned dumps like this exist.

Agriculture

Fertilizers and pesticides are the two most significant groundwater contaminants that result from agricultural activities. The impact of agricultural practices wherein fertilizers and pesticides are normally used is dependent upon local soil conditions. If, for example, the soil is sandy, nitrates from fertilizers are easily carried through the porous soil into the groundwater, contaminating private wells.

Pesticide contamination of groundwater is a subject of national importance because groundwater is used for drinking water by about 50 percent of the nation's population. This especially concerns people living in the agricultural areas where pesticides are most often used, as about 95 percent of that population relies upon groundwater for drinking water. Before the mid-1970s, the common thought was that soil acted as a protective filter, one that stopped pesticides from reaching groundwater. Studies have now shown that this is not the case. Pesticides can reach water-bearing aquifers below ground from applications onto crop fields, seepage of contaminated surface water, accidental spills and leaks, improper disposal, and even through injection of waste material into wells.

Pesticides are mostly modern chemicals. Many hundreds of these compounds are used, and extensive tests and studies of their effect on humans have not been completed. That leads us to ask, "Just how concerned should we be about their presence in our drinking water?" Certainly, treating pesticides as potentially dangerous, and thus handling them with care would be wise. We can say they pose a potential danger if they are consumed in large quantities, but as any experienced scientist knows, you cannot draw factual conclusions unless scientific tests have been done. Some pesticides have had a designated Maximum Contaminant Limit (MCL) in drinking water set by the USEPA, but many have not. Another serious point to consider is the potential effect of combining more than one pesticide in drinking water, which might be different than the effects of each individual pesticide alone. This is another situation where we don't have sufficient scientific data to draw reliable conclusions—in other words, again, we don't know what we don't know.

Did You Know?

Ciguatera fish poisoning (or ciguatera) is an illness caused by eating fish that contain toxins produced by a marine microalgae called *Gambierdiscus toxicus*. Barracuda, black grouper, blackfin snapper, cubera snapper, dog snapper, greater amberjack, hogfish, horse-eye jack, king mackerel, and yellowfin grouper have been known to carry ciguatoxins. People who have ciguatera may experience nausea, vomiting, and neurologic symptoms such as tingling fingers or toes. They also may find that cold things feel hot and hot things feel cold. Ciguatera has no cure. Symptoms usually go away in days or weeks but can last for years. People who have ciguatera can be treated for their symptoms (CDC, 2012).

Saltwater Intrusion

In many coastal cities and towns, as well as in island locations, the intrusion of salty seawater presents a serious water-quality problem. Because fresh water is lighter than saltwater (the specific gravity of seawater is about 1.025), it will usually float above a layer of saltwater. When an aquifer in a coastal area is pumped, the original equilibrium is disturbed and saltwater replaces the fresh water (Viessman and Hammer, 1998). The problem is compounded by increasing population, urbanization, and industrialization, which increases use of groundwater supplies. In such areas, while groundwater is heavily drawn upon, the quantity of natural groundwater recharge is decreased because of the construction of roads, tarmac, and parking lots, which prevent rainwater from infiltrating, decreasing the groundwater table elevation.

In coastal areas, the natural interface between the fresh groundwater flowing from upland areas and the saline water from the sea is constantly under attack by human activities. Since seawater is approximately 2.5 times more dense than freshwater, a high pressure head of seawater occurs (in relationship to freshwater), which results in a significant rise in the seawater boundary. Potable water wells close to this rise in sea level may have to be abandoned because of saltwater intrusion.

Other Sources of Groundwater Contamination

To this point, we have discussed only a few of the many sources of groundwater contamination. For example, we have not discussed mining and petroleum activities that lead to contamination of groundwater, or contamination caused by activities in urban areas. Both of these are important sources. Urban activities (including spreading salt on roads to keep them ice-free during winter) eventually contributes to contamination of groundwater supplies. Underground injection wells used to dispose of hazardous materials can lead to groundwater contamination. As we've discussed, underground storage tanks (USTs) are also significant contributors to groundwater pollution. Other sources of groundwater contamination include the items on a short list:

- Waste tailings
- Residential disposal
- Urban runoff
- Hog wastes
- Biosolids
- Land-applied wastewater
- Graveyards
- Deicing salts
- Surface impoundments
- Waste piles
- Animal feeding operations
- Natural leaching

- Animal burial
- Mine drainage
- Pipelines
- Open dumps
- Open burning
- Atmospheric pollutants

Raw sewage is not listed, because for the most part, raw sewage is no longer routinely dumped into our nation's wells or into our soil. Sewage treatment plants effectively treat wastewater so that it can be safely discharged to local water bodies. In fact, the amount of pollution being discharged from these plants has been cut by over one-third during the past twenty years, even as the number of people served has doubled.

Yet in some areas, raw sewage spills still occur, sometimes because a underground sewer line is blocked, broken, or too small, or because periods of heavy rainfall overload the capacity of the sewer line or sewage treatment plant, so that overflows into city streets or streams occur. Some of this sewage finds its way to groundwater supplies.

The best way to prevent groundwater pollution is to stop it from occurring in the first place. Unfortunately, a perception held by many is that natural purification of chemically contaminated groundwater can take place on its own—without the aid of human intervention. To a degree this is true—however, natural purification functions on its own time, not on human time. Natural purification could take decades and perhaps centuries. The alternative? Remediation. But remediation and mitigation don't come cheap. When groundwater is contaminated, the cleanup efforts are sometimes much too expensive to be practical.

The USEPA has established the Groundwater Guardian Program. The Program is a voluntary way to improve drinking water safety. Established and managed by a nonprofit organization in the Midwest and strongly promoted by the USEPA, this program focuses on communities that rely on groundwater for their drinking water. It provides special recognition and technical assistance to help communities protect their groundwater from contamination. Since beginning in 1994, Groundwater Guardian Programs have been established in nearly 100 communities in thirty-one states (USEPA, 1996).

Self-Purification of Streams

Hercules, that great mythical giant and arguably the globe's first environmental engineer, pointed out that the solution to stream pollution is dilution—that is, dilution is the solution. In reality, today's humans depend on various human-made water treatment processes to restore water to potable and palatable condition. However, it should be pointed out that nature, as Hercules noted, is not defenseless in its fight against water pollution. For example, when a river or stream is contaminated, natural processes (including dilution) immediately kick in to restore the water body and its contents back to its natural state. If the level of contamination is not excessive, the stream or river can restore itself to normal conditions in a relatively short period. In this section, nature's ability to purify and restore typical river systems to normal conditions is discussed.

In terms of practical usefulness the waste assimilation capacity of streams as a water resource has its basis in the complex phenomenon termed stream self-purification. This is a dynamic phenomenon reflecting hydrologic and biologic variations, and the interrelations are not yet fully understood in precise terms. However, this does not preclude applying what is known. Sufficient knowledge is available to permit quantitative definition of resultant stream conditions under expected ranges of variation to serve as practical guides in decisions dealing with water resource use, development, and management (Velz, 1970).

Balancing the "Aquarium"

An outdoor excursion to the local stream can be a relaxing and enjoyable undertaking. On the other hand, when you arrive at the local stream, spread your blanket on the stream bank, and then look out upon the stream's flowing mass and discover a parade of waste and discarded rubble bobbing along the stream's course and cluttering the adjacent shoreline and downstream areas, any feeling of relaxation or enjoyment is quickly extinguished. Further, the sickening sensation the observer feels is not lessened, but made worse as he gains closer scrutiny of the putrid flow. He recognizes the rainbow-colored shimmer of an oil slick, interrupted here and there by dead fish and floating refuse, and the slimy fungal growth that prevails. At the same time, the observer's sense of smell is alerted to the noxious conditions. Along with the fouled water and the stench of rot-filled air, the observer notices the ultimate insult and tragedy: The signs warn "DANGER—NO SWIMMING or FISHING." The observer soon realizes that the stream before him is not a stream at all; it is little more than an unsightly drainage ditch. The observer has discovered what ecologists have known and warned about for years. That is, contrary to popular belief, rivers and streams do not have an infinite capacity for pollution.

Before the early 1970s, such disgusting occurrences such as the one just described were common along the rivers and streams near main metropolitan areas throughout most of the United States. Many aquatic habitats were fouled during the past because of industrialization. However, our streams and rivers were not always in such deplorable condition.

Before the Industrial Revolution of the 1800s, metropolitan areas were small and sparsely populated. Thus, river and stream systems within or next to early communities received insignificant quantities of discarded waste. Early on, these river and stream systems were able to compensate for the small amount of wastes they received; when wounded (polluted), nature has a way of fighting back. In the case of rivers and streams, nature provides flowing waters with the ability to restore themselves through a self-purification process. It was only when humans gathered in great numbers to form great cities that the stream systems were not always able to recover from having received great quantities of refuse and other wastes.

What exactly is it that humans do to rivers and streams? As stated earlier, Halsam pointed out that humans' actions are determined by their expediency. In addition, what most people do not realize is that we have the same amount of water as we did millions of years ago, and through the water cycle, we continually reuse that same water—water that was used by the ancient Romans and Greeks is the same water we are using today. Increased demand has put enormous stress on our water supply. Human are the cause of this stress. Thus, what humans do to rivers and streams is to upset the delicate balance between pollution and the purification process. That is, we tend to unbalance the aquarium.

As mentioned, with the advent of industrialization, local rivers and streams became deplorable cesspools that worsened with time. During the Industrial Revolution, the removal of horse manure and garbage from city streets became a pressing concern, for example, Moran et al. point out that "none too frequently, garbage collectors cleaned the streets and dumped the refuse into the nearest river" (Moran, Moran, and Wiersma, 1986). Halsam (1990) reports that as late as 1887, river keepers gained full employment by removing a constant flow of dead animals from a river in London. Moreover, the prevailing attitude of that day was "I don't want it anymore, throw it into the river."

As pointed out, as early as the 1970s, any threat to the quality of water destined for use for drinking and recreation has quickly angered those affected. Fortunately, since the 1970s we have moved to correct the stream pollution problem. Through scientific study and incorporation of wastewater treatment technology, we have started to restore streams to their natural condition.

Fortunately, we are aided in this effort to restore a stream's natural water quality by the stream itself through the phenomenon of self-purification.

A balance of biological organisms is normal for all streams. Clean, healthy streams have certain characteristics in common. For example, as mentioned, one property of streams is their ability to dispose of small amounts of pollution. However, if streams receive unusually large amounts of waste, the stream life will change and attempt to stabilize such pollutants; that is, the biota will attempt to balance the aquarium. However, if stream biota are not capable of self-purifying, then the stream may become a lifeless body.

Important Point: The self-purification process discussed here relates to the purification of organic matter only.

Sources of Stream Pollution

Sources of stream pollution are normally classified as point or non-point sources. A point source (PS) is a source that discharges effluent, such as wastewater from sewage treatment and industrial plants. Simply put, a point source is usually easily identified as "end of the pipe" pollution; that is, it emanates from a concentrated source or sources. In addition to organic pollution received from the effluents of sewage treatment plants, other sources of organic pollution include runoffs and dissolution of minerals throughout an area and are not from one or more concentrated sources.

Non-concentrated sources are known as non-point sources (see Figure 7.2). Non-point source (NPS) pollution, unlike pollution from industrial and sewage treatment plants, comes from many diffuse sources. NPS pollution is caused by rainfall or snowmelt moving over and through the ground. As the runoff moves, it picks up and carries away natural and man-made pollutants, finally depositing them into streams, lakes, wetlands, rivers, coastal waters, and even our underground sources of drinking water. These pollutants include:

- excess fertilizers, herbicides, and insecticides from agricultural lands and residential areas
- oil, grease, and toxic chemicals from urban runoff and energy production
- sediment from improperly managed construction sites, crop and forest lands, and eroding streambanks
- salt from irrigation practices and acid drainage from abandoned mines
- bacteria and nutrients from livestock, pet waste, and faulty septic systems

Atmospheric deposition and hydromodification are also sources of non-point source pollution (USEPA, 1994).

As mentioned, specific examples of non-point sources include runoff from agricultural fields and also cleared forest areas, construction sites, and roadways. Of particular interest to environmentalists in recent years has been agricultural effluents. As a case in point, take for example farm silage effluent, which has been estimated to be more than 200 times as potent (in terms of BOD) as treated sewage (USEPA, 1994).

Nutrients are organic and inorganic substances that provide food for microorganisms such as bacteria, fungi, and algae. Nutrients are supplemented by the discharge of sewage. The bacteria, fungi, and algae are consumed by the higher trophic levels in the community. Each stream, due to a limited amount of dissolved oxygen (DO), has a limited capacity for aerobic decomposition of organic matter without becoming anaerobic. If the organic load received is above that capacity, the stream becomes unfit for normal aquatic life, and it is not able to support organisms sensitive to oxygen depletion (Mason, 1991).

Effluent from a sewage treatment plant is most commonly disposed of in a nearby waterway. At the point of entry of the discharge, there is a sharp decline in the concentration of DO in the stream. This phenomenon is known as the oxygen sag. Unfortunately (for the organisms that normally occupy a clean, healthy stream), when the DO is decreased, there is a concurrent massive increase in BOD as microorganisms utilize the DO as they break down the organic matter. When the organic matter is depleted, the microbial population and BOD decline, while the DO concentration increases, assisted by stream flow (in the form of turbulence) and by the photosynthesis of aquatic plants. This self-purification process is very efficient, and the stream will suffer no permanent damage as long as the quantity of waste is not too high. Obviously, an understanding of this self-purification process is important to prevent overloading the stream ecosystem.

As urban and industrial centers continue to grow, waste disposal problems also grow. Because wastes have increased in volume and are much more concentrated than earlier, natural waterways must have help in the purification process. This help is provided by wastewater treatment plants. A wastewater treatment plant functions to reduce the organic loading that raw sewage would impose on discharge into streams. Wastewater treatment plants utilize three stages of treatment: primary, secondary, and tertiary treatment. In breaking down the wastes, a secondary wastewater treatment plant uses the same type of self-purification process found in any stream ecosystem. Small bacteria and protozoans (one-celled organisms) begin breaking down the organic material. Aquatic insects and rotifers are then able to continue the purification process. Eventually, the stream will recover and show little or no effects of the sewage discharge. Again, this phenomenon is known as natural stream purification (Spellman and Whiting, 1999).

Discussion Questions

1. Briefly explain the differences in water quality problems.
2. How is water pollution related to agricultural activities?
3. What is thermal pollution?
4. What is acid mine drainage?
5. Explain soil salinity from irrigation.

References and Recommended Reading

CDC, 2012. CDC-Health Studies Program—Harmful Algal Blooms (HABs). Accessed 03/19/12 at http://www.cdc.gov/nceh/hsb/hab/default.htm.

Chilton, J., 1988. *Dry or Drowning*. Beson.

Halsam, S.M., 1990. River pollution: An ecological perspective. New York: Belhaven Press.

Hill, M.K., 1997. *Understanding Environmental Pollution*. Cambridge, UK: Cambridge University Press.

Kay, J., 1996. Chemicals used to cleanse water can also cause problems. *San Francisco Examiner*, October 3.

Lewis, S.A., 1996. *The Sierra Club Guide to Safe Drinking Water*. San Francisco: The Sierra Book Club.

Mason, C.F., 1990. Biological aspects of freshwater pollution. In: Harrison, R.M., ed. *Pollution: Causes, Effects, and Control*. Cambridge: Great Britain. The Royal Society of Chemistry.

Outwater, A., 1996. *Water: A Natural History*. New York: Basic Books.

Rail, C.D., 1985. Groundwater Monitoring Within an Aquifer—A Protocol, *Journal of Environmental Health* 48(3):128–132.

Squillace, P.J., J.F. Pankow, N.E. Korte, and J.S. Zogorski, 1998. Environmental Behavior and Fate of Methyl Tertiary-Butyl Ether. In: Water Online at www.wateronline.com, November 4.

Smith, R.L., 1974. *Ecology and Field Biology*. New York: Harper & Row.

Spellman, F.R., and N.E. Whiting, 1999. *Water Pollution Control Technology*. Rockville, MD: Government Institutes.

Spellman, F.R., 2007. *The Science of Water*, 2nd ed. Boca Raton, FL: CRC Press.

USA Today, 1999. Pollution is top environmental concern, August 29.

USEPA, 1987. *Proposed Regulations for Underground Storage Tanks: What's in the Pipeline?* Washington, D.C.: Office of Underground Storage Tanks.

USEPA, 1994. *What is Nonpoint Source Pollution?* Washington, D.C.: United States Environmental Protection Agency, EPA-F-94-005.

USEPA, 1996. Targeting High Priority Problems, at epamail.epa.gov.

USEPA, 1998. *Drinking Water Priority Rulemaking: Microbial and Disinfection By-products Rules*. Washington, D.C.: United States Environmental Protection Agency, EPA 815-F-95-0014.

USEPA, 2012. Water: Basic information about Regulated Drinking Water Contaminants. Accessed 03/19/12 at http://water.erp.gov/drink/contamiants/basicinformation/disinetecitonbyproducts.cfm.
Velz, F.J., 1970. Applied stream sanitation. New York: Wiley-Interscience.
Viessman, W., Jr., and M.J. Hammer, 1998. *Water Supply and Pollution Control*, 6th ed., Menlo Park, CA: Addison-Wesley.

9

Radiation

> Not only will atomic power be released, but someday we will harness the rise and fall of the tides and imprison the rays of the sun.
>
> —Thomas Edison

IN TODAY'S WORLD, radioactive materials are used somewhat regularly. In the nuclear power industry, for instance, either uranium or plutonium is required to generate electrical energy, albeit indirectly. In the field of medicine, various radioactive materials are used to help diagnose certain diseases, as well as to aid in their treatment. Some radioactive materials are employed industrially as sources of radiation for the sterilization of products against the possible presence of bacteria. Others are used to detect the presence of buried pipelines and to gauge the thickness of plastic film. In research, results derived from the use of specific radioactive materials have greatly influenced our understanding of phenomena in agriculture, medicine, biology, and such diverse fields as astrophysics, art, and archaeology.

Notwithstanding these beneficial features, certain radioactive materials may also be used in ways that could adversely affect life on our entire planet. One such way is associated with the use of certain radioactive materials in nuclear weapons.

In contemporary times, many people throughout the world fear that nuclear weapons may again be used to maintain the current worldwide balance of political power. Such fear of exposure to radioactive materials is not confined to war zones. The operation of nuclear reactors, for instance, involves our primary peaceful use of radioactive materials.

When compared to the other classes of hazardous materials, radioactive materials are generally regarded as having the very highest degree of hazard.

—E. Meyer

Radon—An Environmental Health Concern

(Information in this section is taken from CDC's 2012 ToxTown Radon. Accessed 03/20/12 at http://toxtown.nlm,nih.gov/text_version/chemicals.php?id=27.) Radon is a radioactive gas that is formed naturally from the radioactive decay of uranium in rocks and soil and is found in some homes. It is colorless, odorless, and tasteless, but extremely toxic. When cooled below the freezing point, radon becomes phosphorescent, glowing in yellow and orange-red tones. The chemical symbol for radon is Rn. It can remain in the soil, move to the soil surface and enter the air, or enter groundwater. It is more common in some areas of the country than others. Radon was previously used to treat cancer, arthritis, diabetes, and ulcers. It is currently used to predict earthquakes, study atmospheric transport, and explore for petroleum and uranium. It is also used to initiate and influence chemical reactions in the study of surface reactions and as a tracer in leak detection. You can be exposed to radon through breathing or swallowing it, either as a gas or as particles of radon that attach to dust. Radon is a listed carcinogen. You can test for radon in your home.

In the chapter opening, Eugene Meyer describes the type of radiation that most of us are familiar with, *ionizing radiation*. Very few people have difficulty in recognizing the potential destructive power of this type of radiation. However, fewer individuals are aware of another type of radiation, *non-ionizing radiation*, which we are exposed to each day. Even fewer people can differentiate between the two types. Environmental health practitioners must be familiar with the nature of radiation and understand the detection of radiation, permissible exposure limits, biological effects of radiation, monitoring techniques, control measures, and procedures.

Radiation Safety Program Acronyms and Definitions

This list of acronyms and definitions is typically included in radiation workplace safety programs and is adapted from the U.S. Department of Health and Human Services Public Health Service Centers for Disease Control and Prevention, Atlanta, GA, August 1999, *Radiation Safety Manual*, Spellman

F. R., and N. Whiting, 2010, *The Handbook of Safety Engineering*. Lanham, MD: Government Institutes Press.

Abbreviations Typically Used in Radiation Safety Programs

ALARA—As Low as Reasonably Achievable
ALI—Annual Limit on Intake
AU—Authorized User
CDC—Centers for Disease Control and Prevention
Ci—Curie
cm2—square centimeters
cpm—counts per minute
DAC—Derived Air Concentration
dpm—disintegrations per minute
GM—Geiger-Muller
Nal—Sodium Iodide
Kg—kilogram
lfm—linear feet per minute
LSC—Liquid Scintillation Counter
mCi—milliCurie
ml—milliliters
MeV—mega electron-volts
mrem—millirem (0.001 rem)
NRC—Nuclear Regulatory Commission
OHC—Occupational Health Clinic
OHS—Office of Health and Safety
PSA—Physical Security Activity
RIA—Radioimmunoassay
RSC—Radiation Safety Committee
RSO—Radiation Safety Officer
TLD—Thermoluminescent Dosimeter
^{3}H—Tritium (hydrogen-3)
^{14}C—Carbon-14
^{32}P—Phosphorous-32
^{33}P—Phosphorous-33
^{35}S—Sulfur-35
^{51}Cr—Chromium-51
^{60}Co—Cobalt-60
^{125}I—Iodine-125
^{129}I—Iodine-129
^{131}I—Iodine-131

^{137}CS—Iodine-137
10 CFR 19—NRC's Title 10, Chapter 1, Code of Federal Regulations, Part 19
10 CFR 20—NRC's Title 10, Chapter 1, Code of Federal Regulations, Part 20

Typical Program Definitions

Absorbed Dose is the energy imparted by ionizing radiation per unit mass of irradiated material. The units of absorbed dose are the rad and the gray (Gy).

Activity is the rate of disintegration (transformation) or decay of radioactive material. The units of activity are the curie (Ci) and the Becquerel (Bq).

Alpha Particle is a strongly ionizing particle emitted from the nucleus of an atom during radioactive decay, containing two protons and neutrons and having a double positive charge.

Alternate Authorized User serves in the absence of the authorized user and can assume any duties as assigned.

Authorized User is an employee who is approved by the RSO and RSC and is ultimately responsible for the safety of those who use radioisotopes under his/her supervision.

Beta Particle is an ionizing charge particle emitted from the nucleus of an atom during radioactive decay, equal in mass and charge to an electron.

Bioassay means the determination of kinds, quantities or concentrations, and, in some cases, the locations of radioactive material in the human body, whether by direct measurement (in vivo counting) or by analysis and evaluation of materials excreted or removed from the human body.

Biological Half-Life is the length of time required for one-half of a radioactive substance to be biologically eliminated from the body.

Bremsstrahlung is electromagnetic (x-ray) radiation associated with the deceleration of charged particles passing through matter.

Contamination is the deposition of radioactive material in any place where it is not wanted.

Controlled Area means an area, outside of a restricted area but inside the site boundary, access to which can be limited by the licensee for any reason.

Counts per Minute (cpm) is the number of nuclear transformations from radioactive decay able to be detected by a counting instrument in a one-minute time interval.

Curie (Ci) is a unit of activity equal to 37 billion disintegrations per second.

Declared Pregnant Woman means a woman who has voluntarily informed her employer, in writing, of her pregnancy and the estimated date of conception.

Disintegrations per Minute (dpm) is the number of nuclear transformation from radioactive decay in a one-minute time interval.

Dose Equivalent is a quantity of radiation dose expressing all radiation on a common scale for calculating the effective absorbed dose. The units of dose equivalent are the rem and sievert (SV).

Dosimeter is a device used to determine the external radiation dose a person has received.

Effective Half-Life is the length of time required for a radioactive substance in the body to lose one-half of its activity present through a combination of biological elimination and radioactive decay.

Exposure means the amount of ionization in air from x-rays and gamma rays.

Extremity means hand, elbow, or arm below the elbow, or foot, knee, or leg below the knee.

Gamma Rays are very penetrating electromagnetic radiations emitted from a nucleus and an atom during radioactive decay.

Half-Life is the length of time required for a radioactive substance to lose one-half of its activity by radioactive decay.

Limits (dose limits) means the permissible upper bounds of radiation doses.

Permitted Worker is a laboratory worker who does not work with radioactive materials but works in a radiation laboratory.

Photon means a type of radiation in the form of an electromagnetic wave.

Rad is a unit of radiation-absorbed dose. One rad is equal to 100 ergs per gram.

Radioactive Decay is the spontaneous process of unstable nuclei in an atom disintegrating into stable nuclei, releasing radiation in the process.

Radiation (ionizing radiation) means alpha particles, beta particles, gamma rays, x-rays, neutrons, high-speed electrons, high-speed protons, and other particles capable of producing ions.

Radiation Workers are those personnel listed on the Authorized User Form of the supervisor to conduct work with radioactive materials.

Radioisotope is a radioactive nuclide of a particular element.

Rem is a unit of dose equivalent. One rem is approximately equal to one rad of beta, gamma, or x-ray radiation, or 1/20 of alpha radiation.

Restricted Area means an area, access to which is limited by the licensee for the purpose of protecting individuals against undue risks from exposure to radiation and radioactive materials.

Roentgen is a unit of radiation exposure. One roentgen is equal to 0.00025 Coulombs of electrical charge per kilogram of air.

Thermoluminescent Dosimeter (TLD) is a dosimeter worn by radiation workers to measure their radiation dose. The TLD contains crystalline material, which stores a fraction of the absorbed ionizing radiation and releases this energy in the form of light photons when heated.

Total Effective Dose Equivalent (TEDE) means the sum of the deep-dose equivalent (for external exposures) and the committed effective dose equivalent (for internal exposures).

Unrestricted Area means an area, access to which is neither limited nor controlled by the licensee.

X-rays are a penetrating type of photon radiation emitted from outside the nucleus of a target atom during bombardment of a metal with fast electrons.

Ionizing Radiation

Ionization is the process by which atoms are made into ions by the removal or addition of one or more electrons; they produce this effect by the high kinetic energies of the quanta (discrete pulses) they emit. Simply, ionizing radiation is any radiation capable of producing ions by interaction with matter. Direct ionizing particles are charged particles (e.g., electrons, protons, alpha particles, etc.) having sufficient kinetic energy to produce ionization by collision. Indirect ionizing particles are uncharged particles (e.g., photons, neutrons, etc.) that can liberate direct ionizing particles. Ionizing radiation sources can be found in a wide range of occupational settings, including health-care facilities, research institutions, nuclear reactors and their support facilities, nuclear weapon production facilities, and other various manufacturing settings, just to name a few.

These ionizing radiation sources can pose a considerable health risk to affected workers if not properly controlled. Ionization of cellular components can lead to functional changes in the tissues of the body. Alpha, beta, neutral particles, x-rays, gamma rays, and cosmic rays are ionizing radiation.

Three mechanisms for external radiation protection include time, distance, and shielding. A shorter time in a radiation field means less dose. From a point source, dose rate is reduced by the square of the distance and expressed by the inverse square law:

$$I_1(d_1)^2 = I_2(d_2)^2 \tag{9.1}$$

where

I_1 = dose rate or radiation intensity at distance d_1
I_2 = dose rate or radiation intensity at distance d_2

Radiation is reduced exponentially by thickness of shielding material.

Effective Half-Life

The half-life is the length of time required for one-half of a radioactive substance to disintegrate. The formula depicted below is used when the industrial hygienist is interested in determining how much radiation is left in a worker's stomach after a period of time. Effective half-life is a combination of radiological and biological half-lives and is expressed as:

$$T_{eff} = \frac{(T_b)(T_r)}{T_b + T_r} \tag{9.2}$$

where

T_b = biological half-life
T_r = radiological half-life

It is important to point out that T_{eff} will always be shorter than either T_b or T_r. T_b may be modified by diet and physical activity.

Alpha Radiation

Alpha radiation is used for air ionization—elimination of static electricity (Po-210), clean room applications, and smoke detectors (Am-241). It is also used in air density measurement, moisture meters, non-destructive testing, and oil well logging. Naturally occurring alpha particles are also used for physical and chemical properties, including uranium (coloring of ceramic glaze, shielding) and thorium (high temperature materials). The characteristics of alpha radiation are listed below.

- Alpha (α) radiation is a particle composed of two protons and neutrons with source: Ra-226 → Rn 222 → Accelerators.
- Alpha radiation is not able to penetrate skin.
- Alpha-emitting materials can be harmful to humans if the materials are inhaled, swallowed, or absorbed through open wounds.
- A variety of instruments have been designed to measure alpha radiation. Special training in use of these instruments is essential for making accurate measurements.
- A civil defense instrument (CD V-700) cannot detect the presence of radioactive materials that produce alpha radiation unless the radioactive materials also produce beta and/or gamma radiation.

- Instruments cannot detect alpha radiation through even a thin layer of water, blood, dust, paper, or other material, because alpha radiation is not penetrating.
- Alpha radiation travels a very short distance through air.
- Alpha radiation is not able to penetrate turnout gear, clothing, or a cover on a probe. Turnout gear and dry clothing can keep alpha emitters off of the skin.

The types of high-sensitivity portable equipment used to evaluate alpha radiation in the workplace include:

- Geiger-Mueller counter
- Scintillators
- Solid-state analysis
- Gas proportional devices

Beta Radiation

Beta radiation is used for thickness measurements for coating operations; radioluminous signs; tracers for research; and for air ionization (gas chromatograph, nebulizers). Characteristics of beta radiation are listed below.

- Beta (β) is a high energy electron particle with source: Sr-90 → Y-90 → Electron beam machine.
- Beta radiation may travel meters in air and is moderately penetrating.
- Beta radiation can penetrate human skin to the "germinal layer," where new skin cells are produced. If beta-emitting contaminants are allowed to remain on the skin for a prolonged period of time, they may cause skin injury.
- Beta-emitting contaminants may be harmful if deposited internally.
- Most beta emitters can be detected with a survey instrument (such as a DC V-700), provided the metal probe cover is open. Some beta emitters, however, produce very low energy, poorly penetrating radiation that may be difficult or impossible to detect. Examples of these are carbon-14, tritium, and sulfur-35.
- Beta radiation cannot be detected with an ionization chamber such as CD V-715.
- Clothing and turnout gear provide some protection against most beta radiation. Turnout gear and dry clothing can keep beta emitters off of the skin.

- Beta radiation presents two potential exposure methods, external and internal. External beta radiation hazards are primarily skin burns. Internal beta radiation hazard are similar to alpha emitters.

The types of equipment used to evaluate beta radiation in the workplace include:

- Geiger-Mueller counter
- Gas proportional devices
- Scintillators
- Ion chambers
- Dosimeters

Shielding for beta radiation is best accomplished by using materials with a low atomic number (low z materials) to reduce Bremsstrahlung radiation (i.e., secondary x-radiation produced when a beta particle is slowed down or stopped by a high-density surface). The thickness is critical to stop maximum energy range, and varies with the type of material used. Typical shielding materials include lead, water, wood, plastics, cement, Plexiglas, and wax.

Gamma Radiation and X-Rays

Gamma radiation and x-rays are used for sterilization of food and medical products; radiography of welds, castings, and assemblies; gauging of liquid levels and material density; and oil well logging and material analysis. The characteristics of gamma radiation and x-rays are listed below.

- Gamma (γ) is not a particle (electromagnetic wave) composed of high energy electron with source: Tc-99.
- X-ray is composed of photons (generated by electrons leaving an orbit) with source: most radioactive materials, x-ray machines, secondary to b.
- Gamma radiation and x-rays are electromagnetic radiation like visible light, radio waves, and ultraviolet light. These electromagnetic radiations differ only in the amount of energy they have. Gamma rays and x-rays are the most energetic of these.
- Gamma radiation is able to travel many meters in air and many centimeters in human tissue. It readily penetrates most materials and is sometimes called "penetrating radiation."
- X-rays are like gamma rays. They, too, are penetrating radiation.

- Radioactive materials that emit gamma radiation and x-rays constitute both an external and internal hazard to humans.
- Dense materials are needed for shielding from gamma radiation. Clothing and turnout gear provide little shielding from penetrating radiation but will prevent contamination of the skin by radioactive materials.
- Gamma radiation is detected with survey instruments, including civil defense instruments. Low levels can be measured with a standard Geiger counter, such as the CD V-700. High levels can be measured with an ionization chamber, such as a CD V-715.
- Gamma radiation or x-rays frequently accompany the emission of alpha and beta radiation.
- Instruments designed solely for alpha detection (such as an alpha scintillation counter) will not detect gamma radiation.
- Pocket chamber (pencil) dosimeters, film badges, thermoluminescent, and other types of dosimeters can be used to measure accumulated exposure to gamma radiation.
- The principal health concern associated with gamma radiation is external exposure by penetrating radiation and physically strong source housing. Sensitive organs include the lens of the eye, the gonads, and the bone marrow.

The types of equipment used to evaluate gamma radiation in the workplace include:

- Ion chamber
- Gas proportional
- Geiger Mueller

Shielding gamma and x-radiation depends on energy level. Protection follows an exponential function of shield thickness. At low energies, absorption can be achieved with millimeters of lead. At high energies, shielding can attenuate gamma radiation.

Radiation Dose

In the United States, radiation *absorbed dose*, *dose equivalent*, and *exposure* are often measured and stated in the traditional units called *rad*, *rem*, or *roentgen (R)*. For practical purposes with gamma and x-rays, these units of measure for exposure or dose are considered equal. This exposure can be from an external source irradiating the whole body, an extremity, or other organ or tissue resulting in an *external radiation dose*. Alternately, internally deposited radioactive material may cause an *internal radiation dose* to the whole body or other organ or tissue.

A prefix is often used for smaller measured fractional quantities; for example, milli (m) means 1/1,000. For example, 1 rad = 1,000 mrad. Micro (μ) means 1/1,000,000. So, 1,000,000 μrad = 1 rad, or 10 μR = 0.000010 R.

The SI system (System International) for radiation measurement is now the official system of measurement and uses the "gray" (Gy) and "sievert" (Sv) for absorbed dose and equivalent dose respectively. Conversions are as follows:

1 Gy = 100 rad
1 mGy = 100 mrad
1 Sv = 100 rem
1 mSv = 100 mrem

Radioactive transformation event (radiation counting systems) can be measured in units of "disintegrations per minute" (dpm) and, because instruments are not 100 percent efficient, "counts per minute" (cpm). Background radiation levels are typically less than 10 μR per hour, but due to differences in detector size and efficiency, the cpm reading on fixed monitors and various handheld survey meters will vary considerably.

Uranium Tailings—An Environmental Health Concern

(Information in this section is taken from CDC 2012. ToxTown-*Uranium Tailings*. Accessed 03/20/12 at http://toxtown.nlm.nihg.gov/text_version/ locations.php?id=149.) Uranium is a natural, radioactive element that is mined from the earth. It is extracted from ore by a process called milling. Uranium tailings are the radioactive, sand-like materials left over from uranium milling.

Uranium tailings are placed in mounds called tailings piles, which are located close to uranium mills. There is one abandoned tailings pile in Pennsylvania. All the other tailing piles in the United States are in the West or Southwest.

Demand for uranium has declined due to a lack of demand for new nuclear power plants and increased uranium imports from other countries. Most of the twenty-six licensed U.S. uranium mills no longer process uranium, and another twenty-four sites are abandoned.

Uranium tailings and uranium waste contain radium, which stays radioactive for thousands of years. Many of the radioactive products in uranium tailings and waste produce gamma radiation. The radioactive decay products in tailing pose a health hazard to people in the immediate area of the tailings. Tailings can also contaminate surface water of groundwater that may be used for drinking water. Tailings can also contain selenium and thorium.

Radium decays to produce radon. As mentioned, radon is an invisible and odorless gas. Radon is a carcinogen that can cause lung cancer. Radon for uranium tailings and uranium waste can threaten human health in several ways. People can be exposed to radon gas indoors if tailings are misused as construction material or backfill around homes or buildings. Piles of uranium tailings and uranium waste also can release radon gas into outdoor air or small particles from tailings can be blown into the air for people to inhale or ingest.

Because uranium occurs naturally, small amounts of uranium are found everywhere including in air, food, and water. Some uranium is released from the erosion of uranium tailings.

People who live near uranium mines may have an increased risk of kidney disease. Large amounts of uranium in the drinking water can react with the tissues in the body and damage the kidneys.

Non-Ionizing Radiation

Non-ionizing radiation is described as a series of energy waves composed of oscillating electric and magnetic fields traveling at the speed of light. Non-ionizing radiation includes those electromagnetic regions extending from ultraviolet to radio waves—and usually refers to the portion of the spectrum commonly known as the radio frequency range. Non-ionizing radiation does not cause ionization. In this text we are concerned with the kinds of non-ionizing radiation that can cause injury: ultraviolet light, infrared, laser, microwave, and radiofrequency radiation. Adverse effects on humans range from ultraviolet radiation causing problems that range from serious sunburns (sometimes ultimately causing skin cancers) to photochemical damage to the eyes; high-intensity visible light damaging the eyes; infrared radiation leading to skin burns, dehydration, and eye damage; and microwave radiation, causing thermal damage to body tissues and internal organs, and leading to cataracts or other eye injury.

In comparison to ionizing radiation, non-ionizing radiation is incapable of dislodging orbital electrons, but may leave the atom in an "excited state." All lower energy (frequency) radiation is non-ionizing. Non-ionizing radiation is expressed as a relationship of frequency, wavelength, and the speed of light. The higher the frequency, the higher the energy.

Non-ionizing radiation is found in a wide range of occupational settings and can pose a considerable health risk to potentially exposed workers if not properly controlled. The various types of non-ionizing radiation sources are listed below.

Extremely Low Frequency (ELF) Radiation—ELF radiation at 60 Hz is produced by power lines, electrical wiring, and electrical equipment. Common sources of intense exposure include ELF induction furnaces and high-voltage power lines. Wavelength is in the 50 to 60 Hz range. ACFIH exposure standards are based on understood, verifiable health effects (e.g., magnetophosphenes, induced currents, and potential interference with electronic devices, like pacemakers). ELF radiation applied to surfaces of the body induces electric currents and fields inside the body and excite cells.

Radiofrequency (RF)/Microwave (MW) Radiation—MW radiation is absorbed near the skin, while RF radiation may be absorbed throughout the body. At high enough intensities, both will damage tissue through heating. Sources of RF and MW radiation include radio emitters and cell phones. MW and RF radiation includes frequencies ranging from 0.1 cm to 300 meters or 1 mHz to 300,000 MHz. Microwaves create heat by causing water molecules to vibrate, get agitated, and heat up. Microwaves are reflected by metal but pass through glass, paper, and plastic. Materials containing water absorb them.

Infrared Radiation (IR)—All objects emit IR to other objects that have lower surface temperature. IR has a wavelength of from 700 nanometers to 1 millimeter. The skin and eyes absorb infrared radiation as heat. Workers normally notice excessive exposure through heat sensation and pain. Sources of IR radiation include furnaces, glass blowing, heat lamps, and IR lasers. Infrared light is heat. Exposure standards can be found in the ACGIH TLV booklet. To use this information, the wavelength, geometry of source, and length of exposure must be known.

Visible Light Radiation—The different visible frequencies of the electromagnetic (EM) spectrum are "seen" by our eyes as different colors. Good lighting is conducive to increased production, and can help prevent incidents related to poor lighting conditions. Excessive visible radiation can damage the eyes and skin. Visible light wavelength ranges from 400 nanometers to 700 nanometers. Lasers, compact arc lamps, quartz-iodide-tungsten lamps, gas and vapor discharge tubes, and flash lamps are all sources of visible light. Visible light exposure standards are outlined in the ACGIH TLV booklet. They depend on wavelength and exposure duration.

Ultraviolet (UV) Radiation—UV radiation has a high photon energy range and is particularly hazardous because there are usually no immediate symptoms of excessive exposure. Sources of UV radiation include the sun, black lights, fluorescent lamps, welding arcs, and UV lasers. The wavelength range of UV extends from 100 nanometers to 400 nanometers. The ozone layer only allows wavelengths greater than 290 nanometers to reach the earth. Exposure standards for UV are wavelength dependent. UV-A: 1 mW/cm^2 for 10^3 seconds measuring UV-A at the source. UV-B & C: wavelength dependent on action spectrum, most active at 200 nm. Sunglasses, clothing, sunblock, and enclosing the source provides the best protection against UV.

Laser Hazards—LASER is an acronym for Light Amplification by Stimulated Emission of Radiation. The photon of one atom can cause an excited electron of a neighboring atom to drop to the same energy level, thus causing the emission of another identical photon. Lasers typically emit optical (UV, visible, IR) radiations and are primarily an eye and skin hazard. Common lasers include carbon dioxide IR laser; helium—neon, neodymium YAG, and ruby visible lasers, and the nitrogen UV laser. ANSI has classified lasers into specific categories. The categories range from I to IV. Class I is less hazardous than Class IV.

- Class I lasers are considered to be incapable of producing damaging radiation levels, such as laser printers, and are therefore exempt from most control measures or other forms of surveillance.
- Class II lasers emit radiation in the visible portion of the spectrum, and protection is normally afforded by the normal human aversion response (blink reflex) to bright radiant sources. They may be hazardous if viewed directly for extended periods of time.
- Class IIIa lasers are those that normally would not produce injury if viewed only momentarily with the unaided eye. They may present a hazard if viewed using collecting optics, e.g., telescopes, microscopes, or binoculars. Example: HeNe lasers above 1 milliwatt but not exceeding 5 milliwatts radiant power.
- Class IIIb lasers can cause severe eye injuries if beams are viewed directly or specular reflections are viewed. A Class III laser is not normally fire hazard. Example: visible HeNe lasers above 5 milliwatts but not exceeding 500 milliwatts radiant power. Class IIIa and IIIb lasers require "Caution" signs and well-lighted areas to decrease pupil size.
- Class IV lasers are a hazard to the eye from the direct beam and specular reflections and sometimes even from diffused reflections. Class IV lasers can also start fires and damage skin. Class IV lasers require "Danger" signs.

Optical Density (OD)

Optical density (OD) is a parameter for specifying the attenuation afforded by a given thickness of any transmitting medium. Since laser beam intensities may be a factor of a thousand or a million above safe exposure levels, percent transmission notation can be unwieldy and is not used. As a result, laser protective eyewear filters are specified in terms of the logarithmic units of optical density.

Because of the logarithmic factor, a filter attenuating a beam by a factor of 1,000 (or 10(3)) has an optical density of 3, and attenuating a beam by 1,000,000 or (10(6)) has an optical density of 6. The required optical density is determined by the maximum laser beam intensity to which the individual could be exposed. The optical density of two highly absorbing filters when stacked together is essentially the linear sum of two individual optical densities. The optical density for welding goggles may be 14. A pair of specific protective goggles may have an OD of 7. The formula for calculating optical density is shown below.

$$OD = \mathrm{Log}\,(I_o/I) \tag{9.3}$$

where:

OD = optical density
I_o or I_1 = initial beam intensity
I or I_2 = final beam intensity

Finally, as with ionizing radiation (and all other workplace hazards), industrial hygienists must understand the principles of electromagnetic radiation, its uses in the workplace, its hazards, and effective control measures. The IH will usually find him or herself responsible for the radiation safety program, if one is needed in the organization.

OSHA's Radiation Safety Requirements

OSHA has standards for both ionizing radiation (29 CFR 1910.96) and non-ionizing radiation (29 CFR 1910.97). In order to understand the hazards associated with radiation, environmental health practitioners need to understand the basic terms and concepts summarized in the following paragraphs, adapted from 29 CFR 1910.96.

Radiation—consists of energetic nuclear particles and includes alpha rays, beta rays, gamma rays, x-rays, neutrons, high-speed electrons, and high-speed protons.

Radioactive Material—is material that emits corpuscular or electromagnetic emanations as the result of spontaneous nuclear disintegration.

Restricted Area—is any area to which access is restricted in an attempt to protect employees from exposure to radiation or radioactive materials.

Unrestricted Area—is any area to which access is not controlled because there is no radioactivity hazard present.

Dose—is the amount of ionizing radiation absorbed per unit of mass by part of the body or the whole body.

Rad—is a measure of the dose of ionizing radiation absorbed by body tissues stated in terms of the amount of energy absorbed per unit of mass of tissue. One rad equals the absorption of 100 ergs per gram of tissue.

Rem—is a measure of the dose of ionizing radiation to body tissue stated in terms of its estimated biological effect relative to a dose of one roentgen (r) to x-rays.

Air Dose—means that the dose is measured by an instrument in air at or near the surface of the body in the area that has received the highest dosage.

Personal Monitoring Devices—are devices worn or carried by an individual to measure radiation doses received. Widely used devices include film badges, pocket chambers, pocket dosimeters, and film rings.

Radiation Area—is any accessible area in which radiation hazards exist that could deliver doses as follows: (1) within one hour a major portion of the body could receive more than 5 millirem; or (2) with five consecutive days a major portion of the body could receive more than 100 millirem.

High-Radiation Area—is any accessible area in which radiation hazards exist that could deliver a dose in excess of 100 millirem within one hour.

OSHA's requirements for *ionizing* radiation (according to 29 CFR 1910.96) include the following:

- The employer must ensure that no individual in a restricted area receives higher levels of radiation than those summarized in Table 9.1.
- The employer is responsible for ensuring that no employee under eighteen years of age receives, in one calendar year, a dose of ionizing radiation in excess of 10 percent of the values shown in Table 9.2.

TABLE 9.1

Levels of Radiation (The Office of the Federal Register, Code of Federal Regulations Title 29 Parts 1900–1910, Office of Federal Register, Washington, D.C.: 1985)

Part of Body	*Dose, Rems/Quarter*
Whole body; head and trunk, active, blood-forming organs, lens of eyes, or gonads	1.25
Hands and forearms; feet & ankles	8.75
Skin of whole body	0.5

Source: The Office of the Federal Register, *Code of Federal Regulations Title 29 Parts 1900–1910*, Office of Federal Register, Washington, D.C.: 1985.

TABLE 9.2
Controls for Ionizing Radiation (29 CFR 1910.96, OSHA Standard)

Types of Controls	*Accomplished by*
Limiting radiation emissions at the source	Limiting the quantity of ionizing material.
Limiting time exposure	Limiting employees' time of exposure; preventing access to locations where radiation sources exist; written procedures to limit exposures.
Extending the distance from a source	Increased distance tends to dilute airborne particulates and gases; radiation levels decrease with the square of the distance—the inverse square law.
Shielding	Reducing radiation levels with shielding made of concrete, lead, steel, or water.
Barriers	Walls or fences will keep people out who should not be near or around radiation sources.
Warnings	Radiation areas should be clearly marked.
Evacuation	If a significant release of radioactive material occurs, the site should have a well thought out evacuation plan that employees are familiar with.
Security	Physical monitoring & security procedures can be used.
Training	Employees who work with or around radiation must be trained on the hazards of ionizing radiation.

- The employer is responsible for the provision and use of radiation, and the use of radiation-monitoring devices such as film badges.
- Where a potential for exposure to radioactive materials exists, appropriate warning signs must be posted.

For normal environmental conditions, OSHA requirements for non-ionizing radiation (according to 29 CFR 1910.97) include guidelines for electromagnetic energy of frequencies between 10 MHz and 100 GHz: Power density—10 mW/cm2 for periods of 0.1 hour or more; energy density—1 Mw-hr/cm2 (milliwatt hour per square centimeter) during any 0.1-hour period. Note that this guide applies whether the radiation is continuous or intermittent. Appropriate warning signs must also be posted.

TABLE 9.3
Controls for Non-ionizing Radiation (29 CFR 1910.97, OSHA Standard)

Source	*Controls*
Microwaves	Limiting the intensity of microwaves (frequency or wavelength one is exposed to) or limiting the duration of exposure. Increasing the distance from a source and shielding can also limit intensity of exposure. Signs to warn about radiation hazard or dangers. Employees should handle equipment near microwave sources with insulated gloves to minimize shock and burn hazards. Microwave equipment must be properly grounded to reduce hazards.
Ultraviolet radiation	Limit exposure to most harmful wavelengths. Use absorbing materials to shield skin and eyes.
Infrared radiation	Limit duration of exposure and the intensity of exposure. Looking into infrared sources must be avoided. Shielding (eyewear that absorbs and reflects the impact of infrared radiation on the eyes) reduces the intensity of exposure.
Lasers	Depends on the class of Laser (The Food and Drug Administration [FDA] has standards for the classification and safety design features of lasers). Controls may include enclosure of the laser source, control of potentially reflective surfaces, interlocks on doors to location where lasers are used, fail-safe pulsing controls to prevent accidental actuation, remote firing room and controls, use of baffles to limit location of beams and wearing suitable protective eyewear and clothing.

Source: 29 CFR 1910.97, OSHA Standard.

Radiation Exposure Control

Controls, both engineering and administrative, are an important element in any radiation safety program. The industrial hygienist can employ some controls (depending upon the situation) to protect employees and the public. Again, as we have stated throughout this text, engineering controls are the preferred methodology, when they are appropriate and possible. Tables 9.2 and 9.3 list the kinds of engineering and other control methods that can be employed to protect people from ionizing radiation, as well as the controls for non-ionizing radiation. The information contained in these tables comes primarily from publications by the American National Standards Institute (ANSI), New York; readers should refer to a complete listing of ANSI standards.

Discussion Questions

1. Discuss the difference between ionizing and non-ionizing radiation.
2. Discuss the difference between "rad" and "rem."
3. What is the ionizing process?
4. Are OSHA's radiation exposure control methods adequate?
5. Should we increase our reliance on nuclear power?

References and Recommended Reading

Bond, C., 2002. Statement on S. 2579. *Congressional Record*, daily edition.

DHS, 2003. *The National Strategy for the Physical Protection of Critical Infrastructure and Key Assets*. Washington, D.C.

(DHS) Department of Homeland Security, 2008. Homeland Security Presidential Directives. Accessed at www.dhs.gov/xabout/laws/editorial_0607.shtm.

(DHS) Department of Homeland Security, 2007. *Energy: Critical infrastructure and key resources sector-specific plan as input to the national infrastructure protection plan*. Washington, D.C.

DOL, 2008. *Communication*. Washington, D.C.: U.S. Department of Labor.

(EIA) Energy Information Administration, 2002. *Nuclear Glossary*. Accessed 10/18/11 at www.Eia.doe.gov/cneaf/nuclear/page/glossary.html.

(EIA) Energy Information Administration, 2003. Carbon Dioxide Emissions from Generation of Electric Power. Accessed 10/19/09 at www.eia.doe.gov/cneaf/electricity.html.

(EIA) Energy Information Administration, 2009a. Official energy statistics. Accessed 04/26/09 at www.eia.doe.gov/fuelelectric.html.

(EIA) Energy Information Administration, 2009b. Introduction to Nuclear Power. Accessed 10/21/09 at http://www.eia.doe.gov/cneaf/nuclear/page/into.html.

USDOE, 2009. Science and Technology. Retrieved 04/25/09 at http://www.energy.gov/sciencetech/index.htm.

U.S. International Trade Commission, 1991. *Global Competitiveness of U.S. Advanced-Technology Manufacturing Industries: Pharmaceuticals*. Report to the Committee of Finance, U.S. Senate, on Investigation No. 332-302 under Section 332(g) of the Tariff Act of 1930.

United States Regulatory Commission. *NRC Regulations Title 10, Code of Federal Regulations*. Accessed 11/8/2011 at http://www.nrc.gov/reading-rm/doc-collections/cfr/.

10

Occupational Health

Jurgis talked lightly about work, because he was young. They told him stories about the breaking down of men, there in the stockyards of Chicago, and of what had happened to them afterward—stories to make your flesh creep, but Jurgis would only laugh. He had only been there four months, and he was young, and a giant besides. There was too much health in him. He could not even imagine how it would feel to be beaten. "That is well enough for men like you," he would say, "silpnas, puny fellows—but my back is broad."

—Upton Sinclair, *The Jungle*

[With regard to occupational health,] we are dealing here with the tip of a treacherous ecological iceberg. Few exact studies have been made to measure the full dimension of occupational illness, occupation pollution, [and] occupational exposures. It is, unhappily, mostly guesswork. We do know, for instance, how certain occupations lead to a high rate of specific kinds of cancer. But there are only scant and fragmentary epidemiological studies of the health effects of chemical or noise pollution on whole groups of workers. Most of the data consists of braided hints of a far flung, unfathomed problem yet to be accurately measured.

—Wallick, 1972

Much of the information in this chapter is taken from F. R. Spellman's (2000–2009) Occupational Health and Safety Lectures presented during ENVH 401/501 classes at Old Dominion University (ODU) Norfolk, VA.

Environmental Health versus Occupational Health

In a nutshell, *occupational health* is the multidisciplinary discipline or approach that recognizes, diagnoses, treats, prevents, and controls workplace diseases, injuries, and other conditions. It is the preventive medicine aspect of public health—it is an integral part of many clinical disciplines.

Occupational health and safety is part of preventive medicine. It is distinguished by its focus on environmental determinants of diseases and methods of disease prevention.

Environmental health comprises those aspects of human health, including quality of life, that are determined by physical, chemical, biological, social, and psychosocial processes in the environment. It also refers to the theory and practice of assessing, correcting, controlling, and preventing those factors in the environment that can potentially affect adversely the health of present and future generations.

Simply put, environmental health encompasses a wide array of determinants that can affect a person's health, including occupational exposures—that is, in a sense, the workplace can be viewed as a subset of the environment, an environment unique to those who are employed at that workplace and very much dependent on their specific tasks and protections provided therein.

Environmental health encompasses a wide array of agents that may cause acute or chronic health effects in the population. For example:

- Lead-based paint in housing
- Ambient air pollutants (e.g., ozone, particulate matter, and toxic chemicals)
- Indoor air pollutants (e.g., molds, formaldehyde, carbon monoxide, and tobacco smoke)
- Pathogens in food and drinking water (e.g., Cryptosporidia and E. coli)
- Pesticide residues in food (e.g., organo-phosphates and pesticides that are suspected carcinogens)
- Disinfection by-products in drinking water
- Stressors that cause injury (e.g., automobiles and firearms in the home)
- Hazards in the work environment

Occupational hazards are the most preventable causes of disease, disability, and death. Why? Because unlike many other public health problems, we often know who is exposed to what. When the effect of an exposure can be measured, it is often possible to construct an exposure-effect curve.

Thus, good epidemiology can describe causal relationships. The bottom line: *because the exposures are in the workplace, they are by definition preventable.*

Although the science of occupational medicine is universal, the economic, political, and social environments determine whether and how we make progress and prevent occupational disease. To be an effective occupational health professional, one must understand five basic points concerning occupational health in the United States today:

1. The occupational health laws are strong. (They may often be diluted by various interpretations, but they are strong.)
2. The ethics of the field need strengthening.
3. Prevention must be the underlying objective of all activities.
4. Control technology and substitution are the critical strategies for prevention.
5. To prevent disease effectively, we must rely on lab tests for the data on which to base public health policy rather than waiting for epidemiology to count human bodies.

The effectiveness of occupational physicians hinges almost entirely on their ability to prevent disease. As an environmental health practitioner, effectiveness in occupational health requires an awareness of production technology. Think about it! If you do not understand the basics of industrial work, it is not possible to understand the hazards the worker has the potential to be exposed to.

All health professionals need to be able to deal with challenges effectively. They therefore need to be able to *recognize*, *manage*, and *prevent* work-related injuries and diseases.

According to the NIOSH National Traumatic Occupational Fatality database, approximately 7,000-plus traumatic occupational fatalities occur in the United States each year; currently, the highest rates are in agriculture, construction, and mining.

Estimates of the Frequency of Work-Related Diseases

The difficulty in obtaining accurate estimates of the frequency of work-related diseases is due to several factors:

1. Many problems do not come to the attention of health professionals and employers and, therefore, are not included in data collection systems.
2. Many occupational medical problems that do come to the attention of physicians and employers are not recognized as work related.
3. Some medical problems recognized by health professionals or employers as work related are not reported because the association with work is equivocal and because reporting requirements are not strict.

4. Because many occupational medical problems are preventable, their very persistence implies that some individual group is legally and economically responsible for creating or perpetuating them.

Two important aspects of occupational disease distinguish it from other medical problems:

1. Recognition of work-related medical problems is almost totally dependent on obtaining occupational information in the medical history.
2. In contrast to many nonoccupational diseases, occupational diseases can almost always be prevented.

Current Concerns with Occupational Health Issues

Governmental Role—OSHA/NIOSH greatly expanded epidemiologic and lab research into the causes of occupational diseases and injuries and the methods of preventing them. They began to strengthen the training of occupational health and safety professionals.

Occupational Safety and Health Education—This includes unions, worker right-to-know laws, and public awareness via mass media (e.g., lead, asbestos, pesticides, and ionizing radiation).

Concern with Social and Ethical Questions—The Supreme Court has upheld a worker's right to refuse hazardous work; it ruled that he or she could not be discharged or discriminated against for exercising the right not to work under conditions he or she reasonably believed to be very dangerous.

Workplace-Related Health and Medical Programs—Recently, there has been an increased emphasis on prevention programs focused on education and screening. Some programs deal not only with specific occupational medical problems but also with problems such as hypertension and smoking that may be a function of the personal lifestyles of workers.

Liability—The fear of liability suits has driven many employers to preventive activity.

Advances in Technology—Recent advances in technology have facilitated the identification of workplace hazards and potential hazards. Most notable are improvements in ways of determining the presence and measuring the levels of workplace hazards and new methods of monitoring concentrations of hazardous substances in body fluids and the physiological impairments they cause.

The Environmental Movement—Public concern for a safe and healthful environment, from protecting air, soil, and water from contamination to ensuring safety in consumer products, extends to the workplace.

Occupational Health: The Social Perspective

As mentioned, to be effective in recognizing and preventing work-related diseases and injuries, environmental health professionals need an understanding of workplaces and an appreciation for the nature of work. They need to know what work is.

Work versus Labor

To truly understand what work is, we must understand a few basic precepts:

- We must work to survive.
- The experience of work is very different from person to person and from situation to situation.
- The word *work* blurs important distinctions that are not often noticed but are acknowledged all the time in our use of the terms *work* and *labor*.
- Hard work is thought of as healthy; hard labor is punitive.
- Workers are a general category (artists).
- Laborers are those who engage in the most backbreaking and mindless chores (slaves).

Bottom Line: We may sometimes say *work* when we are referring to labor; we never say *labor* when we are referring to work.

Profit Motive versus Individual Fulfillment

The single most important fact about work in America is that there is little of it. That is, in the current work environment, there seems to be an almost even split between those who work and those who are or feel they are entitled. Americans both work and labor, but most, though, labor. They find what jobs they can, by and large, and do what they need to do to keep them. They do not choose what they will make, under what conditions they will make it, or what will happen to it afterward. Their employers, the sales and labor markets, and the workings of the economy as a whole, make these choices for them (Levenstein, Wooding, Rosenberg, 2000).

Whatever control most Americans have over how much they receive in return for their labor, how long they labor, how hard they labor, and the quality of the workplace environment is acquired in a contractual situation in which the workers' desire for comfort, income, safety, and leisure is continually counterbalanced by the employers' need for *profit* (Levenstein, Wooding, Rosenberg, 2000).

Studies indicate that most Americans would continue to work even if they did not need the money. Why? Americans would continue to work even if not necessary because of their need for social interactions, their desire to feel needed, and their desire to contribute to the common good.

Ironically, many employers see increased craftsmanship as of little economic value compared to increased productivity. Why? We will leave the answer to this to the reader.

The Drive for Greater Productivity

The capital expense involved in modern production leads employers to seek the highest possible rates of productivity. Their bottom line is *the bottom line*. Normal social reactions among workers, which in a less mechanized and fragmented work process appear as part of the rhythm of work itself, are seen as disruptive of production, and attempts on the part of workers to establish some level of control and sociability in the workplace are often misconstrued.

Managers who see such acts as threats to productivity and efficiency consider them to be indications of disloyalty and laziness (Levenstein, Wooding, Rosenberg, 2000).

Unemployment

Whiteside (1988) points out how striking it is that, even given the unsatisfying nature of most jobs and the hostility that this produces in many workers, almost all workers would rather have a job than no job at all. Or at least this has been the norm in the past—general entitlement programs skew views about working instead of receiving and vice versa.

The fact is that unemployment (along with long-term entitlements) is more destructive to physical and mental health than all but the most dangerous jobs (Leeflang, Klein-Hesselink, Spruit, 1992). Lack of a job is often equated with personal worthlessness.

The changing work environment compounds worker problems. Manufacturing jobs are disappearing. Lower paying service jobs are increasing. Worker benefits are things of the past. Years of building up skills and experience have evaporated and impacted the "middle-class" lifestyle. In the past, the meaning of work has been more frequently controlled by the workers themselves rather than imposed on them by others.

Because work-related disease is often a by-product of the conditions in which many workers now labor, only by changing those conditions can occupational disease be minimized. A preventive approach to work-related disease requires, in part, restoring control over the labor process and work conditions to those who are at risk.

Occupational Health Practice: Recognize, Evaluate, and Control

(The sections that follow are based on Spellman, F. R., and N. Whiting, 2005. *Safety Engineering: Principles and Practice*, 2nd ed. Lanham, MD: Government Institutes Press.) Occupational health practice, in coordination with and cooperation with safety engineers, begins with recognition. It ends with implementation of a control for a hazard taken from one of several options. In between recognition and control are many steps, which include evaluation. This entire process—actually a paradigm (recognition-evaluation-control)—requires the application of several principles that are both germane and important to the outcome.

What outcome? For the environmental health practitioner who works in occupational health, the outcome is quite simple: the elimination or the control of workplace hazards.

To what purpose? To protect from harm, ensure personal health, and to save lives.

In this section, we discuss the substance (the meat) of occupational health practice: recognizing a hazard, evaluating the hazard, and then controlling (or eliminating) the hazard. When we think in terms of maintaining the good occupational health of the worker, these three vital steps or processes make sense. By adding the term *occupational health* we simply specify the target of our practice.

Often, when first hired, the novice (but well equipped and highly energetic) environmental health professional who works in occupational health has a vision—that he or she is going to make the working world safer for all workers (a lofty goal, with much merit, but laced with small probability). Notwithstanding the obstacles placed in the way (namely, human beings and the laws of physics), our occupational health novice is determined to accomplish that goal . . . and woe be it to the element, object, rule, proceeding, previous practice, or engineering principle that stands in the way—our novice is determined to succeed. And many do. Unfortunately, just as many fail.

But why? Good question—one with many answers, and many of these answers aren't pleasant to comprehend or consider—but they are there, and they affect how occupational health specialists function.

To be a successful occupational health practitioner, one must be a lot of things to a lot of different people—including oneself. Any baggage the novice occupational health professional brings to the job should include a sense of humor, the habit of listening, and the ability to see things from all directions. Let's take a look at this "baggage," one "bag" at a time.

Humor? Absolutely. The quirks and foibles of the people around you, from the least-skilled worker to the most senior member of management, will affect what you do continuously, and if you do not possess enough of a sense of humor to be able to laugh at your situation and yourself, the battle is ultimately lost. What you really need is a mind open to the possibility of your own limitations, and the sure knowledge that perfection is in the eye of the beholder—and remember, in the eye of the beholder, we only have a 40 percent field of vision. We cannot see behind our heads, unless we turn our heads and actually look.

Let's get back to this humor thing. What are we really getting at here?

The best way to illustrate our point is to look at a few occupational health corollaries from A. Block in *Murphy's Law and Other Reasons Why Things Go Wrong*, and *Murphy's Law Book Two*, 1977, 1980. When you begin to take yourself too seriously, remember that you—and the workers you are trying to protect (sometimes you'll think they are deliberately trying to work against you)—are subject to Murphy's Law as well as to OSHA's.

- Most projects require three hands.
- Hindsight is an exact science.
- When all else fails, read the instructions.
- A fail-safe circuit will destroy others.
- If a test installation functions perfectly, all subsequent systems will malfunction.
- Only God can make a random selection.
- A failure will not appear until a unit has passed final inspection.
- Any system that depends on human reliability is unreliable.

This is what you're up against. For your own sanity, you'll need to be able to see the funny side of many situations you'll encounter on the job.

In addition to humor, the occupational health professional should also have the habit (a well-established one) of listening. *Listening* . . . this one word is powerful in meaning, essential to success, vital to anything we do, yet at the same time, listening is something we often fail to do—a tool we fail to use. When you get right down to it, this common tendency not to listen, not to hear what others say, is rather odd and certainly a waste. Why? Simply because one of the most powerful tools any manager or practitioner in any profession has is to listen. For the occupational health professional, listening is vital—absolutely vital.

The occupational health professional must be able to see things from all directions. But what does this mean?

What we mean is that most of us never quite know everything to look at—particularly for the occupational health professional during the hazard evaluation phase. We do use other senses to make judgments or evaluations (such as taste, smell, etc.), but the visual experience is what we rely on most, and that visual experience primarily affects the decisions that we end up making—the ones that affect the outcome. To make solid evaluations takes focus, concentration, and the imagination to see possibilities. To say that the evaluation process is a tricky task is like saying going to the moon and back is a leisurely undertaking.

Based on personal experience, we add one more piece of luggage that should be part of any occupational health practitioner's baggage: the ability to persuade. Persuasion, while simple enough in concept, connotes so many different meanings and methodologies of implementation, especially when it comes to persuading certain managers.

Whether managing a widget or computer factory, the manager, in regards to expertise, must be a well-rounded, highly skilled individual. No one questions the need for incorporation of these highly trained practitioners—well-versed in the disciplines of engineering, chemistry, computer science, environmental science, safety principles, widgetology, accounting, auditing, technical aspects, and operations. In our experience, however, engineers, chemists, scientists, and widgetologists, and others with no formal management training, are often hindered (limited) in their ability to solve the complex management problems currently facing both industries.

There are those who will view this opinion with some disdain. However, in the current environment where privatization, the need for upgrading security, and other pressing concerns are present, skilled management professionals are needed to manage and mitigate these problems. One final point: when dealing with technical managers versus professional managers, we are reminded of another quote: "Common sense is not very common" (Mark Twain).

In the sections that follow, we look at recognition, evaluation, and control—all critical to the control or elimination of workplace hazards.

Hazard Recognition

Logically, only after hazards are recognized can suitable controls be identified and selected. In safety engineering, we must acknowledge two points about hazard recognition right up front: (1) recognition relates to the identification of hazards, and (2) no one individual can be fully knowledgeable about all hazards. What we are saying here is what we say throughout this text: the occupational health professional must have knowledge of hazards and potential controls across many engineering and scientific disciplines (the Jack- or Jill-of-all-trades syndrome).

The occupational health professional is both a proactive and reactive analyst. Obviously, we favor the proactive mode (find the hazard before it causes a problem) versus the reactive mode (find out what went wrong and prevent it from occurring again).

Analytical ability is first tempered and then honed by training and experience. Coupled with analytical ability is observation; a learned skill, like analytical ability, is also improved by training and experience.

The analytical process used by the occupational health professional is paradoxical, because the methods he or she uses to recognize hazards involves viewing normal (the acceptable norm) processes, equipment and machinery, and/or situations to seek primarily for the abnormal. Simply put, the abnormal is what the occupational health professional is attempting to recognize—to eliminate.

Obviously, recognition employs the use of the senses (sight, hearing, smell, taste, and touch), but it also requires an investment of time. Unless sufficient time is allowed for observation, the view is cursory at best—an unsatisfactory condition. For example, if you observe a person working, your presence may affect how they work. To really see them in action as they normally work, you must be there long enough to not provide a distractive element.

The primary proactive technique used by the safety professional in the recognition phase of hazard control is the safety and health audit or inspection process. Inspecting for safety and health compliance is a structured and detailed approach to reducing and controlling the seriousness of accidents—before they occur.

Important Note: Any occupational health professional (or other designated official armed with OSHA regulations) can conduct a safety and health audit of any operating facility. He or she might even do a credible job, insofar as recognizing and reporting discrepancies from the written rules. While you don't have to understand the physical science of a process (nor be competent at job safety analysis) to determine whether guardrails are in compliance with 42" nominal height, a 21" midrail and 4" toeboard; if flammable storage cabinets are properly labeled; or if a guard on a mechanical gear box is in place, the process just described is not true inspecting—and definitely not recognition.

The bottom line? Recognition can't be accomplished unless the safety professional understands the processes and technology of an operation. Without process and technology information, the occupational health professional is likely to fail in his or her recognition of latent hazards, failures, or errors that can cause accidents. You must understand the tasks a job requires, the materials, and the equipment involved to be able to recognize the hazards.

We have said, "No one individual can be fully knowledgeable about all the hazards." Common sense helps us understand the logic and validity of this statement. However, if the occupational health practitioner is to do a credible job (and using a book of regulations will only allow him or her to ensure compliance and not really recognize hazards), how is the effective recognition of potential hazards generated? Although they should never be used as substitute for sound understanding, preparation, and planning (or for keeping one's eyes and ears open while inspecting a process), the most common (and most effective) tool for hazard recognition is the checklist.

Checklists? Absolutely. An occupational health professional who does not incorporate the use of a comprehensive checklist is like the traveler who does not use a map or GPS. Checklists serve many purposes, but the primary purpose is to "jog" the memory—to make the safety professional much more aware of—and knowledgeable about—potential hazards.

What type of checklist is advised? It depends on the intended objective. In some cases, a checklist that focuses on a narrow field of hazards and problems (one specifically tailored for the operation under inspection) may be advised. At other times the checklist might need to be comprehensive or general, especially when you are looking at the entire plant site.

The recognition phase must examine three things to be effective: unsafe conditions, unsafe operation, and unsafe acts. Recognition is not only accomplished by the supervisor as a routine function of his or her job (he or she notices something wrong or abnormal with the system he or she is responsible for), but also by workers who operate and maintain a system. The occupational health professional routinely audits the system, operation, and/or process with the main intention of looking for flaws in the system that allow unsafe conditions to arise. These audits and other inspections must become a permanent part of the occupational health professional's role.

We pointed out earlier that the occupational health professional must understand the processes, systems, or operations. This is a crucial element in the auditing process. If the inspector (any inspector) is not familiar with the operation and the operating parameters of a process, system, or operation, then he or she would find conducting an effective audit difficult. This is especially the case in energy exchange. The occupational health professional must understand that free energy transfer in any of its forms (thermal, electrical, kinetic, potential, chemical, and radiation), in any uncontrolled manner generates damage (which can be drastic if energy is of high magnitudes), in processes—and in people.

Hazard Evaluation

An integral part of occupational health is the hazard evaluation process. In the evaluation process, the goal is to be forward-looking and predictive, with the result of the evaluation process applied *prior* to realization of any losses.

In formal college training programs, the hazard evaluation phase of environmental health and safety engineering curriculums include large amounts of training time expended on modeling. In the computer age, models have become the basic tools of all science, business, and engineering students. Predictive modeling (a forward-looking hazard analysis technique) is greatly needed in certain areas of occupational health. Occupation health professionals not only need to be taught how to use these models, but also to understand their application.

We point out, however, that unless the occupational health student is destined for employment in a manufacturing firm that produces products, and where there is a need to predict certain safety aspects of the product (product hazard risk assessment), and where the product is engineered for safety, his or her expertise in computer predictive modeling has little application in the "real world." We define "real world" as the deckplates of a factory—where the occupational health professional is responsible for the safety and health of all workers.

Some professors will dispute this point. Some occupational health practitioners will also dispute this point—we understand this. However, in this text, we stress the occupational health practitioner who is employed in the "field"—on the job in whatever activity, with the primary function as the company safety and health person. Experience has shown us that many college-level training programs emphasize hazard risk assessment on the "widget," or product, concept, instead of on practical application in the working world. Simply put, hiring a newly graduated occupational health professional who is extremely well-versed in computer modeling, but is lacking in on-the-spot recognition and evaluation of hazards in the field is not at all unusual. In this text, we attempt to "fix" this glaring deficiency by providing grounding in "real world" information.

To gain better appreciation for the kinds of hazard recognition and evaluation activities the field occupational health professional is likely to actually perform and get involved with, we provide a simple but important example.

When a company decides to add on to its present infrastructure for whatever reason, often this includes the construction of new structures to house complex process equipment. If the company has its own internal engineering staff, the project will most likely be assigned inhouse, instead of being put out for bid to engineering firms. The actual construction, of course, is placed out on bid to outside construction companies.

The inhouse engineering department will be responsible for the entire project: planning, design, construction, testing, etc. In the preliminary design phase, a prudent course of action for the engineering department would include the company's occupational health professional in the process.

Why? For several reasons. For instance, the occupational health professional should be in on the ground floor whenever such a project is undertaken, if for no other reason than his or her need to be familiar with the new process or equipment to be installed (the company occupational health professional must be familiar with *all* company processes).

Another reason is because he or she normally is cognizant of any new regulations or requirements covered under the National Fire Protection Association's (NFPA) Life Safety Code NFPA 101, local safety codes, and new OSHA requirements. Incorporating any new regulatory requirements into a system or process is better in the design phase than it is after construction is completed.

Probably the single most important reason that the occupational health professional should be included in the design phase is for his or her evaluation of (scrutiny of) the systems to be installed.

What other tools (besides common sense) does the occupational health professional have at his or her disposal to aid them in the evaluation phase—to identify hazards and their probability? Good question—with several answers.

Familiarity with Similar Units or Processes

A good example is a chemical process. Some chemical processes have an inherent hazard, such as high temperature or pressure; they could create a "runaway" (exothermic) type of reaction, or use hazardous materials, or produce hazardous or toxic substances.

Company Accident Reports

Written plant accident reports are normally included in those "must-have and must-keep" recordkeeping requirements. These reports are valuable for many reasons. In this instance, they are valuable because they may provide a good reference for anticipating hazards and estimating probability. Many occupational health professionals computerize these accident reports to facilitate easy access and to compile probability indices—very valuable statistics.

HAZOP Review Process

HAZOP (Hazard and Operability Study) is a concept that is gaining importance in environmental health, occupational health, industrial hygiene, and safety engineering (especially with the advent of OSHA's Process Safety Management Standard [PSM]). The purpose of HAZOP is to identify hazard and operability problems. Though it can be used for existing facilities, it is most successful when applied to new plants at the point where the design is nearly firm and documented or to existing plants where a major redesign is planned. The HAZOP is a team function; the industrial hygienist, along with a process engineer and operating personnel, performs the HAZOP to determine hazards and operating problems, make recommendations in design change, procedures, and so forth (primarily to improve health and safety), and recommend follow-up studies where no conclusion was possible from lack of information. The results are qualitative and require detailed plant descriptions (e.g., drawings, procedures, and flow charts). A HAZOP also requires considerable knowledge of the process, instrumentation, and operation, and this information is usually provided by team members expert in these areas (American Institute of Chemical Engineers, 1985).

Training (Attending Seminars and Meetings)

Training or information received while attending seminars and meetings conducted by professional societies and trade organizations (such as the American Society of Safety Engineers [ASSE], the National Safety Council [NSC], the American Petroleum Institute, the American Society of Mechanical Engineers, the American Institute of Chemical Engineers, and the Society of Fire Protection Engineers) can go a long way toward increasing the occupational health professional's hazard identification/evaluation knowledge.

Published Materials

The occupational health professional who does not review papers published in journals misses the opportunity to read about problems others have identified, which may allow the occupational health professional to correct problems without having to experience them personally.

Insurance Companies

The company's insurance vendors can provide valuable information to the occupational health professional. Insurance vendors normally compile a history of losses, with documentation of events causing these losses. The insurer's experience may point out similar problems that occurred in other similar companies (Miller et al., 1987).

Hazard Control

Recognition leads to evaluation, which leads to the last phase—hazard control. Seems simple enough, doesn't it? But it's simple only in the sense that it doesn't take rocket science to recognize a potential hazard, to evaluate it and determine that indeed it is a hazard, and then to control the hazard. No, NASA occupational health professionals excepted, rocket science is not normally part of the occupational health professional's general knowledge. Instead, in hazard control, an understanding of risk is part of the general knowledge the industrial hygienist needs.

What is risk? ASSE (1988) defines *risk* as a measure of both the probability and the consequence of all hazards of an activity or condition; a subjective evaluation of relative failure potential; in insurance, a person or thing insured. Seems simple enough, doesn't it?

But is risk really that simple? No, not really. Why? Let's take a look.

Standard occupational health and safety practice dictates that after a hazard has been identified and evaluated, the potential results should be assessed to determine which risks are acceptable, and which are not. "Acceptable" risks? Is there such a thing as accepting a risk? Yes, there is, and throughout your classes in college, you may spend many countless, boring hours studying this elusive topic.

What are acceptable risks? It depends. Let's try to answer this question by stating what an unacceptable risk is. Simply put, any risk involving human life and health is not acceptable. However, risk involving only property may be acceptable—to a limited degree.

Wait a minute. We know what some of you out there might be thinking—"Well, if risk involving property may be acceptable, then why should I provide expensive fire protection for my factory? The economics are against fire protection—look at the money I could save the company." We certainly hope this view is not shared by anyone reading this text. Providing no fire protection for a factory (or any other business operation), and letting it burn down on the premise that it is fully insured and can be rebuilt (the "let the insurance company sort-it-out" syndrome) is *not* an acceptable risk.

Several methods are available to control or eliminate risks. Standard occupational health and safety practice to control hazards has been prioritized over the years. The priorities, in order of importance, are:

Eliminate the hazard (do whatever you are doing in a safer way or simply stop doing it).
Reduce the hazard level (overdesign to reduce risk).

Provide safety devices (failsafe design, install barriers, isolate the machine, equipment, process, pressure relief valves).
Provide warnings (caution or warning labels and alarms).
Provide safety procedures (training, written operating procedures, personal protective equipment).
Justify residual risk (do nothing).

Did You Know?

Workers who are exposed to extreme cold or work in cold environments may be at risk of cold stress. Extreme cold weather is a dangerous situation that can bring on health emergencies in susceptible people, such as those without shelter, those who work outdoors, and those who work in an area that is poorly insulated or without heat.

Whatever control method you choose to employ, the engineering approach—the prudent approach—is to think "multiple" controls. Multiple controls should be of different design to eliminate a common cause defeating all redundant controls. Note: Redundancy means providing more than one means to accomplish something, where each means is independent of the other.

Industrial Hygiene

Most safety professionals are already involved in some aspects of industrial hygiene. They study work operations, look for potential hazards, and make recommendations to minimize these hazards. The industrial hygienist, through specialized study and training, has the expertise to deal with these complex problems. If the safety professional carries on the day-to-day safety functions involving immediate decisions, he or she must know when and where to get help on industrial hygiene problems.

After the industrial hygienist surveys the plant, makes recommendations, and suggests certain control measures, it may become the safety professional's responsibility to see that the control measures are being applied and followed. Or such responsibility may be vested in an individual whose education and training is in the combined disciplines of safety and health (Olishifski, J.B., and B.A. Plog, 1988, p. 4).

What is industrial hygiene? Most countries of the world identify industrial hygiene as the practice of occupational hygiene. The exception is the United States, where it is identified as industrial hygiene (Moeller, 2011).

The American Industrial Hygiene Association (AIHA) defines *industrial hygiene* as "that science and art devoted to the anticipation, recognition, evaluation, and control of those environmental factors or stresses—arising in or from the workplace—which may cause sickness, impaired health and well-being, or significant discomfort and inefficiency among workers or among citizens of the community."

What is an industrial hygienist? A well-trained, well-prepared industrial hygienist (IH) is equipped to deal with virtually any situation that arises (Ogle, 2003).

Is the industrial hygienist also an occupational health professional? Maybe. It depends. Like safety engineering, occupational health and industrial hygiene have commonly been thought to be separate entities (especially by safety professionals, occupational health professionals, and industrial hygienists). In fact, over the years, a considerable amount of debate and argument has risen between those in the safety, occupational health, and industrial hygiene professions on many areas concerning safety and health issues in the workplace—and on exactly who is best qualified to administer a workplace safety and health program.

Historically, the safety professional had the upper hand in this argument—that is, prior to the enactment of the OSH Act. Until OSHA went into effect, occupational health and industrial hygiene were not topics that many thought about, cared about, or had any understanding of. Safety was safety, job safety included health and industrial hygiene, and that was that.

After the OSH Act, however, things changed. In particular, people began to look at work injuries and work diseases differently. In the past, they were regarded as separate problems. Why?

The primary reason for this view was obvious—and not so obvious. The obvious was work-related injury. Work injuries occurred suddenly, and their agent (i.e., the electrical source, chemical, machine, tool, work or walking surface, or whatever unsafe element caused the injury) usually was readily obvious.

Not so readily obvious were the workplace agents (occupational diseases) that caused illnesses. Why? Because most occupational diseases develop rather slowly, over time. In asbestos exposure, for example, workers who abate (remove) asbestos-containing materials without the proper training (awareness) and personal protective equipment are subject to exposure. Typically, asbestos exposure may be a one-time exposure event (the silver bullet syndrome) or the exposure may go on for years. No matter the length of exposure, one thing is certain, with asbestos contamination, pathological change occurs slowly—some time will pass before the worker notices a difference in his or her pulmonary function. Disease from asbestos exposure has a latency period that may be as long as twenty to thirty years before the effects are realized. The point? Any exposure to asbestos, short-term or long-term, may eventually lead to a chronic disease (in this case, restrictive lung disease) that is irreversible (e.g., asbestosis). Of course, many other types of workplace toxic exposures can affect workers' health. The prevention, evaluation, and control of such occurrences is the role of the industrial hygienist.

Thus, because of the OSH Act, and also because of increasing public awareness and involvement by unions in industrial health matters, the role of the industrial hygienist has continued to grow over the years. Certain colleges and universities have incorporated industrial hygiene majors into environmental health programs. For example, Old Dominion University in Norfolk, Virginia, has an industrial hygiene major or specialty as part of its environmental health program.

The other offshoot of the OSH Act has been, in effect (though many practitioners in the field will disagree with this point of view), a continuing tendency toward uniting safety, environmental health, and industrial hygiene into one entity.

This presents a problem with definition. When we combine the three entities, do we combine them into a "safety" or an occupational health or an "industrial hygiene" title or profession? The debate on this issue continues. What is the solution to this problem—how do we end the debate? To attempt an answer, we need to look at several factors, and at the actual experience gained from practice in the real world of safety, occupational health, and industrial hygiene.

Again, the occupational health professional must be a generalist—a Jack or Jill of all trades. You should already have a pretty good feel for what the occupational health professional is required to do, and what he or she is required to know to be effective in the workplace.

How about the industrial hygienist? What are industrial hygienists required to do and know to be effective in the workplace? Let's take a look at what a typical industrial hygienist does (you should be able to determine the level of knowledge they should have).

The primary mission of the industrial hygienist is to examine the workplace environment and its environs by studying work operations and processes. From these studies, he or she is able to obtain details related to the nature of the work, materials and equipment used, products and by-products, number of employees, hours of work, and so on. At the same time, appropriate measurements are made to determine the magnitude of exposure or nuisance (if any) to workers and the public. The hygienist's next step is to interpret the results of the examination of the workplace environment and environs, in terms of ability to impair worker health. (For example, is there a health hazard in the workplace that must be mitigated?) With examination results in hand, the industrial hygienist then presents specific conclusions to the appropriate managerial authorities.

Is the process described above completed? Yes and no. In many organizations, the industrial hygienist's involvement stops there. But remember, discovering a problem is only half the safety battle. Knowing that a problem exists, but not taking any steps to mitigate it is leaving the job half-done. In this light, the industrial hygienist normally will make specific recommendations for control measures—an important part of industrial hygiene's anticipate-recognize-evaluate-control paradigm.

Any further involvement in working toward permanently changing the hazard depends on the industrial hygienist's role in a particular organization. Organization size is one of the two most important factors that determine the industrial hygienist's role within an organization. Obviously, an organization that consists of several hundred (or thousands) of workers will increase the work requirements for the industrial hygienist.

The second factor has to do with what the organization produces. Does it produce computerized accounting records? Does it perform a sales or telemarketing function? Does it provide office supplies to those businesses requiring such services? If the organization accomplishes any of these functions (and many similar functions) the organization probably does not require the services of a full-time staff industrial hygienist. On the other hand, if the organization handles, stores, or produces hazardous materials, if the organization is a petroleum refinery, or if the organization is a large environmental laboratory, then there might be a real need for the services performed by a full-time staff industrial hygienist.

Let's look at these factors in combination, and assume the organization employs 5,000 full-time workers in the production of chemical products. In this situation, not only are the services normally performed by an industrial hygienist required, but also the company probably needs more than just one full-time industrial hygienist—perhaps several. In this case, each industrial hygienist might be assigned duties that are narrow in scope, with limited responsibility—in short, each hygienist must specialize.

What this all means, of course, is that the size of the organization and the type of work performed dictates the need for an industrial hygienist, and each organization's needs sets the extent of their work requirements.

An important area of responsibility for the industrial hygienist that we have not mentioned here (and one often overlooked in the real work world) is the industrial hygienist's responsibility to conduct training. If the industrial hygienist examines a workplace environment and environs for occupational hazards and discovers one (or more), he or she will perform the anticipate-recognize-evaluate-control actions. However, an important part of the "control" area of the industrial hygiene paradigm is *informing* those exposed to the hazards about the hazards—to train them. In the field of industrial hygiene in general, preparation in this area and emphasis on its importance on-the-job has been weak.

What does all this mean? Good question—one answered most simply by pointing out that occupational health professionals are (or should be) generalists. Industrial hygienists are specialists. This simple statement sums up the main difference between the two professions.

Which one is best? Another good question. Actually which one is "best" is not the issue—rather, which one is better suited to perform the functions of the occupational health practitioner? Based on our own personal experience, we profess the need for generalization (we have stated this throughout the text and will continue to do so to the end). The occupational health professional needs to be well versed in all aspects of industrial hygiene—no question about this. We also hold no doubt about the industrial hygienist being efficacious to safety and health.

The problem lies in present perception and use. The safety engineer views his role as all-encompassing (as he or she rightly should), and the occupational health professional and industrial hygienist views him or herself as a step above safety, occupying that high pinnacle position known as "specialist." Which one is right, and which one is wrong? Neither. Again, it is a matter of perception. On the use side of the issue, the safety professional is usually employed to run or manage an organization's safety program (as safety manager, safety engineer, safety professional, safety director, safety coordinator, occupational health director/manager, or another similar title), while the occupational health practitioner and industrial hygienist are usually hired to fill only a position of organizational occupational health specialist or industrial hygienist.

Can you see the difference? Believe us when we say that practicing safety professionals, occupational health professionals, and industrial hygienists not only see the difference, they feel the difference—and they learn to live and work with the difference—sometimes as colleagues and sometimes as competitors.

Workplace Stressors

As mentioned in chapter 1, stressors can cause occupational sickness. The industrial hygienist focuses on evaluating the healthfulness of the workplace environment, either for short periods or for a work-life of exposure. When required, the industrial hygienist recommends corrective procedures to protect health, based on solid quantitative data, experience, and knowledge. The control measures he or she often recommends include: isolation of a work process, substitution of a less harmful chemical or material, and/or other measures designed solely to increase the healthfulness of the work environment.

Important Note: Woe be it to the rookie industrial hygienist, occupational health professional, or safety engineer who has the audacity (and downright stupidity) to walk up to any supervisor (or any worker with experience at task) and announce that he or she is going to find out everything that is unhealthy (and thus injurious to workers) about the work process—without having any idea how the process operates, what it does, or what it is all about. We do not recommend this scenario—as an industrial hygienist (or as a safety engineer), you must first understand the work operations and processes to the point that you could almost operate the system efficiently and safely yourself.

What are the workplace stressors the industrial hygienist should be concerned with? According to Pierce (1984), the industrial hygienist should be concerned with those workplace stressors that are likely to accelerate the aging process, cause significant discomfort and inefficiency, or may be immediately dangerous to life and health. Several stressors fall into these categories; the most important ones include:

Chemical stressors—gases, dusts, fumes, mists, liquids, or vapors.

Physical stressors—noise, vibration, extremes of pressure and temperature, and electromagnetic and ionizing radiation.

Biological stressors—bacteria, fungi, molds, yeasts, insects, mites, and viruses.

Ergonomic stressors—repetitive motion, work pressure, fatigue, body position in relation to work activity, monotony/boredom, and worry.

Areas of Concern: Industrial Toxicology

From the list of stressors above, you can see that the industrial hygienist has many areas of concern related to protecting the health of workers on the job. In this section, we focus on the major areas that the industrial hygienist typically is concerned with in the workplace. We also discuss the important areas of industrial toxicology, industrial health hazards, industrial noise, vibration, and environmental control. All of these areas are important to the industrial hygienist (and to the worker, of course), but they are not all-inclusive; the industrial hygienist also is concerned with other areas—ionizing and nonionizing radiation, for example, and many others.

Normally, we give little thought to the materials (chemical substances, for example) we are exposed to on a daily, almost constant basis, unless they interfere with our lifestyles, irritate us, or noticeably physically affect us. Most of these chemical substances do not present a hazard—under ordinary conditions. However, keep in mind that all chemical substances have the potential for being injurious at some sufficiently high concentration and level of exposure. The industrial hygienist understands this, and to prevent the lethal effects of overexposure for workers must have an adequate understanding and knowledge of general toxicology.

To provide a basic review, we ask: What is toxicology? *Toxicology* is a very broad science that studies the adverse effects of chemicals on living organisms. It deals with chemicals used in industry, drugs, food, and cosmetics, as well as those occurring naturally in the environment. Toxicology is the science that deals with the poisonous or toxic properties of substances. The primary objective of industrial toxicology is the prevention of adverse health effects in workers exposed to chemicals in the workplace. The industrial hygienist's responsibility is to consider all types of exposure and the subsequent effects on living organisms. Following the prescribed precautionary measures and limitations placed on exposure to certain chemical substances by the industrial toxicologist is the worker's responsibility. The industrial hygienist uses toxicity information to prescribe safety measures for protecting workers.

To gain a better appreciation for what industrial toxicology is all about, you must understand some basic terms and factors—many of which contribute to determining the degree of hazard particular chemicals present. You must also differentiate between *toxicity* and *hazard. Toxicity* is the intrinsic ability of a substance to produce an unwanted effect on humans and other living organisms when the chemical has reached a sufficient concentration at a certain site in the body. *Hazard* is the probability that a substance will produce harm under specific conditions. The industrial hygienist and other safety professionals employ the opposite of hazard—*safety*—that is, the probability that harm will not occur under specific conditions. A toxic chemical—used under safe conditions—may not be hazardous.

Basically, all toxicological considerations are based on the *dose-response relationship*, another toxicological concept important to the industrial hygienist. In its simplest terms, the dose of a chemical to the body resulting from exposure is directly related to the degree of harm. This relationship means that the toxicologist is able to determine a *threshold level* of exposure for a given chemical—the highest amount of a chemical substance to which one can be exposed with no resulting adverse health effect. Stated another way, chemicals present a threshold of effect, or a no-effect level.

Threshold levels are critically important parameters. For instance, under the OSH Act, threshold limits have been established for the air contaminants most frequently found in the workplace. The contaminants are listed in three tables in 29 CFR 1910 subpart Z—Toxic and Hazardous Substances. The threshold limit values listed in these tables are drawn from values published by the American Conference of Governmental Industrial Hygienists (ACGIH) and from the "Standards of Acceptable Concentrations of Toxic Dusts and Gases," issued by the American National Standards Institute (ANSI).

An important and necessary consideration when determining levels of safety for exposure to contaminants is their effect over a period of time. For example, during an eight-hour work shift, a worker may be exposed to a concentration of Substance A (with a 10 ppm [parts per million—analogous to a full shot glass in a swimming pool] TWA [time-weighted average], 25 ppm ceiling and 50 ppm peak) above 25 ppm (but never above 50 ppm) only for a maximum period of 10 minutes. Such exposure must be compensated by exposures to concentrations less than 10 ppm, so that the cumulative exposure for the entire eight-hour work shift does not exceed a weighted average of 10 ppm. Formulas are provided in the regulations for computing the cumulative effects of exposures in such instances. Note that the computed cumulative exposure to a contaminant may not exceed the limit value specified for it.

Although air contaminant values are useful as a guide for determining conditions that may be hazardous and may demand improved control measures, the industrial hygienist must recognize that the susceptibility of workers varies.

Even though it is essential not to permit exposures to exceed the stated values for substances, note that even careful adherence to the suggested values for any substance will not assure an absolutely harmless exposure. Thus, the air contaminant concentration values should only serve as a tool for indicating harmful exposures, rather than the absolute reference on which to base control measures.

For a chemical substance to cause or produce a harmful effect, it must reach the appropriate site in the body (usually via the bloodstream) at a concentration (and for a length of time) sufficient to produce an adverse effect. Toxic injury can occur at the first point of contact between the toxicant and the body, or in later, systemic injuries to various organs deep in the body (Hammer, 1989). Common routes of entry are ingestion, injection, skin absorption, and inhalation. However, entry into the body can occur by more than one route (e.g., inhalation of a substance that can also be absorbed through the skin).

Ingestion of toxic substances is not a common problem in industry—most workers do not deliberately swallow substances they handle in the workplace. However, ingestion does sometimes occur either directly or indirectly. Industrial exposure to harmful substances through ingestion may occur when workers eat lunch, drink coffee, chew tobacco, apply cosmetics, or smoke in a contaminated work area. The substances may exert their toxic effect on the intestinal tract or at specific organ sites.

Injection of toxic substances may occur just about anywhere in the body where a needle can be inserted, but is a rare event in the industrial workplace.

Skin absorption or contact is an important route of entry in terms of occupational exposure. While the skin may act as a barrier to some harmful agents, other materials may irritate or sensitize the skin and eyes, or travel through the skin into the bloodstream, thereby impacting specific organs.

Inhalation is the most common route of entry for harmful substances in industrial exposures. Nearly all substances that are airborne can be inhaled. Dusts, fumes, mists, gases, vapors, and other airborne substances may enter the body via the lungs and may produce local effects on the lungs, or may be transported by the blood to specific organs in the body.

Upon finding a route of entry into the body, chemicals and other substances may exert their harmful effects on specific organs of the body, such as the lungs, liver, kidneys, central nervous system, and skin. These specific organs are termed *target organs* and will vary with the chemical of concern.

The toxic action of a substance can be divided into *short-term (acute)* and *long-term (chronic)* effects. Short-term adverse (acute) effects are usually related to an accident where exposure symptoms (effects) may occur within a short time period following either a single exposure or multiple exposures to a chemical. Long-term adverse (chronic) effects usually occur slowly after a long period of time, following exposures to small quantities of a substance (as lung disease may follow cigarette smoking). Chronic effects may sometimes occur following short-term exposures to certain substances.

Table 10.1 shows the harmful effects that can result from overexposure to some chemical agents.

Industrial Health Hazards

NIOSH and OSHA's *Occupational Health Guidelines for Chemical Hazards*, DHHS (NIOSH) Publication No. 81-123 (Washington, D.C.: Superintendent of Documents, U.S. Government Printing Office, current edition) illustrates quite clearly that the number of known industrial poisons is quite large, and also that their effects and means of control are generally understood. Generally, determining if a substance is hazardous or not is simple, if the following is known: (1) what the agent is and what form it is in; (2) the concentration; and (3) the duration and form of exposure.

However, practicing safety engineers and industrial hygienists come face-to-face with one problem rather quickly. Many new compounds of somewhat

TABLE 10.1
Comparison of Selected Chemical Agents and Their Harmful Effects Resulting from Overexposure

Agent Type	*Major Route of Entry*	*Acute/Chronic Effects*
Asbestos	Inhalation	*Chronic:* asbestosis, mesothelioma, lung cancer
Arsenic	Skin absorption	*Acute:* skin irritation, conjunctivitis, sensitization dermatitis *Chronic:* possible epidermal cancer
Cadmium	Inhalation	*Acute:* chest pain, shortness of breath, pulmonary edema, digestive effects
Lead	Inhalation	*Chronic:* gastrointestinal disturbance, anemia due to red blood cell effects, kidney disease and reproductive effects
Aromatic Solvent	Inhalation	*Acute:* central nervous system effects, depression, narcotic effects *Chronic:* liver, blood system disorders
Sulfur dioxide	Skin absorption, Inhalation	*Dermatitis* (chronic or acute) *Acute:* eye & respiratory irritation *Chronic:* bronchitis

uncertain toxicity are introduced into the workplace each year. Another related problem occurs when manufacturers develop chemical products with unfamiliar trade names, and do not properly label them to indicate the chemical constituents of the compounds (of course, under OSHA's 29 1910.1200 Hazard Communication Standard, this practice is illegal).

One of the primary categories of industrial health hazards that the occupational health professional must deal with is airborne contaminants. Two main forms of airborne contaminants are of chief concern: particulates, and gases or vapors. Particulates include dusts, fumes, smoke, aerosols, and mists, classified additionally by size and chemical makeup, and sometimes by shape.

Dusts are solid particles of matter produced by grinding, crushing, handling, detonation, rapid impact, etc. Size may range from 0.5 to 100 mm (micron: 1 mm = 1/25000 inch), with most (over 90 percent) airborne dust in the 0.5-5-mm range. Dusts do not tend to flocculate except under electrostatic forces.

Fumes are solid particles of matter formed by condensation of vapors. Heating or volatilizing metals (welding) or other solids usually produces them. Size usually ranges from 0.01 to 0.5 microns. Fumes flocculate and sometimes coalesce.

Gases are normally formless fluids (a state of matter separate from solids and liquids) that occupy the space of an enclosure and that can change to liquid or solid states only by the combined effects of increased pressure and decreased temperatures. Gases diffuse.

Mists are fine liquid droplets suspended in or falling through air. Mist is generated by condensation from the gaseous to liquid state, or by breaking up liquid into fine particles through atomizing, spraying, mechanized agitation, splashing, or foaming.

Smoke is the visible carbon or soot particles (generally less than 0.1 micron in size) resulting from the incomplete combustion of carbonaceous materials such as oil, tobacco, coal, and tar.

Vapor is the gaseous phase of a substance that is liquid or solid at normal temperature and pressure. Vapors diffuse.

Industrial atmospheric contaminants exist in virtually every workplace. Sometimes they are readily apparent to workers, because of their odor, or because they can actually be seen. Occupational health professionals, however, cannot rely on odor or vision to detect or measure airborne contaminants. They must rely on measurements taken by detection and sampling devices. Many different commercially available instruments permit the detection and concentration evaluation of many different contaminants. Some of these instruments are so simple that nearly any worker can learn to properly operate them. A note of caution, however: the untrained worker may receive an instrument reading that seems to indicate a higher degree of safety than may actually exist. Thus, the qualitative and quantitative measurement of atmospheric contaminants generally is the job of the occupational health professional. Any samples taken should also be representative—samples should be taken of the actual air the workers breathe, at the point they breathe them, in their breathing zone—between the top of the head and the shoulders.

Did You Know?

What constitutes cold stress and its effects can vary across different areas of the country. In regions relatively unaccustomed to winter weather, near freezing temperatures are considered factors for "cold stress." Whenever temperatures drop decidedly below normal and as wind speed increases, heat can more rapidly leave your body. These weather-related conditions may lead to serious health problems.

Workplace exposure to toxic materials can be reduced or controlled by a variety of individual control methods, or by a combination of methods. Various control methods available to safety engineers are broken down into three categories: engineering controls, administrative controls, and personal protective equipment.

Engineering Controls

Engineering controls are methods of environmental control whereby the hazard is "engineered out," either by initial design specifications, or by applying methods of substitution (e.g., replacing toxic chlorine used in disinfection processes with relatively non-toxic sodium hypochlorite). Engineering control may entail utilization of isolation methods. For example, a diesel generator when operating produces noise levels in excess of 120 decibels (dBA); however, it could be controlled by enclosing it inside a soundproofed enclosure—effectively isolating the noise hazard. Another example of hazard isolation can be seen in the use of tightly closed enclosures that isolate an abrasive blasting operation. This method of isolation is typically used in conjunction with local exhaust ventilation. Ventilation is one of the most widely used and effective engineering controls (because it is so crucial in controlling workplace atmospheric hazards) and is discussed in detail in the following section.

Ventilation

Simply put, ventilation is "the" classic method, and the most powerful tool of control used in occupational health and safety engineering to control environmental airborne hazards. Experience (much experience) has shown that the proper use of ventilation as a control mechanism can assure that workplace air remains free of potentially hazardous levels of airborne contaminants. In accomplishing this, ventilation works in two ways: (1) by physically removing the contaminated air from the workplace, or (2) by diluting the workplace atmospheric environment to a safe level by the addition of fresh air.

A ventilation system is all very well and good (virtually essential, actually), but an improperly designed ventilation system can make the hazard worse. This essential point cannot be overemphasized. At the heart of a proper ventilation system are proper design, proper maintenance, and proper monitoring. The industrial hygienist plays a critical role in ensuring that installed ventilation systems are operating at their optimum level.

Because of the importance of ventilation in the workplace, the industrial hygienist must be well versed in the general concepts of ventilation, the principles of air movement, and monitoring practices. The industrial hygienist must be properly prepared (through training and experience) to evaluate existing systems and design new systems for control of the workplace environment. In the next few sections, we present the general principles and concepts of ventilation system design and evaluation. This material should provide the basic concepts and principles necessary for the proper application of industrial ventilation systems. This material also serves to refresh the knowledge of the practitioner in the field. Probably the best source of information on ventilation is the ACGIH's *Industrial Ventilation: A Manual of Recommended Practice* (current edition)—this text is a must-have reference for every occupational health professional.

The purpose of industrial ventilation is essentially to (under control) recreate what occurs in natural ventilation. Natural ventilation results from differences in pressure. Air moves from high-pressure areas to low-pressure areas. This difference in pressure is the result of thermal conditions. We know that hot air rises, which (for example) allows smoke to escape from the smokestack in an industrial process, rather than disperse into areas where workers operate the process. Hot air rises because air expands as it is heated, becoming lighter. The same principle is in effect when air in the atmosphere becomes heated. The air rises and is replaced by air from a higher-pressure area. Thus, convection currents cause a natural ventilation effect through the resulting winds.

What does all of this have to do with industrial ventilation? Actually, quite a lot. Simply put, industrial ventilation is installed in a workplace to circulate the air within, in order to provide a supply of fresh air and replace air with undesirable characteristics.

Could this be accomplished simply by natural workplace ventilation? That is, couldn't we just heat the air in the workplace so that it will rise and escape through natural ports—windows, doors, cracks in walls, or mechanical ventilators in the roof (installed wind-powered turbines, for example)? Yes, we could design a natural system like this, but in such a system, air does not circulate fast enough to remove contaminants before a hazardous level is reached, which defeats our purpose in providing a ventilation system in the first place. Thus, we use fans to provide an artificial, mechanical means of moving the air.

Along with controlling or removing toxic airborne contaminants from the air, installed ventilation systems perform several other functions within the workplace. These functions include the following:

1. Ventilation is often used to maintain an adequate oxygen supply in an area. In most workplaces, this is not a problem because natural ventilation usually provides an adequate volume of oxygen; however, in some work environments (deep mining and thermal processes that use copious amounts of oxygen for combustion) the need for oxygen is the major reason for an installed ventilation system.
2. An installed ventilation system can remove odors from a given area. This type of system (as you might guess) has applications in such places

as athletic locker rooms, restrooms, and kitchens. In performing this function, the noxious air may be replaced with fresh air, or odors may be masked with a chemical-masking agent.

3. One of the primary uses of installed ventilation is one that we are familiar with—providing heat, cooling, and humidity control.
4. A ventilation system can remove undesirable contaminants at their source, before they enter the workplace air (e.g., from a chemical dipping or stripping tank). Obviously, this technique is an effective way to ensure that certain contaminants never enter the breathing zone of the worker—exactly the kind of function occupational health and safety engineering is intended to accomplish.

Earlier we stated that installed ventilation is able to perform its designed function via the use of a mechanical fan. Actually, a mechanical fan is the heart of any ventilation system, but like the human heart, certain ancillaries are required to make it function as a system. Ventilation is no different. Four major components make up a ventilation system: (1) The fan forces the air to move; (2) an inlet or some type of opening allows air to enter the system; (3) an outlet must be provided for air to leave the system; and (4) a conduit or pathway (ducting) not only directs the air in the right direction, but also limits the amount of flow to a predetermined level.

An important concept regarding ventilation systems is the difference between exhaust and supply ventilation. An *exhaust ventilation system* removes air and airborne contaminants from the workplace. Such a system may be designed to exhaust an entire work area, or it may be placed at the source to remove the contaminant prior to its release into the workplace air. The second type of ventilation system is the *supply system,* which (as the name implies) adds air to the work area, usually to dilute work area contaminants to lower the concentration of these contaminants. However, a supplied-air system does much more; it also provides movement to air within the space (especially when an area is equipped with both an exhaust and supply system—a usual practice, because it allows movement of air from inlet to outlet and is important in replenishing exhausted air with fresh air).

Air movement in a ventilation system is a result of differences in pressure. Note that pressures in a ventilation system are measured in relation to atmospheric pressure. In the workplace, the existing atmospheric pressure is assumed to be the zero point. In the supply system, the pressure created by the system is *in addition to* the atmospheric pressure that exists in the workplace (i.e., a positive pressure). In an exhaust system, the objective is to lower the pressure in the system below the atmospheric pressure (i.e., a negative pressure).

When we speak of increasing and decreasing pressure levels within a ventilation system, what we are really talking about is creating small differences in pressure—small when compared to the atmospheric pressure of the work area. For this reason, these differences are measured in terms of *inches of water* or *water gauge*, which results in the desired sensitivity of measurement. Air can be assumed to be incompressible, because of the small-scale differences in pressure.

Let's get back to the water gauge or inches of water. Since one pound per square inch of pressure is equal to twenty-seven inches of water, one inch of water is equal to 0.036 pounds pressure, or 0.24 percent of standard atmospheric pressure. Remember the potential for error introduced by considering air to be incompressible is very small at the pressure that exists with a ventilation system.

The industrial hygienist must be familiar with the three pressures important in ventilation: velocity pressure, static pressure, and the total pressure. To understand these three pressures and their function in ventilation systems, you must first to be familiar with pressure itself. In fluid mechanics, the energy of a fluid (air) that is flowing is termed *head*. Head is measured in terms of unit weight of the fluid or in foot-pounds/pound of fluid flowing. Note: The usual convention is to describe head in terms of feet of fluid that is flowing.

So what is pressure? *Pressure* is the force per unit area exerted by the fluid. In the English system of measurement, this force is measured in lbs/ft^2. Since we have stated that the fluid in a ventilation system is incompressible, the pressure of the fluid is equal to the head.

Velocity pressure (VP) is created as air travels at a given velocity through a ventilation system. Velocity pressure is only exerted in the direction of airflow and is *always* positive (i.e., above atmospheric pressure). When you think about it, velocity pressure has to be positive, and obviously the force or pressure that causes it also must be positive.

Note that the velocity of the air moving within a ventilation system is directly related to the velocity pressure of the system. This relationship can be derived into the standard equation for determining velocity (and clearly demonstrates the relationship between velocity of moving air and the velocity pressure):

$$v = 4005/\sqrt{VP} \tag{10.1}$$

Static pressure (SP) is the pressure that is exerted in all directions by the air within the system, which tends to burst or collapse the duct. It is expressed in inches of water gauge (wg). A simple example may help you grasp the concept of static pressure. Consider the balloon that is inflated at a given pressure. The pressure within the balloon is exerted equally on all sides of the balloon. No air velocity exits within the balloon itself. The pressure in the balloon is totally the result of static pressure. Note that static pressure can be both negative and positive with respect to the local atmospheric pressure.

Total pressure (TP) is defined as the algebraic sum of the static and velocity pressures, or the following:

$$\mathrm{TP} = \mathrm{SP} + \mathrm{VP} \tag{10.2}$$

The total pressure of a ventilation system can be either positive or negative (i.e., above or below atmospheric pressure). Generally, the total pressure is positive for a supply system, and negative for an exhaust system.

For the industrial hygienist to evaluate the performance of any installed ventilation system, he or she must make measurements of pressures in the ventilation system. Measurements are normally made using instruments such as a manometer or a Pitot tube.

The *manometer* is often used to measure the static pressure in the ventilation system. The manometer is a simple, U-shaped tube, open at both ends, and usually constructed of clear glass or plastic so that the fluid level within can be observed. To facilitate measurement, a graduated scale is usually present on the surface of the manometer. The manometer is filled with a liquid (water, oil, or mercury). When pressure is exerted on the liquid within the manometer, the pressure causes the level of liquid to change as it relates to the atmospheric pressure external to the ventilation system. The pressure measured, therefore, is relative to atmospheric pressure as the zero point.

When manometer measurements are used to obtain positive pressure readings in a ventilation system, the leg of the manometer that opens to the atmosphere will contain the higher level of fluid. When a negative pressure is being read, the leg of the tube open to the atmosphere will be lower, thus indicating the difference between the atmospheric pressure and the pressure within the system.

The *Pitot tube* is another device used to measure static pressure in ventilation systems. The Pitot tube is constructed of two concentric tubes. The inner tube forms the impact portion, while the outer tube is closed at the end and has static pressure holes normal to the surface of the tube. When the inner and outer tubes are connected to opposite legs of a single manometer, the velocity pressure is obtained directly. If the engineer wishes to measure static pressure separately, two manometers can be used. Positive and negative pressure measurements are indicated on the manometer as above.

Local Exhaust Ventilation--(the most predominant method of controlling workplace air) is used to control air contaminants by trapping and removing them near the source. In contrast to dilution ventilation (which lets the contamination spread throughout the workplace, later to be diluted by exhausting quantities of air from the workspace), local exhaust ventilation surrounds the point of emission with an enclosure, and attempts to capture and remove the emissions before they are released into the worker's breathing zone. The contaminated air is usually drawn through a system of ducting to a collector, where it is cleaned and delivered to the outside through the discharge end of the exhauster. A typical local exhaust system consists of a hood, ducting, an air-cleaning device, fan, and a stack. Hazard (1988) points out that a local exhaust system is usually the proper method of contaminant control if:

- the contaminant in the workplace atmosphere constitutes a health, fire, or explosion hazard.
- national or local codes require local exhaust ventilation at a particular process.
- maintenance of production machinery would otherwise be difficult.
- housekeeping or employee comfort will be improved.
- emission sources are large, few, fixed, and/or widely dispersed.
- emission rates vary widely by time.
- emission sources are near the worker-breathing zone.

The industrial hygienist must remember that determining beforehand precisely the effectiveness of a particular system is often difficult. Thus, measuring exposures and evaluating how much control has been achieved after a system is installed is essential. A good system may collect 80 to 90+ percent, but a poor system may capture only 50 percent or less.

Once the system is installed, and has demonstrated that it is suitable for the task at hand, the system must be well maintained. Careful maintenance is a must. In dealing with ventilation problems, the industrial hygienist soon finds out that his or her worst headache in maintaining the system is poor—or no—maintenance.

A phenomenon that many practitioners in the occupational health and safety field forget (or never knew in the first place) is that ventilation, when properly designed, installed, and maintained, can go a long way to ensure a healthy working environment. However, ventilation does have limitations. For example, the effects of blowing air from a supply system and removing air through an exhaust system are different. To better understand the difference and its significance, let's take an example of air supplied through a standard exhaust duct.

When air is exhausted through an opening, it is gathered equally from all directions around the opening. This includes the area behind the opening itself. Thus, the cross-sectional area of airflow approximates a spherical form, rather than the conical form that is typical when air is blown out of a supply system. To correct this problem, a flange is usually placed around the exhaust opening, which reduces the air contour, from the large spherical contour to that of a hemisphere. As a result, this increases the velocity of air at a given distance from the opening. This basic principle is used in designing exhaust hoods. Remember that the closer the exhaust hood is to the source, and the less uncontaminated air it gathers, the more efficient the hood's percentage of capture will be. Simply put, it is easier for a ventilation system to blow air than it is for one to exhaust it. Keep this in mind whenever you are dealing with ventilation systems and/or problems.

General and Dilution Ventilation—Along with local exhaust ventilation are two other major categories of ventilation systems: general and dilution ventilation. Each of these systems has a specific purpose, and finding all three types of systems present in a given workplace location is not uncommon.

General ventilation systems (sometimes referred to as heat control ventilation systems) are used to control indoor atmospheric conditions associated with hot industrial environments (such as those found in foundries, laundries, bakeries, and other workplaces that generate excess heat) for the purpose of preventing acute discomfort or injury. General ventilation also functions to control the comfort level of the worker in just about any indoor working environment. Along with the removal of air that has become process-heated beyond a desired temperature level, a general ventilation system supplies air to the work area to condition (by heating or cooling) the air, or to make up for the air that has been exhausted by dilution ventilation in a local exhaust ventilation system.

A *dilution ventilation system* dilutes contaminated air with uncontaminated air, to reduce the concentration below a given level (usually the threshold limit value of the contaminant) to control potential airborne health hazards, fire and explosive conditions, odors, and nuisance-type contaminants. This is accomplished by removing or supplying air, to cause the air in the workplace to move, and as a result, mix the contaminated with incoming uncontaminated air.

This mixing operation is essential. To mix the air there must be, of course, air movement. Air movement can be accomplished by natural draft caused by prevailing winds moving through open doors and windows of the work area.

Thermal draft can also move air. Whether the thermal draft is the result of natural causes or is generated from process heat, the heated air rises, carrying any contaminant present upward with it. Vents in the roof allow this air to escape into the atmosphere. Makeup air is supplied to the work area through doors and windows.

A mechanical air-moving device provides the most reliable source for air movement in a dilution ventilation system. Such a system is rather simple. It requires a source of exhaust for contaminated air, a source of air supply to replace the air mixture that has been removed with uncontaminated air, and a duct system to supply or remove air throughout the workplace. Dilution ventilation systems often are equipped with filtering systems to clean and temper the incoming air.

Industrial Noise Control

Only recently has noise has been recognized as a significant industrial health problem. In fact, now, workers' compensation laws in all states recognize hearing losses due to industrial noise as an occupational disease.

What Is Noise?—The obvious question is "What is noise?" Simply put, *noise* is any unwanted sound. The industrial hygienist is concerned about noise (or any workplace sound) that exceeds OSHA regulated levels and may be injurious to workers—that cause hearing damage.

Hearing damage risk criteria for exposure to noise are found in OSHA's 29 CFR 1910.95 (Hearing Conservation Standard) and are stated in table 10.2.

Determining Workplace Noise Levels—The industrial hygienist's primary concern when starting a noise reduction or control program is first to determine if any "noisemakers" in the facility exceed the OSHA limits for worker exposure—exactly which machines or processes produce noise at unacceptable levels. Making this determination is accomplished by conducting a noise level survey of the plant or facility.

When conducting the noise level survey, the industrial hygienist should use an ANSI-approved *sound-level meter* (a device used most commonly to measure sound pressure). Sound is measured in decibels. One decibel is one-tenth of a bel (a unit of measure in electrical-communication engineering) and is the minimum difference in loudness that is usually perceptible.

The sound level meter consists of a microphone, an amplifier and an indicating meter, which responds to noise in the audible frequency range of about 20 to 20,000 Hz. Sound level meters usually contain "weighting" networks designated "A," "B," or "C." Some meters have only one weighting network; others are equipped with all three. The A-network approximates the equal loudness curves at low sound pressure levels, the B-network is used for medium sound pressure levels, and the C-network is used for high levels.

In conducting a routine workplace sound level survey, using the A-weighted network (referenced dBA) in the assessment of the overall noise hazard has

TABLE 10.2
Permissible Noise Exposures (U.S. Department of Labor, Part 1910. Occupational Safety and Health Standards, subpart G., Dection 1910.95. Washington, D.C.: Occupational Safety and Health Administration, 1995)

Duration per Day, Hours	*Sound Level dBA*
	Slow Response
8	90
6	92
4	95
3	97
2	100
1.5	102
1	105
0.5	110
0.25	115

Notes: When the daily noise exposure is composed of two or more periods of noise exposure of different levels, their combined effect should be considered, rather than the individual effect of each. If the sum of the following fractions $C_1/T_1 + C_2/T_2 + C_n/T_n$ exceeds unity, then the mixed exposure should be considered to exceed the limit value. C_n indicates the total time of exposure at a specified noise level, and T_n indicates the total time of exposure permitted at that level. Exposure to impulsive or impact noise should not exceed 140–dB peak sound pressure level. The numbers in this table are measured on the A weighting scale of a standard sound level meter is slow response mode.

Source: U.S. Department of Labor, Part 1910. *Occupational Safety and Health Standards*, subpart G., section 1910.95. Washington, D.C.: Occupational Safety and Health Administration, 1995.

become common practice. The A-weighted network is the preferred choice because it is thought to provide a rating of industrial noises that indicates the injurious effects such noise has on the human ear (gives a frequency response similar to that of the human ear at relatively low sound pressure levels).

With an approved and freshly calibrated (always calibrate test equipment prior to use) sound level meter in hand, the industrial hygienist is ready to begin the sound level survey. In doing so, the industrial hygienist is primarily interested in answering the following questions: (1) What is the noise level in each work area, (2) what equipment or process is generating the noise, (3) which employees are exposed to the noise, and (4) how long are they exposed to the noise?

In answering these questions, industrial hygienists record their findings as they move from workstation to workstation, following a logical step-by-step procedure. The first step involves using the sound level meter set for A-scale slow response mode to measure an entire work area. When making such measurements, restrict the size of the space being measured to under 1,000 square feet. If the maximum sound level does not exceed 80 dBA, it can be assumed that all workers in this work area are working in an environment with a satisfactory noise level. However, a note of caution is advised here: The key words in the preceding statement are "maximum sound level." To assure an accurate measurement, the industrial hygienist must ensure that all "noisemakers" are actually in operation when measurements are taken. Measuring an entire work area does little good when only a small percentage of the noisemakers are actually in operation.

The next step depends on the readings recorded when the entire work area was measured. For example, if the measurements indicate sound levels greater than 80 dBA, and then another set of measurements needs to be taken at each worker's workstation. The purpose here, of course, is to determine two things: which machine or process is making noise above acceptable levels (i.e., >80 dBA), and which workers are exposed to these levels. Remember that the worker who operates the machine or process might not be the only worker exposed to the noisemaker. You need to inquire about other workers who might, from time to time, spend time working in or around the machine or process. Our experience in conducting workstation measurements has shown us noise levels usually fluctuate. If this is the case, you must record the minimum and maximum noise levels. If you discover that the noise level is above 90 (dBA) (and it remains above this level), you have found a noisemaker that exceeds the legal limit (90 dBA). However, if your measurements indicate that the noise level is never greater than 85 (dBA) (OSHA's action level), the noise exposure can be regarded as satisfactory.

If workstation measurements indicate readings that exceed the 85 dBA level, you must perform another step. This step involves determining the length of time of exposure for workers. The easiest, most practical way to make this determination is to have the worker wear a noise dosimeter, which records the noise energy to which the worker was exposed during the workshift.

What happens next? You must then determine if the worker is exposed to noise levels that exceed the permissible noise exposure levels listed in table 10.2. The key point to remember is that your findings must be based on a time-weighted average (TWA). For example, from table 10.2 you will notice that a noise level of 95 dBA is allowed up to four hours per day.

Important Note: This parameter assumes that the worker has good hearing acuity with no loss. If the worker has documented hearing loss, then exposure to 95 dBA or higher may be unacceptable under any circumstances without proper hearing protection.

So exactly what does four-hour maximum exposure per day mean? It means that, cumulatively, a worker cannot be exposed for more than four hours of noise at the 95-dBA level. Cumulative maximum exposures are used because all noisemakers are not necessarily *continuous*; instead, they may be *intermittent* or *impact-type* noisemakers. Consider this—a worker who runs a machine operates the machine eight hours each day. When the machine is running, it continuously produces 95 dBA. Obviously, this worker must be protected from the 95-dBA noisemaker, because his or her exposure will be over an eight-hour period—which is not allowed under OSHA. Another worker operates a machine that produces 95-dBA noise, but the operator only operates it for a few minutes at a time, with several minutes without the machine running in between operations. The worker is exposed to noise from other workstations, at varying levels. This is considered intermittent operation, with intermittent noise generation—and possibly intermittent exposure (depending upon the level of noise). Is this worker exposed to noise levels above the Permissible Exposure Limit of four hours maximum (i.e., without hearing protection)?

It depends. To make this determination we must calculate the daily noise dose. We can accomplish this by using equation 10.3:

$$E_m = C_1/T_1 + C_2/T_2 + C_3/T_3 + \ldots C_n/T_n \quad (10.3)$$

where:

E^m = mixed exposure
C = total time of exposure at a specified noise level
T = total time of exposure permitted at that level

For purposes of illustration let's assume that the worker's intermittent noise levels expose him or her to the following noise levels during the workday:

85 dBA for 2.75 hours
90 dBA for 1 hour
95 dBA for 2.25 hours
100 dBA for 2 hours

The question is, has the worker received an excessive exposure during the workday?

To answer this question we use equation 10.3 and plug in the parameters. From our calculation, if we find that the sum of the fractions equals or exceeds 1, then the mixed exposure is considered to exceed the limit value. Daily noise dose (D) is expressed as a percentage of Em = 1, the mixed exposure is equivalent to a noise dose of 100 percent. Keep in mind that noise levels below 90 dBA are not considered in the calculation of daily noise.

So, again, has our worker received an excessive exposure during her workday?

Let's find out:

$$\text{Dose} = \frac{1}{0} + \frac{2.2}{8} + \frac{5}{4} + \frac{2}{2} = 169\%$$

The sum exceeds 1, therefore, obviously, the results indicate that the employee has received an excessive exposure during her workday.

Engineering Controls for Industrial Noise—When the industrial hygienist investigates the possibility of using engineering controls to control noise, the first thing he or she recognizes is that reducing and/or eliminating all noise is virtually impossible. And this should not be the focus in the first place . . . eliminating or reducing the "hazard" is the goal. While the primary hazard may be the possibility of hearing loss, the distractive effect (or its interference with communication) must also be considered. The distractive effect of excessive noise can certainly be classified as hazardous whenever the distraction might affect the attention of the worker. The obvious implication of noise levels that interfere with communications is emergency response. If ambient noise is at such a high level that workers can't hear fire or other emergency alarms, this is obviously a hazardous situation.

So what does all this mean? The industrial hygienist must determine the "acceptable" level of noise. Then he or she can look into applying the various noise control measures. These include making alterations in engineering design (obviously this can only be accomplished in the design phase) or making modifications after installation. Unfortunately, this latter method is the one the industrial hygienist is usually forced to apply—and also the most difficult, depending upon circumstances.

Let's assume that the safety engineer is trying to reduce noise levels generated by an installed air compressor to a safe level. Obviously, the first place to start is at the source: the air compressor. Several options are available for the industrial hygienist to employ at the source. First, the safety engineer would look at the possibility of modifying the air compressor to reduce its noise output. One option might be to install resilient vibration mounting devices. Another might be to change the coupling between the motor and the compressor.

If the options described for use at the source of the noise are not feasible or are only partially effective, the next component the industrial hygienist would look at is the path along which the sound energy travels. Increasing the distance between the air compressor and the workers could be a possibility. (Note: Sound levels decrease with distance.) Another option might be to install acoustical treatments on ceilings, floors, and walls. The best option available (in this case) probably is to enclose the air compressor, so that the dangerous noise levels are contained within the enclosure, and the sound leaving the space is attenuated to a lower, safer level. If total enclosure of the air compressor is not practical, then erecting a barrier or baffle system between the compressor and the open work area might be an option.

The final engineering control component the industrial hygienist might incorporate to reduce the air compressor's noise problem is to consider the receiver (the worker). An attempt should be made to isolate the operator by providing a noise reduction or soundproof enclosure or booth for the operator.

Industrial Vibration Control

Vibration is often closely associated with noise, but is frequently overlooked as a potential occupational health hazard. Vibration is defined as the oscillatory motion of a system around an equilibrium position. The system can be in a solid, liquid, or gaseous state, and the oscillation of the system can be periodic or random, steady state or transient, continuous or intermittent (NIOSH, 1973). Vibrations of the human body (or parts of the human body) are not only annoying, they also affect worker performance, and sometimes causes blurred vision and loss of motor control. Excessive vibration can cause trauma, which results when external vibrating forces accelerate the body or some part so that amplitudes and restraining capacities by tissues are exceeded.

Vibration results in the mechanical shaking of the body or parts of the body. These two types of vibration are called *whole-body vibration* (affects vehicle operators) and *segmental vibration* (occurs in foundry operations, mining, stonecutting, and a variety of assembly operations, for example). Vibration originates from mechanical motion, generally occurring at some machine or series of machines. This mechanical vibration can be transmitted directly to the body or body part, or it may be transmitted through solid objects to a worker located at some distance away from the actual vibration.

The effect of vibration on the human body is not totally understood; however, we do know that vibration of the chest may create breathing difficulties, and that an inhibition of tendon reflexes is a result of vibration. Excessive vibration can cause reduced ability on the part of the worker to perform complex tasks, and indications of potential damage to other systems of the body also exist.

More is known about the results of segmental vibration (typically transmitted through hand to arm), and a common example is the vibration received when using a pneumatic hammer—a jackhammer. One recognized indication of the effect of segmental vibration is impaired circulation to the appendage, a condition known as *Raynaud's Syndrome*, also known as "dead fingers" or "white fingers." Segmental vibration can also result in the loss of the sense of touch in the affected area. Some indications that decalcification of the bones in the hand can result from vibration transmitted to that part of the body. In addition, muscle atrophy has been identified as a result of segmental vibration.

As with noise, the human body can withstand short-term vibration, even though this vibration might be extreme. The dangers of vibration are related to certain frequencies that are resonant with various parts of the body. Vibration outside these frequencies is not nearly so dangerous as vibration that results in resonance.

Control measures for vibration include substituting some other device (one that does not cause vibration) for the mechanical device that causes the vibration. An important corrective measure (often overlooked) that helps in reducing vibration is proper maintenance of tools, or support mechanisms for tools, including coating the tools with materials that attenuate vibrations. Another engineering control often employed to reduce vibration is the application of balancers, isolators, and damping devices/materials that help to reduce vibration.

Administrative Controls

After the design, construction, and installation phase, installing engineering controls to control a workplace hazard or hazards often becomes difficult. Some exceptions were mentioned in the previous section. A question occupational health professionals face on almost a continuous basis is, "If I can't engineer out the hazard, what can I do?"

This question would not arise, of course, if the safety engineer had been allowed to participate in the design, construction, and installation phases. However, our experience has shown us that more often than not, the occupational health professional is excluded from such preliminary construction phases. This certainly is not "good engineering practice"—but it happens. And thus the questions arise on how best to reduce or remove hazards after they have been installed. The occupational health professional is tasked with finding the answers.

As a second line of defense, after engineering controls are determined to be impossible, not practicable, not feasible, or cannot be accomplished for technological reasons—or for any reasons—*administrative controls* might be an alternative.

What are administrative controls? Administrative controls attempt to limit the worker's exposure to the hazard. Normally accomplished by arranging work schedules and related duration of exposures so that employees are minimally exposed to health hazards, another procedure transfers workers who have reached their upper permissible limits of exposure to an environment where no additional exposure will be experienced. Both control procedures are often used to limit worker exposure to air contaminants or noise. For example, a worker who is required to work in an extremely high noise area where engineering controls are not possible would be rotated from the high noise area to a quiet area when the daily permissible noise exposure is reached.

Reducing exposures by *limiting the duration of exposure* (basically by modifying the work schedule) must be carefully managed (most managers soon find that attempting to properly manage this procedure takes a considerable amount of time, effort, and "imagination"). When practiced, reducing worker exposure is based on limiting the amount of time a worker is exposed, ensuring that OSHA Permissible Exposure Limits (PELs) are not exceeded.

Let's pause right here and talk about Permissible Exposure Limits (PELs) and Threshold Limit Values (TLVs). You should know what they are and what significance they play in the safety engineer's daily activities. Let's begin with TLVs.

Threshold Limit Values (TLVs) are published by the American Conference of Governmental Industrial Hygienists (ACGIH) (an organization made up of physicians, toxicologists, chemists, epidemiologists, and industrial hygienists) in its *Threshold Limit Values for Chemical Substances and Physical Agents in the Work Environment*. These values are useful in assessing the risk of a worker exposed to a hazardous chemical vapor; concentrations in the workplace can often be maintained below these levels with proper controls. The substances listed by ACGIH are evaluated annually, limits are revised as needed, and new substances are added to the list, as information becomes available. The values are established from the experience of many groups in industry, academia, and medicine, and from laboratory research.

The chemical substance exposure limits listed under both ACGIH and OSHA are based strictly on airborne concentrations of chemical substances in terms of milligrams per cubic meter (mg/m^3), parts per million (ppm), and fibers per cubic centimeters ($fibers/cm^3$). Allowable limits are based on three different time periods of average exposure: (1) eight-hour work shifts known as TWA (time weighted average), (2) short terms of fifteen minutes or STEL (short-term exposure limit), and (3) instantaneous exposure of "C" (ceiling). Unlike OSHA's PELs, TLVs are recommended levels only, and do not have the force of regulation to back them up.

OSHA has promulgated limits for personnel exposure in workplace air for approximately 400 chemicals listed in tables Z1, Z2, and Z3 in part 1910.1000 of the Federal Occupational Safety and Health Standard. These limits are defined as permissible exposure limits (PEL), and like TLVs are based on eight-hour time weighted averages (or ceiling limits when preceded by a "C"). Keeping within the limits in the Z Tables is the only requirement specified by OSHA for these chemicals. The significance of OSHA's PELs is that they have the force of regulatory law behind them to back them up—compliance with OSHA's PELs is the law.

Evaluation of personnel exposure to physical and chemical stresses in the industrial workplace requires the use of the guidelines provided by TLVs and the regulatory guidelines of PELs. For the occupational health practitioner to carry out the goals of recognizing, measuring, and effecting controls (of any type) of workplace stresses, such limits are a necessity, and have become the ultimate guidelines in occupational health practice. A word of caution is advised, however. These values are set only as guides for the best practice, and are not to be considered absolute values. What are we saying here? These values provide reasonable assurance that occupational disease will not occur, if exposures are kept below these levels. On the other hand, occupational disease is likely to develop in some people—if the recommended levels are exceeded on a consistent basis.

Let's get back to administrative controls.

We stated that one option available to the occupational health professional in controlling workplace hazards is the use of an administrative control that involves modifying workers' work schedules to limit the time of their exposure so that the PEL/TLV is not exceeded. We also said (or at least implied) that this procedure is a manager's nightmare to implement and manage. Practicing safety engineers don't particularly like it, either; they feel that such a strategy merely spreads the exposure out, and does nothing to control the source. Experience has shown that in many instances this statement is correct. Nevertheless, work schedule modification is commonly used for exposures to such stressors as noise and lead.

Another method of reducing worker exposure to hazards is by ensuring good *housekeeping practices*. Housekeeping practices? Absolutely. Think about it. If dust and spilled chemicals are allowed to accumulate in the work area, workers will be exposed to these substances. This is of particular importance for flammable and toxic materials. Ensuring that housekeeping practices do not allow toxic or hazardous materials to disperse into the air is also an important concern.

Administrative controls can also reach beyond the workplace. For example, if workers work to abate asbestos eight hours a day, they should only wear approved Tyvek protective suits, and other required personal protective equipment (PPE). After the work assignment is completed, these workers must decontaminate following standard protocol. The last thing these workers should be allowed to do is to wear their personal clothing for such work, to avoid decontamination procedures, then take their contaminated clothing with them when they leave for home. The idea is to leave any contaminated clothing at work (safely stored or properly disposed of).

Implementation of standardized *materials handling* or *transferring procedures* are another administrative control often used to protect workers. In handling chemicals, any transfer operations taken should be closed-system, or should have adequate exhaust systems to prevent worker exposure or contamination of the workplace air. This practice should also include the use of spill trays to collect overfill spills or leaking materials between transfer points.

Programs that involve visual inspection and automatic sensor devices (*leak detection programs*) allow not only for quick detection, but also for quick repair and minimal exposure. When automatic system sensors and alarms are deployed as administrative controls, tying the alarm system into an automatic shutdown system (close a valve, open a circuit, etc.), allows the sensor to detect a leak, sound the alarm, and initiate corrective action (for example, immediate shutdown of the system).

Finally, two other administrative control practices that go hand-in-hand are *training* and *personal hygiene*. For workers to best protect themselves from workplace hazards (to reduce the risk of injury or illness), they must be made aware of the hazards. OSHA puts great emphasis on the worker-training requirement. This emphasis is well placed. No worker can be expected to know the entire workplace, process, or equipment hazards, unless he or she has been properly trained on the hazards.

An important part of the training process is worker awareness. Legally (and morally) workers have the right to know what they are working with, what they are exposed to while on the job; they must be made aware of the hazards. They must also be trained on what actions to take when they are exposed to specific hazards. Personal hygiene practices are an important part of worker protection. The safety engineer must ensure that appropriate cleaning agents and facilities such as emergency eyewashes, showers, and changing rooms are available and conveniently located for worker use.

Did You Know?

When exposed to cold temperatures, your body begins to lose heat faster than it can be produced. Prolonged exposure to cold will eventually use up your body's stored energy. The result is hypothermia, or abnormally low body temperature. A body temperature that is too low affects the brain, making the victim unable to think clearly or move well. This makes hypothermia particularly dangerous—because a person may not know it is happening and will not be able to do anything about it (CDC, 2012).

Personal Protective Equipment (PPE)

As a hazard control method, *personal protective equipment* (PPE) should only be used when other methods fail to reduce or eliminate the hazard; PPE is the safety engineer's last line (last resort) of defense against hazard exposure in the workplace. Briefly, the types commonly used to control materials-related hazards include:

Respiratory Protection—When engineering controls are not feasible or are in the process of being instituted, appropriate respirators should be used to control exposures to airborne hazardous materials. Note that under OSHA regulations (specifically 29 CFR 1910.134), the employer is required to provide such equipment whenever it is necessary.

Protective Clothing—This includes chemical, thermal, and/or electrical clothing such as gloves, aprons, coveralls, suits, etc. Many materials and types are available, to suit different applications and needs.

Head, Eye, Hand, Foot Protection—This type of PPE is required in any situation that presents a reasonable probability of injury. These items are worn for protection from physical injury and include hard hats, safety glasses, goggles, leather gloves, laboratory gloves, and steel-toed safety shoes.

A final word on PPE as a method of environmental control: PPE has one serious drawback—it does nothing to reduce or eliminate the hazard. This critical point is often ignored or overlooked. What PPE really does is afford the wearer a barrier between him/herself and the hazard—and that is all. Sometimes workers gain a false sense of security when they don PPE, thinking that somehow the PPE is the element that makes them safe, not working safely. An electrician wears the proper type of electrical insulating gloves and stands on a rubber mat while she services a high-voltage electrical switchgear. If she performs her work in a haphazard manner, will she be safe? Will the gloves and rubber mat protect her from electrocution? Maybe. Maybe not. PPE provides some personal protection, but it is no substitute for safe work practices.

Another problem with PPE is that often PPE offers the temptation to employ its use without first attempting to investigate thoroughly the possible methods of correcting the unsafe physical conditions. This results in substituting PPE in place of safety engineering methods to correct the hazardous environment (Grimaldi and Simonds, 1989).

The occupational health professional also learns (rather quickly) that employees often resist using it. We see a constant struggle between the occupational health professional and the worker about ensuring that the worker wears his or her PPE. We hear their excuses: "Those safety glasses get in my way." "That hardhat is too heavy for my head." "I can't do my work properly with those clumsy gloves on my hands." "Gee, I forgot my safety shoes. I think I left them at my girlfriend/boyfriend's house." Like homework assignment excuses, these statements are common, frequent, often irritating, and never-ending (though some of the more original ones are even quite entertaining). But one thing is certain—workers who do not wear PPE when required are leaving themselves wide open to injury or death. For the novice occupational health professional, we can only add: "Welcome to the challenging field of occupational health."

Discussion Questions

1. Discuss the similarities and differences between the job functions of safety engineers, industrial hygienists, and occupational health professionals.
2. What are the three parts that make up the industrial hygiene paradigm?
3. Define *stressors.*
4. How much detailed knowledge should an industrial hygienist have concerning workplace processes and operations?
5. What stressors are of greatest concern to industrial hygienists?
6. What principle areas should the industrial hygienist be concerned with?
7. Define *toxicology.*
8. What's the difference between toxic and hazard?
9. Discuss dose-response relationship and threshold level. What effect does time have on these?
10. What is the difference between acute and chronic exposure?
11. Describe how ventilation can control airborne hazards.
12. Three elements ensure ventilation systems are safe and effective. What are they, and why is each important?
13. How does natural ventilation occur?
14. What is the difference between exhaust and supply ventilation?
15. What is a water gauge and how does it work?
16. Define *noise.* How is it measured?
17. Discuss time-weighted average.
18. How and why are checklists important?

References and Recommended Reading

American Institute of Chemical Engineers, 1985. *Guidelines for Hazard Evaluation Procedures*. New York: American Institute of Chemical Engineers.

ASSE, 1988. *The Dictionary of Terms Used in the Safety Profession*, 3rd ed. Des Plaines, IL: American Society of Safety Engineers.

Best's Loss Control Engineering Manual. Oldwick, NJ: A.M. Best Co., Inc., annual.

Burgess, W.A., 1981. *Recognition of Health Hazards in Industry*. New York: Wiley.

Block, A., 1977. *Murphy's Law and Other Reasons Why Things Go Wrong*. Los Angeles: Price/Stern/Sloan.

Bureau of National Affairs, 1989. *Occupational Safety and Health: 7 Critical Issues for the 1990s* (SSP-136), July.

CDC, 2012. Cold Stress. Accessed 03/22/12 at http://www.cdc.gov/niosh/topics/coldstress/.

Eisma, T.L., 1990. Demand for Trained Professionals Increases Educational Interests. *In: Occupational Health and Safety*. Waco, TX: Stevens, 52–56.

Ferry, T., 1990. *Safety and Health Management Planning*. New York: Van Nostrand Reinhold.

Gloss, D.S., and M.G. Wardle, 1984. *Introduction to Safety Engineering*, New York: Wiley.

Grimaldi, J.V., and R.H. Simonds, 1989. *Safety Management*, 5th ed., Homewood, IL: Irwin.

Hardie, D.T., 1981. The Safety Professional of the Future. *Professional Safety*, May: 17–21.

Hazard, W.G., 1988. Industrial Ventilation. Plog, B., ed. *Fundamentals of Industrial Hygiene*, 3rd ed., Chicago, IL: National Safety Council.

Hammer, W., 1989. *Occupational Safety Management and Engineering*, 4th ed., Englewood Cliffs, NJ: Prentice Hall.

Hoover, R.L., R.L. Hancock, K.L. Hylton, O.B. Dickerson, and G.E. Harris, 1989. *Health, Safety, and Environmental Control*, New York: Van Nostrand Reinhold.

Leeflang, R.L., D.J. Klein-Hesselink, and I.P. Spruit, 1992. Heath effects of unemployment. *Soc Sci Med* 34:151–163.

Levenstein, C., J. Wooding, and B. Rosenberg, 2000. Occupational Health: A Social Perspective, 4th ed. In: Levy, F.S., and D.H. Wegman, eds. Philadelphia: Lippincott, Williams and Wilkins.

Levy, B.S., and D.H. Wegman, eds., 1988. *Occupational Health: Recognizing and Preventing Work-Related Disease*, 2nd ed. Boston: Little, Brown, and Company.

Lewis, R., 1992. *Sax's Dangerous Properties of Industrial Materials*, 8th ed., New York: Van Nostrand Reinhold.

Miller, D.E., R.J. Petrucco, and A.A. Corona Jr., 1987. How to Establish Industrial Loss Prevention and Fire Control. *Handbook of Occupational Safety and Health*. New York: Wiley, 1987.

Moeller, D.W., 2011. *Environmental Health*, 4th ed. Cambridge, MA: Harvard University Press.

NFPA, 1991. *Life Safety Code Handbook*, 5th ed. Battery Park, MA: National Fire Protection Association.

NIOSH, 1973. *The Industrial Environment, Its Evaluation and Control.* Cincinnati, OH: NIOSH.

NIOSH, 1994. *Pocket Guide to Chemical Hazards*, Washington, D.C.: U.S. Department of Health and Human Services.

Ogle, B.R., 2003. Introduction to Industrial Hygiene. In: *Fundamentals of Occupational Safety and Health*, 3rd ed. Rockville, MD: Government Institutes.

Olishifski, J.B., and B.A. Plog, 1988. Overview of Industrial Hygiene. In: Plog, B.A., ed. *Fundamentals of Industrial Hygiene*, 3rd ed. Chicago, IL: National Safety Council.

USAA Aide Magazine, 1989. Health and Safety: The Importance of Wearing Helmets, Oct. 25.

U.S. Department of Labor, 1995. *Occupational Safety and Health Standards* (part 1910), subpart G., section 1910.95. Washington, D.C.: OSHA.

Wallick, R., 1972. *The American Worker: An Endangered Species.* New York: Ballantine Books.

Whiteside, N., 1988. Unemployment and health: an historical perspective. *Social Policy* 17:177–194.

Appendix

Units of Measurement

A BASIC KNOWLEDGE OF UNITS OF MEASUREMENT and how to use them is essential for students of environmental science. Environmental science students and practitioners should be familiar both with the U.S. Customary System (USCS) or English System and the International System of Units (SI). We summarize some of the important units here to enable better understanding of material covered later in the text. Table App.1 gives conversion factors between SI and USCS systems for some of the most basic units encountered.

In the study of environmental science, you will commonly encounter both extremely large quantities and extremely small ones. The concentration of some toxic substance may be measured in parts per million or billion (ppm or ppb), for example. PPM may be roughly described as the volume of liquid contained in a shot glass compared to the volume of water contained in a swimming pool. To describe quantities that may take on such large or small values, a system of prefixes that accompanies the units is useful. We present some of the more important prefixes in table App.2.

Units of Mass

Simply defined, *mass* is a quantity of matter and measurement of the amount of inertia that a body possesses. Mass expresses the degree to which an object resists a change in its state of rest or motion, and is proportional to the amount of matter in the object. Another, simpler way to understand mass is to think of it as the quantity of matter an object contains.

TABLE 1
Commonly Used Units and Conversion Factors

Quantity	*SI units*	*SI symbol x*	*Conversion factor =*	*USCS units*
Length	Meter	M	3.2808	Ft
Mass	Kilogram	Kg	2.2046	lb.
Temperature	Celsius	C	1.8 (C) + 32	F
Area	Square meter	m2	10.7639	ft2
Volume	Cubic meter	m3	35.3147	ft3
Energy	Kilojoule	Kj	0.9478	Btu
Power	Watt	W	3.4121	Btu/hr
Velocity	Meter/second	m/s	2.2369	mi/hr

Beginning science students often confuse mass with weight. They are different. *Weight* is the gravitational force action upon an object and is proportional to mass. In the SI system (a modernized metric system) the fundamental unit of mass is the *gram* (g). How does this stack up against weight?

To show the relationship between mass and weight, consider that a pound contains 452.6 grams. In laboratory-scale operations, the gram is a convenient unit of measurement. However, in real-world applications the gram is usually prefixed with one of the prefixes shown in table 2.4. For example, human body mass is expressed in *kilograms* (1 kg = 2.2 pounds). In everyday term, a kilogram is the mass of one liter of water. When dealing with units of measurement pertaining to environmental conditions such as air pollutants and toxic water pollutants, they may be measured in *teragrams* (1×10^{12} grams) and *micrograms* (1×10^{-6} grams) respectively. When dealing with large-scale industrial commodities, the mass units may be measured in units of megagrams (Mg), also known as a metric ton.

TABLE 2
Common Prefixes

Quantity	*Prefix*	*Symbol*
10-12	Pico	P
10-9	Nano	N
10-6	Micro	M
10-3	Milli	M
10-2	Centi	C
10-1	Deci	D
10	Deca	Da
102	Hecto	H
103	Kilo	K
106	Mega	M

Often *mass* and *density* are mistaken as signifying the same thing—they are not. Where mass is the quantity of matter and measurement of the amount of inertia that a body contains, *density* refers to how compacted a substance is with matter; density is the mass per unit volume of an object, and its formula can be written as shown in the following:

$$\text{Density} = \frac{\text{mass}}{\text{volume}}$$

Something with a mass of 25 kg that occupies a volume of 5 m^3 would have a density of 25 kg/5m^3 = 5 kg/m^3. In this example the mass was measured in kilograms and the volume in cubic meters.

Units of Length

In measuring locations and sizes, we use the fundamental property of *length*, defined as the measurement of space in any direction. Space has three dimensions, each of which can be measured by length (see figure App.1). This can be easily seen by considering the rectangular object shown in the figure above. It has length, width, and height, but each of these dimensions is a length.

In the metric system, length is expressed in units based on the *meter* (m), which is 39.37 inches long. A kilometer (km) is equal to 1000m and is used to measure relatively great distances. In practical laboratory applications the centimeter (cm = 0.01 m) is often used. There are 2.540 cm per inch, and the cm is employed to express lengths that would be given in inches in the English system. The micrometer (μm) is also commonly used to express measurements of bacterial cells and wavelengths of infrared radiation by which the

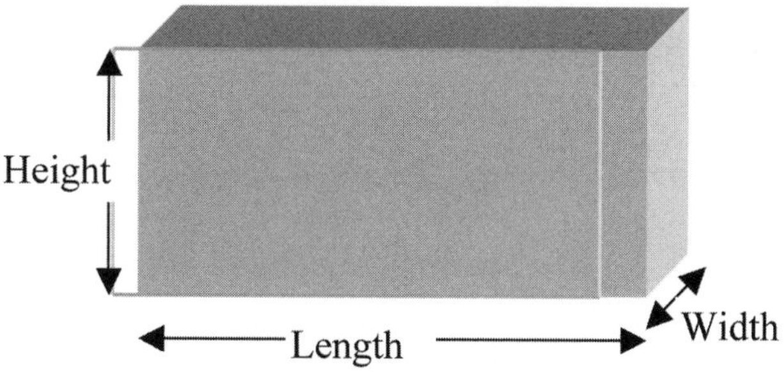

FIGURE App. 1

earth re-radiates solar energy back to outer space. For measuring visible light (400 to 800 nm), the nanometer (nm) (10^{-9}) is often used.

Units of Volume

The easiest way to approach measurements involving volume is to remember that volume is surface area multiplied by a third dimension. The liter is the basic metric unit of volume and is the volume of a decimeter cubed (1 L = 1 dm^3). A milliliter (Ml) is the same volume as a cubic centimeter, cm^3.

Units of Temperature

Temperature is a measure of how "hot" something is—how much thermal energy it contains. Temperature is a fundamental measurement in environmental science, especially in most pollution work. The temperature of a stack gas plume, for example, determines its buoyancy and how far the plume of effluent will rise before attaining the temperature of its surroundings. This in turn determines how much it will be diluted before traces of the pollutant reach ground level.

Temperature is measured on several scales: for example, the centigrade or Celsius and Fahrenheit scales are both measured from a reference point—the freezing point of water—which is taken as 0°C or 32°F. The boiling point of water is taken as 100°C or 212°F respectively. Thermodynamic devices usually work in terms of absolute or "thermodynamic temperature" where the reference point is absolute zero, the lowest possible temperature attainable. For absolute temperature measurement, the thermodynamic unit or Kelvin (K) scale—which uses centigrade divisions for which zero is the lowest attainable measurement—is used. A unit of temperature on this scale is equal to a Celsius degree, and is not called a degree, but is called a Kelvin and designated as K, not °K. The value of absolute zero on the Kelvin scale is -273.15°C, so that the Kelvin temperature is always a number 273 (rounded) higher than the Celsius temperature. Thus water boils at 373 K and freezes at 273 K.

To convert from the Celsius scale to the Kelvin scale, simply add 273 to the Celsius temperature and you have the Kelvin temperature. Mathematically,

$$K = {}^\circ C + 273$$

where K = temperature on the Kelvin scale and °C = temperature on the Celsius scale. Converting from Fahrenheit to Celsius or vice versa is not so easy. The equations used are:

$$°C = 5/9(°F - 32)$$

and

$$°F = 9/5°C + 32$$

where °C = temperature on the Celsius scale and °F = temperature on the Fahrenheit scale. As examples, 15°C = 59°F and 68°F = 20°C. °F or °C or both, of course, can be negative numbers.

Units of Pressure

Pressure is force per unit area and can be expressed in a number of different units, including the atmosphere (atm), the average pressure exerted by air at sea level, or the pascal (Pa), usually expressed in kilopascal (1 kPa = 1000 Pa, and 101.3 kPa = 1 atm). Pressure can also be given as millimeters of mercury (mm Hg), based on the amount of pressure required to hold up a column of mercury in a mercury barometer. 1 mm of mercury is a unit called the torr, and 760 torr equal 1 atm.

Units Often Used in Environmental Studies

In environmental studies, often the concentration of some substance (foreign or otherwise) in air or water is of interest. In either medium, concentrations may be based on volume or weight, or a combination of the two (which may lead to some confusion). To understand how weight and volume are used to determine concentrations when studying liquids or gases/vapors, study the following explanations.

Liquids

Concentrations of substances dissolved in water are usually expressed in terms of weight of substance per unit volume of mixture. In environmental science, a good practical example of this weight per unit volume is best observed whenever a contaminant is dispersed in the atmosphere in solid or liquid form as a mist, dust, or fume. When this occurs, its concentration is usually expressed on a weight-per-volume basis. Outdoor air contaminants and stack effluents are frequently expressed as grams, milligrams, or micrograms per cubic meter, ounces per thousand cubic feet, pounds per thousand pounds of air, and grains per cubic foot. Most measurements are expressed in metric units. However, the use of standard U.S. units is justified for purposes of comparison with existing data, especially those relative to the specifications for air-moving equipment.

Alternatively, concentrations in liquids are expressed as weight of substance per weight of mixture, with the most common units being parts per million (ppm), or parts per billion (ppb).

Since most concentrations of pollutants are very small, one liter of mixture weighs essentially 1000 g, so that for all practical purposes we can write:

$$1 \text{ mg/L} = 1 \text{ g/m}^3 = 1 \text{ ppm (by weight)}$$
$$1 \text{ µg/L} = 1 \text{ mg/m}^3 = 1 \text{ ppb (by weight)}$$

The environmental health practitioner may also be involved with concentrations of liquid wastes that may be so high that the specific gravity (the ratio of an object's or substance's weight to that of an equal volume of water) of the mixture is affected, in which case a correction may be required:

$$\text{mg/L} = \text{ppm (by weight) x specific gravity}$$

Gases/Vapors

For most air pollution work, by custom, we express pollutant concentrations in volumetric terms. For example, the concentration of a gaseous pollutant in parts per million (ppm) is the volume of pollutant per million volumes of the air mixture. Calculations for gas and vapor concentrations are based on the gas laws:

- The volume of gas under constant temperature is inversely proportional to the pressure.
- The volume of a gas under constant pressure is directly proportional to the Kelvin temperature. The Kelvin temperature scale is based on absolute zero (O°C = 372K).
- The pressure of a gas of a constant volume is directly proportional to the Kelvin temperature.

When measuring contaminant concentrations, you must know the atmospheric temperature and pressure under which the samples were taken. At standard temperatures and pressure (STP), 1 gm-mol of an ideal gas occupies 22.4 liters (L). The STP is 0°C and 760 mmHg. If the temperature is increased to 25°C (room temperature) and the pressure remains the same, 1 g-mol of gas occupies 24.45 liters.

Sometimes you'll need to convert milligrams per cubic meter (mg/m^3) (weight-per-volume ratio) into a volume-per-unit-volume ratio. If one gmole of an ideal gas at 25°C occupies 24.45 L, the following relationships can be calculated:

$$\text{ppm} = \frac{24.45\ \text{mg/m}^3}{\text{molecular wt}}$$

$$\text{mg/m}^3 = \frac{\text{molecular wt}}{24.45}\ \text{ppm}$$

Index

About the Authors

Frank R. Spellman is a retired U.S. naval officer with twenty-six years of active duty, a retired environmental safety and health manager for a large wastewater sanitation district in Virginia, and a retired assistant professor of environmental health at Old Dominion University, Norfolk, Virginia. He is the author or coauthor of seventy-eight books and consults on environmental matters with the U.S. Department of Justice and various law firms and environmental entities across the world. He holds a BA in public administration, a BS in business management, and an MBA and PhD in environmental engineering. In 2011, he traced and documented the ancient water distribution system at Machu Pichu, Peru, and surveyed several drinking water resources in Coco and Amazonia, Ecuador.

Melissa L. Stoudt is a radiation controls training instructor at an atomic power laboratory. She previously instructed students and officers in the U.S. Navy in proper radiation controls and handling of radioactive material. She is coauthor of *Nuclear Infrastructure Protection and Homeland Security* (Scarecrow Press, 2011).